DEAD CITIES

Also by Mike Davis

Nonfiction

Prisoners of the American Dream:
Politics and Economy in the History of the U.S. Working Class
(1986, 1999, 2018)

City of Quartz: Excavating the Future in Los Angeles (1990, 2006)

Ecology of Fear: Los Angeles and the Imagination of Disaster (1998)

Casino Zombies: True Stories From the Neon West (1999, German only)

Magical Urbanism: Latinos Reinvent the U.S. Big City (2000)

Late Victorian Holocausts: El Niño Famines and the
Making of the Third World (2001)

The Grit Beneath the Glitter: Tales from the Real Las Vegas,
edited with Hal Rothman (2002)

Under the Perfect Sun: The San Diego Tourists Never See,
with Jim Miller and Kelly Mayhew (2003)

The Monster at Our Door: The Global Threat of Avian Flu (2005)

Planet of Slums: Urban Involution and the Informal Working Class (2006)

No One Is Illegal: Fighting Racism and State Violence
on the U.S.-Mexico Border, with Justin Akers Chacon (2006)

Buda's Wagon: A Brief History of the Car Bomb (2007)

In Praise of Barbarians: Essays against Empire (2007)

Evil Paradises: Dreamworlds of Neoliberalism,
edited with Daniel Bertrand Monk (2007)

Be Realistic: Demand the Impossible (2012)

Old Gods, New Enigmas: Marx's Lost Theory (2018)

Set the Night on Fire: L.A. in the Sixties, coauthored by Jon Wiener (2020)

The Monster Enters: COVID-19, Avian Flu, and the Plagues of Capitalism
(2022)

Fiction

Land of the Lost Mammoths (2003)

Pirates, Bats, and Dragons (2004)

DEAD CITIES

And Other Tales

MIKE DAVIS

with a new foreword by Rebecca Solnit

Haymarket Books
Chicago, IL

First published in 2002 by The New Press, New York

Published in 2024 by
Haymarket Books
P.O. Box 180165
Chicago, IL 60618
773-583-7884
www.haymarketbooks.org
info@haymarketbooks.org

ISBN: 979-8-88890-257-8

Distributed to the trade in the US through Consortium Book Sales and
Distribution (www.cbsd.com) and internationally through Ingram Publisher
Services International (www.ingramcontent.com).

This book was published with the generous support of Lannan Foundation,
Wallace Action Fund, and the Marguerite Casey Foundation.

Special discounts are available for bulk purchases by organizations and
institutions. Please email info@haymarketbooks.org for more information.

Original cover design by Steven Hiatt.
This edition cover design by Eric Kerl.

Printed in the Canada by union labor.

Library of Congress Cataloging-in-Publication data is available.

10 9 8 7 6 5 4 3 2 1

For two old heroes, sorely missed:
Sid Schneck and Mike ("Scotty") Napier

Contents

Foreword by Rebecca Solnit   xi

Preface: The Flames of New York   1

PART I NEON WEST   21

  1   'White People Are Only a Bad Dream ...'   23

  2   Ecocide in Marlboro Country   33

  3   Berlin's Skeleton in Utah's Closet   65

  4   Las Vegas Versus Nature   85

  5   Tsunami Memories   107

PART II HOLY GHOSTS   117

  6   Pentecostal Earthquake   119

  7   Hollywood's Dark Shadow   127

  8   The Infinite Game   143

  9   The Subway That Ate L.A.   183

  10   The New Industrial Peonage   191

Part III Riot City   205

11 'As Bad as the H-Bomb'   207

12 Burning All Illusions   227

13 Who Killed L.A.?: A Political Autopsy   239

14 Fear and Loathing in Compton   275

15 Dante's Choice   285

Part IV Extreme Science   305

16 Cosmic Dancers on History's Stage?   307

17 Dead Cities: A Natural History   361

18 Strange Times Begin   401

Acknowledgments   419

Index   421

But he says, "Get rid of these dark thoughts,"
And he gets rid of these dark thoughts.
And what could he say,
And what could he do
That's any better?

Robert Desnos

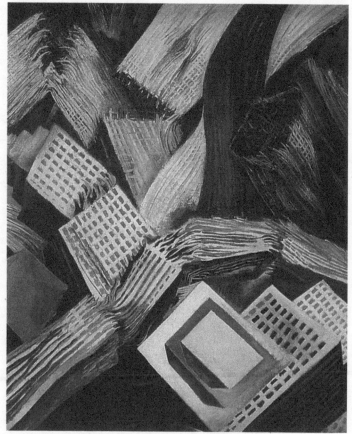

José Clemente Orozco, *Los Muertos* (1931)

Foreword

Rebecca Solnit

People were right that Mike Davis was a prophet but wrong about what a prophet is. There's a debased version in which prophets are oracles, like Nostradamus and witches staring into crystal balls in bad movies, equipped with a supernatural ability to see the future the rest of us can't. Mike had an entirely natural ability, earned through decades of reading, observing, and participating, to see the past and present with depth and breadth and in them to read the consequences of what we were doing and some of the potential futures. "You don't have to be a weatherman to know which way the wind blows," Bob Dylan sang, but if you wanted to know which way it was going to blow, studying meteorology would help. Remembering that we sowed the wind helps us know that we will reap the whirlwind.

The idea that Mike was a prophet seemed to go hand in hand with the idea that things had long been calm and stable but were about to go haywire, that the apocalypse was looming. Those who considered impending chaos, violence, and destruction surprising intrusions on a static reality often had failed to notice that they had been with us all along. The future surprised them because they had forgotten the past and, too often, not even scrutinized the present. A 2003 profile recounts, "As Davis cataloged the natural disruptions in Los Angeles's history—tornadoes, mud slides, fires—he was surprised by three things: disasters occur regularly; the media covers them unequally; and they are always labeled 'catastrophes, unusual, exceptional,' which, in Davis's mind, makes them the opposite of what they really are: to be expected." To be expected if you know the history.

Mike knew apocalypses had been coming at us all along, the expanded us, that included, for example, Native Californians for whom the apocalypse came, in Franciscan robes and then with an obscene greed for gold. It's in the title of his 2000 book *Late Victorian Holocausts*. And, of course, the other meaning of apocalypse, the original one, is revelation, and places are full of revelations for those who study them deeply enough. In this sense, *Dead Cities* is a book of revelations about cities and the places beyond them whose destiny is inseparable from urban life, urban power, and the ravenous urban appetites for raw materials, just as urban life is never separate from the natural world, as water in the overabundance of flood and in the shortage of drought, in storms, weather, and climate chaos.

A prophet is nothing more or less than someone who can speak with a moral voice, who rises high and sees far. It's not just a rhetorical power; it's a capacity to see and passion to do so. Frederick Douglass declared, in lines I imagine Mike might have liked, "For it is not light that is needed, but fire; it is not the gentle shower, but thunder. We need the storm, the whirlwind, and the earthquake. The feeling of the nation must be quickened; the conscience of the nation must be roused; the propriety of the nation must be startled; the hypocrisy of the nation must be exposed."

This stormy book has ancestors; there's an American prophetic tradition that includes (but is not limited to) Thomas Paine, Henry David Thoreau, Abraham Lincoln, Frederick Douglass, Chief Seattle, Wovoka, Black Elk, Dorothy Day, Martin Luther King Jr., Ella Baker, James Baldwin, Gloria Anzaldúa, Octavia Butler, Ursula K. LeGuin, Patti Smith, Reverend William Barber II, Julian Aguon. Prophecy is the ability to move freely in imagination and language, to look across the broad swathe of time with a bird's-eye view, to draw connections across the distance, to see the patterns that can't be seen from up close on the ground.

Ayana Mathis writes, "The prophets in the Hebrew Bible and the Jewish Tanakh are myriad and complex but have in common their proximity to periods of big trouble for the ancient Israelites—the sort of trouble that alters a nation and its people forever." Theologian Walter Bruggeman, in his book *The Prophetic Imagination*, says the task is "to nurture, nourish, and evoke a consciousness and perception alternative to the consciousness of the dominant culture around us." That works. Mike wrote about a prophet in the theological sense in *Dead Cities*, Nevada's Paiute messiah Wovoka, whose visions and teachings led to the Ghost Dance religion among the devastated indigenous survivors of genocide. "The essence of the Ghost Dance is, perhaps, precisely the moral stamina to outlast this great mirage," he writes.

Dead Cities is the third book in an urbanist trilogy that began with *City of Quartz*, the 1990 book that made Mike's reputation as a prophet—but its subtitle was *Excavating the Future in Los Angeles*, an argument that you get to the future through a deep dive into the past. The second book, *Ecology of Fear*, has a passage in its opening essay that sums up some of what prophetic power consists of. He wrote in praise of the essential lushness of the Los Angeles basin, while contradicting the accounts of it as a bleak and arid place, "Southern California has reaped flood, fire, and earthquake tragedies that were as avoidable, as unnatural, as the beating of Rodney King and the ensuing explosion in the streets." Then, a page later, he cites Father Juan Crespi, writing in 1769 about the Los Angeles basin's "admirable" landscape with its "soil...black and loamy." It takes a kind of mental athleticism to transition from Rodney King to the Spanish friars in such a compressed space, from the social to the economic to the ecological with such ease, to see both the grand sweep of history across a place and hold so many details together.

Those brief passages are already violating a lot of rules of academic writing, including the "stay in your lane" notions of expertise, defined narrowly. The use of police brutality as ecological metaphor—I'm not sure that's allowed either, and Mike's prose style had bravura and pyrotechnics that are not normal for that arena in general, any more than was his passionate engagement on behalf of the underdogs and outsiders. Academic writing often strives to sound dispassionate by means of disengagement; Mike floored it in the opposite direction with full-throttle sentences like this one in *Dead Cities*: "Las Vegas, moreover, is a major base camp for the panzer divisions of motorized toys—dune buggies, dirt bikes, speed boats, jet-skis, and the like—that each weekend make war on the fragile desert environment."

Mike seemed to have read everything on political science, geology, geography, environmental science, Western and urban history—it all seemed to be churning into ideas within him, and when he spoke it was as though he just stopped damming up the torrent of ideas coming together from his contemplation of this colossal database. I met him in 1995, after I'd sent him my second book, *Savage Dreams*, about the nuclear wars that weren't supposed to have started, the Indian wars supposedly long over, raging together in an American West most people outside the zones of impact seemed unable to see. I was just starting out and he was endlessly kind and encouraging to me, as he was to a lot of other people. I remember a man whose son had died by suicide whom he called regularly for years to support, a prisoner to whom he was a faithful correspondent, and wonder if I was just another stray lamb he was shepherding. We exchanged letters and then emails, hundreds of them, over the next decade, and hung out in person from time to time in those years, and then, through no ill will, didn't keep up the connection. But it meant a lot to me at a crucial time in my life as a writer and Westerner. Those years of friendship let me see his mind at work, his enthusiasm for the work of others from organizers to artists, and his expansive empathy and solidarity that underlies his writings about the forces of destruction and oppression.

The essays in this book are ferocious. Ferocious but far from hopeless. Some of them feel very much of their time—but that time haunts us and has things to teach us, as a new generation of right-wing pundits pushes fear of crime and immigrants and lobbies for more policing and fewer civil rights. Mike was born in 1946 in Fontana, east of L. A. Octavia Butler was born a year later, in nearby Pasadena, and her visionary science fiction novels—notably *Parable of the Sower*—spring from a world that has a lot in common with the one described in Mike's books. She saw harshness, trouble, racism, environmental destruction, but she, too, wasn't going to surrender. She wrote in an essay, "The very act of trying to look ahead to discern possibilities and offer warnings is in itself an act of hope." That is, it assumes we still have some choices to make about what kind of a future we will have. A prophet, in the glib sense, tells us what will happen as though it's inevitable, encouraging passivity. A warning tells us that there's still time to decide what will happen, that we are shaping the present in the future. Warnings call for action.

Mike himself affirmed hope toward the end of his life when he told an interviewer: "This seems an age of catastrophe, but it's also an age equipped, in an abstract sense, with all the tools it needs. Utopia is available to us. If, like me, you lived through the civil-rights movement, the antiwar movement, you can never discard hope. I've seen social miracles in my life, ones that have stunned me—the courageousness of ordinary people in a struggle." Here, in *Dead Cities*, are the catastrophes, the tools, the movements, and the courage.

Preface

The Flames of New York

> Lower Manhattan was soon a furnace of crimson flames, from which
> there was no escape. Cars, railways, ferries, all had ceased, and never
> a light lit the way of the distracted fugitives in that dusky confusion
> but the light of burning. Dust and black smoke came pouring into the
> street, and were presently shot with red flame.
>
> H. G. Wells, The War in the Air (1908)

This image, part of a long warning note about the "Massacre of New York,"
slumbered for nearly a century on a back shelf of the New York Public Library.
H. G. Wells, that socialist Nostradamus, penned it in 1907. The American edition
of his *War in the Air* includes an extraordinary illustration (is it not from CNN?)
of a firestorm devouring Wall Street, with Trinity Church smoldering in the
background. Wells also offered some shrewd and unfriendly thoughts about New
York's messianic belief in its exemption from the bad side of history:

> For many generations New York had taken no heed of war, save as a thing that
> happened far away, that affected prices and supplied the newspapers with exciting
> headlines and pictures. The New Yorkers felt that war in their own land was an
> impossible thing.... They saw war as they saw history, through an irridescent mist,
> deodorised, scented indeed, with all its essential cruelties tactfully hidden away.

They cheered the flag by habit and tradition, they despised other nations, and whenever there was an international difficulty they were intensely patriotic, that is to say, they were ardently against any native politician who did not say, threaten, and do harsh and uncompromising things to the antagonist people.[1]

When a foreign policy dominated by the Trusts and Monopolies entangles America in a general War of the Powers, New Yorkers, still oblivious to any real danger, rally to flags, confetti, and an imperial Presidency.

And then suddenly, into world peacefully busied for the most part upon armaments and the perfection of explosives, war came.... The immediate effect on New York ... was merely to intensify her normal vehemence.... Great crowds assembled ... to listen to and cheer patriotic speeches, and there was a veritable epidemic of little flags and buttons ... strong men wept at the sight of the national banner ... the trade in small arms was enormously stimulated ... and it was dangerous not to wear a war button.

One of the most striking facts historically about this war, and one that makes complete the separation between the methods of warfare and democracy, was the effectual secrecy of Washington.... They did not bother to confide a single fact of their preparations to the public. They did not even condescend to talk to Congress. They burked and suppressed every inquiry. The war was fought by the President and the Secretary of State in an entirely autocratic manner.[2]

But the Americans, blinded by the solipsistic delusion that they live in a history solely of their own making, are easy targets for that scheming New Assyria: Wilhelmine Germany. Surprise-attacked by the Imperial zeppelin fleet, ragtime New York becomes the first modern city destroyed from the air. In a single day, haughty Manhattanites are demoted to slaughtered natives:

As the airships sailed along they smashed up the city as a child will shatter its cities of brick and card. Below they left ruins and blazing conflagrations and heaped and scattered dead: men, women and children mixed together as though they had been no more than Moors, or Zulus or Chinese.[3]

The Fantastic Mask

> The Mask. Look at the Mask.
> Sand, Crocodile, and Fear above New York.
> *Federico García Lorca*, "Dance of Death" (1929)

If Wells, looking through his Edwardian spyglass, foresaw the end of American exceptionalism in eerily accurate focus, his is only one of myriad visions hurled back at us since the World Trade Center became the womb of all terror. There is, for instance, José Clemente Orozco's fierce 1931 painting *Los Muertos* ("The Dead"), which depicts Manhattan skyscrapers being broken apart like piñatas at a satanic birthday party. Orozco, still an anarcho-syndicalist in temperament, spent eight years in the city, observing the growing armies of "hard, desperate, angry men, with opaque eyes and clenched fists." His cityscapes are dominated by alarming tectonic tensions: an approach to the Queensboro Bridge buckles, Eighth Avenue is shrouded in sulfurous haze, the Elevated is portrayed as the fiery portal of Hell, and so on. Orozco seems to be warning of earthquakes, volcanoes, infernos.[4]

His homesick acquaintance, Federico García Lorca, also augured a Manhattan apocalypse. Lorca's New York poems are so saturated with fear and prophecy that he originally entitled them "Introduction to Death." On the original Black Tuesday in 1929, the Andalusian poet wandered through the canyons of Wall Street, watching in amazement as ruined investors flung themselves from windows of monstrous buildings. "The ambulances collected suicides," he wrote, "whose hands were full of rings." Amid the "merciless silence of money," Lorca "felt the sensation of real death, death without hope, death that is nothing but rottenness." It was easy, then, for him to visualize the inevitable destruction of lower Manhattan by "hurricanes of gold" and "tumults of windows"—a Gypsy intuition, perhaps, of the deadly black cloud that engulfed Wall Street last September.[5] Or maybe the deathcloud was actually that "storm blowing from Paradise ... piling wreckage upon wreckage" that Walter Benjamin warned about.[6] In either case, it was not only "what we call progress" (that is to say, the real history of the American imperium in the Middle East) that has blown back, but also all of our imagined catastrophes, vengeful angels, and days of reckoning.

The walled subdivision on End of History Lane turned out to be only one subway stop from The War of the Worlds. The *fatwa* from a cave in Afghanistan has sent amuck every invader and monster that ever thrilled fans of *Amazing Tales* or Universal Pictures. Wells's zeppelins rain fiery death on Wall Street. King Kong and Godzilla pulverize Fifth Avenue. Extraterrestrials broil Soho in brimstone and pitch. Nightmare spores turn Radio City into a ghost town. Fu Manchu and the evil Ming have a cousin in Afghanistan. Sci-fi happens. Indeed, anything can happen. But the *frisson* is different than we expected.

Fear Studies

Indeed, 9/11 has been societal exorcism in reverse. It is important to recall the already fraught collective condition before Real Terror arrived in a fleet of hijacked airliners. The *X-Files* defined the 1990s in the same way that the *Honeymooners* had defined the 1950s. It was an age of inexplicable anxiety. Although it seems laughable now, millions purportedly trembled before the occult menaces of black helicopters, killer asteroids, maddog teenagers, recovered memories, Lyme disease, Satanic preschools, road rage, Ebola fever, Colombian drug cartels, computer viruses, and Chinese atomic spies. There was a diagnostic consensus among social scientists and culture theorists that Americans were suffering from acute hypochondria. On the eve of the Y2K nonapocalypse, "Fear Studies" — or "Sociophobics" as it is sometimes called—had emerged as the hottest new niche in academia. Dozens of pundits were raving about the "mainstreaming of conspiracy culture," the arrival of "risk society," the "hermeneutic of suspicion," the "plague of paranoia," "the mean world syndrome," or the newly discovered role of the amygdala as the "center of the [brain's] wheel of fear."[7]

In the best of the genre, Barry Glassner systematically debunked some of the more common goblins—young Black men, street drugs, terroristic political correctness, and so on—that deliberately spook the path toward public understanding of such social problems as unemployment, bad schools, racism, and world hunger. He carefully showed how media-conjured scares were guilty "oblique expressions" of the postliberal refusal to reform real conditions of inequality. Fear had become the chief ballast of the rightward shift since 1980. Americans, in his view, "were afraid of the wrong things," and were being hoaxed by the latter-day equivalents of Orson Welles's notorious "War of the Worlds" broadcast. "The Martians," he underscored, "aren't coming."[8]

But, alas, they have come after all, brandishing box-cutters. Although movies, like kites and women's faces, were banned in the Hindu Kush version of utopia, the attacks on New York and Washington, D.C. were organized as epic horror cinema with meticulous attention to *mise-en-scène*. The hijacked planes were aimed to impact precisely at the vulnerable border between fantasy and reality. In contrast to the 1937 radio invasion, thousands of people who turned on their televisions on 9/11 were convinced that the cataclysm was just a broadcast, a hoax. They thought they were watching rushes from the latest Bruce Willis film. Nothing since has thrown cold water on this sense of illusion. The more improbable the event, the more familiar the image. The "Attack on America," and its sequels, "America Fights Back" and "America Freaks Out," has continued to unspool as a succession of celluloid hallucinations, each of which can be rented from the corner video shop: *The Siege, Independence Day, Executive Action, Outbreak, The Sum of All Fears*, and so on. George W. Bush, who has a bigger studio, meanwhile responds to Osama bin Laden as one *auteur* to another with his own fiery wide-angle hyperboles.

Has history, then, simply become a montage of prefabricated horrors crafted in Hollywood writers' huts? Certainly the Pentagon thought so when it secretly conscripted a group of famous screenwriters, including Spike Jonze (*Being John Malkovich*) and Steven De Souza (*Die Hard*), to "brainstorm about terrorist targets and schemes in America and to offer solutions to those threats." The working group is based at the Institute for Creative Technology, an Army joint venture with the University of Southern California that mines Hollywood expertise to develop interactive war games with sophisticated story paths. One of its products is *Real War*, a video game that helps train military leaders to "battle against insurgents in the Middle East." When on 20 September an unidentified "foreign intelligence agency" warned the FBI of a potential attack on a major Hollywood studio, it was the last twist in a Mobius strip weaving simulation into reality and back again.[9]

The Interminable Uncanny

Mere skepticism seems powerless to remove the fantastic mask worn by such events. When hypochondriacs actually contract the plague of their worst fear, their ontologies tend to be thrown out of kilter. Watching the South Tower of the World Trade Center collapsing on its thousands of victims, a friend's child blurted out: "But this isn't real the way that real things are real." Exactly. Nor does it feel real the way real things do. There is a proper name, of course, for this eerie sensation of reality invaded by fantasy. "An *uncanny* effect," Freud wrote, "is often and easily produced when the distinction between imagination and reality is effaced, as when something that we have hitherto regarded as imaginary appears before us in reality."[10]

I am not sure, however, that Freud anticipated such a Walpurgis night of uncanny doubles and repetitions. The Israeli psychoanalyst Yolanda Gampel, an expert on second-generation legacies of the Holocaust, has addressed this more extreme condition, which she calls "interminable uncanniness." It is a sensibility—now perhaps being mass-franchised—that usurps the lives of those who have witnessed an "astounding, unbelievable, and unreal reality" like mass murder. "They no longer fully believed their own eyes: they had difficulty distinguishing between this unreal reality and their own imagination. [Moreover] such an assault on the boundary between fantasy and reality becomes traumatic in itself and leads to great fear of one's thoughts and expectations."[11]

There is also a large and perhaps short-lived dimension of old-fashioned hysteria. When the mayor of Chicago has to go on TV to reassure his citizenry that a glob of guacamole on a sidewalk is not some deadly andromeda strain, then we are back in the realm of familiar panics like Welles's radio Martians in New Jersey or the Japanese "bombing" of Los Angeles in the aftermath of Pearl Harbor. But when hysteria subsides, the uncanny will likely endure, as Gampel explains, "not [as] a symptom, behavior, or neurotic organization," but as "lived experience": a permanent foreboding about urban space as potential Ground Zero.[12]

Black Utopia

The bourgeois' hat flies off his pointed head.
...
Trains fall off bridges.
— *Jakob van Hoddis*, "World's End" (1910)

From a psychoanalytic perspective, of course, there is more to the story. Freud defined the uncanny as always involving some "return of the repressed" as when, "after the collapse of their religion, [a people's] gods turn into demons."[13] (Or their skyscrapers into infernos?) But what is the repressed root of modern urban fear? What is the ultimate psycho-social substrate upon which *politics* (and what else is it?) has deposited layer after layer of spectral dangers: the fear of the poor, fear of crime, fear of Blackness, and now fear of bin Laden?

The most interesting answer, at least within the Marxist tradition, comes from Ernst Bloch. Although primarily known as a dialectician of hope, Bloch was also attentive to the uncanny qualities of the big city. As the one unrepentant Expressionist in the ranks of Western Marxism, he retained that apocalyptic sensibility that had first burst forth in the revolutionary poem "World's End" that Jakob van Hoddis read in the Das Neopathetische Cabaret in late 1910. "Something uncanny was in the air," and *Expressionismus* was the lightning rod that captured urban fear on the brink of the First World War and converted it into a prefigurative vision of the horrors to come.[14] The poems of George Heym and Georg Trakl, and the canvasses of Franz Marc, Ernst Kirchner, Erich Heckel, and, above all, Ludwig Meidner were ablaze with clairvoyant images of murder victims, tumbling tenements, exploding cities, and flying bodies. Indeed, Meidner—who wrote that "the street bears the apocalyptic within itself"—could not look out his window without being shattered by the imminence of disaster. "My brain bled dreadful visions," he wrote of the torrid summer of 1913. "I could see nothing but a thousand skeletons jigging in a row. Many graves and burned cities writhed across the plains."[15]

In the equally ominous year of 1929, Bloch returned to this eschatological nervousness. In "The Anxiety of an Engineer," he explains the "fearful bourgeois," intriguingly, in terms of the contrasting urban ecologies of capitalist and precapitalist cities. In the latter (he uses Naples as an example), there is no delusion of total command over Nature, just constant ecological adaptation. The city is an imperfect and carnivalesque improvisation that yields to the fluxes of a dynamic Mediterranean environment: "Things are allowed to remain in a halfway real condition, and delight is taken in the way things come to their own equilibrium and completion." Although the objective hazards it faces (volcanoes, earthquakes, landslides, and tsunamis) are arguably greater than those of any other large European city, Naples is on familiar terms (*heimlich* in Freud's sense) with the "old dragon" of catastrophic nature. Anxiety does not infuse daily life on the slopes of Vesuvius.[16]

In the "Americanized big city," by contrast, the quest for the bourgeois utopia of a totally calculable and safe environment has paradoxically generated radical insecurity (*unheimlich*). Indeed, "where technology has achieved an apparent victory over the limits of nature ... the coefficient of known and, more significantly, unknown danger has increased proportionately." In part, this is because the metropolis's interdependent technological systems—as Americans discovered in fall 2001—have become "simultaneously so complex and so vulnerable." More profoundly, the capitalist big city is "extremely dangerous" because it dominates rather than cooperates with Nature. (Although Bloch has the old-fashioned centralized industrial city in mind, his argument would presumably apply to the networked and polycentric metroregion as well.)[17]

The Uncanny is precisely that "nothingness [non-integration with Nature] that stands behind the mechanized world." Although Bloch was acutely aware of the imminent dangers of fascism and a new world war, he insisted that the deepest structure of urban fear is not Wells's war in the air, but "detachment and distance from the natural landscape."

> The subject is teetering on the brink of absolute nihilism; and if this mechaniza-
> tion with or without purpose, this universal depletion of meaning, should come to
> fulfillment, then the future void may prove equal to all the death anxieties of late
> antiquity and all the medieval anxieties about hell.[18]

Years later in *The Principle of Hope* (1938–47), Bloch again reflected on the relationship between modern anxiety and urban-technological "perversion." This
time he focused on science fiction and catastrophe. His pretext was Grandville's
bizarre 1844 book, *Another World*, with its images of a monstrously technolo-
gized Nature: giant iron insects, gas lamps as big as the moon, men with amazing
mechanical prostheses, and so on. In Bloch's interpretation, the "schizophrenic
petit-bourgeois" Grandville (who "died three years later in a madhouse") was
the Hieronymus Bosch of the steam age, and his book is a huge anxiety dream
"full of the terror of the technological challenge and of what it is calling." Yet
the landscape of terror is also, as in Bosch, voluptuous and nearly infinite in
irony. Reminding us that hell is full of laughter, Bloch calls this cataclysm where
everything bad is foretold in dark humor, a "black utopia."[19] He might have been
thinking of New York.

Manhattan Nightmares

> All these April nights combing the streets alone a skyscraper
> has obsessed him—a grooved building jutting up with
> uncountable bright windows falling onto him out of a scud-
> ding sky.
> John Dos Passos, Manhattan Transfer (1925)

"Irony," of course, is now an illegal alien in the land of liberty. Even professional
ironists like Christopher Hitchens police the sacred "no irony" zone that surrounds the ruins of the World Trade Center. Otherwise it might be possible to
draw various parallels between Jimmy Herf's nightmare in *Manhattan Transfer*
of a skyscraper falling on him and the hatless bourgeoisie of the Expressionist
apocalypse. Urban anxiety snakes like a 50,000-volt current through Dos Passos's
famed novel (called "expressionist" by many reviewers), written a few years after
Italian anarchists (16 September 1920) had exploded a wagon load of dynamite in
front of J. P. Morgan's offices on Wall Street.

> The horse and wagon were blown to bits. Glass showered down from office win-
> dows, and awnings twelve stories above the street burst into flames. People fled in

terror as a great cloud of dust enveloped the area. In Morgan's office Thomas Joyce of the securities department fell dead on his desk amid a rubble of plaster and glass. Outside scores of bodies littered the streets. Blood was everywhere.[20]

Dos Passos's New York, like Bloch's Berlin, is a great engine roaring down tracks that engineers have yet to build, toward destinations unknown. The sheer out-of-control velocity of the metropolis, including the drunken swaying of its arrogant skyline, is the master theme of *Manhattan Transfer*. It is not surprising that passengers on this runaway train should be more than a little anxious. In the end, Jimmy Herf answers his own rhetorical question—"But what's the use of spending your whole life fleeing the City of Destruction?"—by hitching a ride out of town. ("How fur ye goin?" asks the truckdriver. "I dunno," he answers. "... Pretty far.")[21]

It was the hubris of New York's landowners and cops in the 1990s that ruthless "zero tolerance" could expunge this constitutive anxiety: the "edginess" that generations of twenty-somethings have sought with the desperation of junkies. The Gotham express was shunted into a suburban siding, a national showcase that "big cities were again safe." Ruling from his so-called "bunker" (Emergency Command Center) on the twenty-third floor of the World Trade Center, Mayor Giuliani reshaped Manhattan into "an electric urban theme park as safe and, some said, sterile as a suburban mall."[22]

The Worm in the Apple

In a savage new biography of Giuliani, the *Village Voice*'s Wayne Barrett shows how a police department with a dangerously high testosterone level became the city's urban planning agency. "The bunker was emblematic of an administration that had unconstitutionally closed City Hall Park to all but mayorally sanctioned public spectacle, blockaded bridges to kill a cab protest, barricaded midtown crosswalks to regulate pedestrians and yanked the homeless out of shelter beds on the coldest night of the year to enforce ancient bench warrants for open beer can violations."[23] The media generally viewed the fascistic bullying of squeegee men, panhandlers, cabbies, street vendors, and welfare recipients as a small price

PREFACE   11

to pay for the triumphs of having brought Disney (the ultimate imprimature of suburban safety) to Times Square and tourism back to New York.

Now folks in Iowa watch grisly television footage of the FBI raking the rubble at Fresh Kills for rotting body parts (fireworks are used to keep the landfill's huge turkey vultures away) and thank God that they still live on the farm or, at least, in a gated suburb of Des Moines. However much they may admire the Churchillian pose struck by Rudolph Giuliani or the fortitude of New York's rescue workers, family vacations are not usually envisioned as exercises in "overcoming fear." So they stay at home in droves: as do the myriads of low-wage, largely immigrant hotel and restaurant workers laid off by the tourist depression. Every ancient connotation of the Big City as the sinister abode of danger, death, and infection has been revalorized by the almost weekly "terrorist alerts" and the Bush administration's constant hyping of imminent nuclear or "dirty" bomb threats to New York, Washington, and other metropolises. Big Apple tourist officials must have been especially appalled by Federation of American Scientists' recent warning to Congress of the likely consequences of a low-tech radiological attack on Manhattan:

> ... a bomb made with a single footlong pencil of cobalt from a food irradiation plant and just ten pounds of TNT and detonated at Union Square in a light wind would send a plume of radiation drifting across three states. Much of Manhattan would be as contaminated as the permanently closed area around the Chernobyl nuclear plant. Anyone living in Manhattan would have at least a 1-in-100 chance of dying from cancer caused by the radiation. An area reaching deep into the Hudson Valley would, under current Environmental Protection Agency Standards, have to be decontaminated or destroyed.[24]

Although many surprises undoubtedly lurk down river, it is already clear that the advent of "catastrophic terrorism" in tandem with protracted recession will produce major mutations in the American city. Indeed, it is conceivable that bin Laden et al. have put a silver stake in the heart of the "downtown revival" in New York and elsewhere. The traditional central city where buildings and land values soar toward the sky is not yet dead, but the pulse is weakening. The current globalization of fear will accelerate the high-tech dispersal of centralized organizations, including banks, securities firms, government offices, and tele-communications centers, into regional multisite networks. Terror, in effect, has become the business partner of technology providers like Sun Microsystems and Cisco Systems, which have long argued that distributed processing (sprawling PC networks) mandates a "distributed workplace." In this spatial model (of which the al Qaeda network might be an exemplar), satellite offices, telecommuting, and, if need be, comfortable bunkers will replace most of the functions of that obsolete behemoth, the skyscraper. Very tall buildings have long been fundamen-tally uneconomic; indeed, the absurdly overbuilt World Trade Center—a classic Rockefeller boondoggle—was massively subsidized by public sector tenants.[25] (Will the hijacked airliners someday be seen as having played the same role in the extinction of skyscrapers as the Chicxulub asteroid in the demise of dinosaurs?)

The Fear Economy

Meanwhile, the "Fear Economy," as the business press has labeled the complex of military and security firms rushing to exploit the national nervous breakdown, will grow fat amid the general famine. Fear, of course, has been reshaping Ameri-can city life since at least the late 1960s; but the new terror provides a powerful Keynesian multiplier: "According to *Fortune* magazine, the private sector will spend over $150 billion on homeland security–related expenses such as insur-ance, workplace security, logistics, and information technology—approximately four times the federal government's announced homeland security budget."[26] Thus the already million-strong army of low-wage security guards is expected to increase 50 percent or more in the next decade, while video surveillance, finally beefed up to the British standard with face recognition software, will strip the last privacy from daily routine. The security regime of airport departure lounges will likely provide a template for the regulation of crowds at malls, shopping concourses, sports events, and elsewhere. Americans will be expected to express gratitude as they are scanned, frisked, imaged, taped, and interrogated "for their own protection." Venture capital will flood into avant-garde sectors developing germ warfare sensors, and cyber-security systems. Some pundits predict that the real heroes of the War Against Terrorism will be a "private sector army" of venture capitalists and security-tech startups. ("Its best troops will be regiments of geeks rather than the Special Forces that struck the first blows against the Taliban in Afghanistan. These pocket-protector brigades live on rations of cold pizza and coffee... They take orders not from generals or admirals but from markets and stockholders."[27]) In any event, as the evolution of homeland security already illustrates, the discrete technologies of surveillance, environmental monitoring, data processing, building design, and even entertainment will grow into a single, integrated system. "Security," in other words, will become a full-fledged urban utility like water, electric power, and telecommunications.

Despite massive plans for "hardening" and "terror-proofing" downtown public spaces and monumental buildings,[28] however, most white-collar workers and managers will prefer to consume enhanced security closer to their suburban homes. Physical security retrofits—the reinforcement of building structures, vapor-and-trace detection systems, traffic barricades, bomb mitigation containers, smart doors, metal detectors, bomb-proof trash cans, biometric surveillance portals, reduced surface and underground parking—will impose huge and unavoidable expenses on cities that are trying to shore up their downtown economies, but they are unlikely to stem the new exodus of jobs and tax resources. Massive public sector subsidies to developers and corporate tenants may likewise slow but probably won't reverse the trend toward deconcentration. In addition, as self-advertised "world cities" hunker down for the long siege, urban economists and fiscal analysts must wrestle with the new demon of "deglobalization": the portion of global service production and international tourism that may be lost forever.

Needless to say, all this adds up to a fiscal crisis of a magnitude that may dwarf the notorious municipal meltdown of the mid 1970s. Certainly this is the case in New York City, where Felix Rohatyn, the city's bank-appointed financial overlord from 1973 to 1993, has warned of approaching bankruptcy as City Hall grapples with a projected $6 billion deficit in a $40 billion budget.[29] The city's crisis has been amplified by the post-9/11 flight of 22,000 jobs and their contribution to the tax rolls from Lower Manhattan to New Jersey. Manhattan boosters are terrified that Wall Street giants like American Express, Deutsche Bank, Merrill Lynch, and

Dow Jones will learn to love dispersal into the "bombproof," low-rent districts of the megalopolis.[30] Fiscal crisis and corporate deconcentration are especially bad news for the new immigrant working class already buried under the rubble of the city's fallen tourist and service industries. As Mark Green, the androidlike Democratic candidate for mayor, warned, the reconstruction of Lower Manhattan "may require sacrifice from others." Since Giuliani-era crime control is sacrosanct, as is the goodwill of big business, budget-cutters will hack away at lifeline public services—housing, libraries, sanitation, recreation, job programs, and the like—in New York's neglected Black and Latino neighborhoods. Whatever twin-tower replica or monumental novelty eventually fills the void in Lower Manhattan, it will likely be financed by savage retrenchment in Washington Heights, Mott Haven, and Brownsville. So much for the famous "solidarity"of New Yorkers.[31]

War of the Worlds

> In an immediate and inclusive way suspicion of the Arabs became second nature.
>
> Franz Fanon, "Racist Fury in France" (1959)

Long ago a tourist in New York sent home a postcard. "If all the world became America," wrote the poet Sayyid Qutb, "it would undoubtedly be the disaster of humanity." Seconded by the Egyptian government to study US educational methods, Qutb disembarked at the 42nd Street Pier in autumn 1948 an admirer of liberal modernity. But he was repulsed by Truman's America and underwent a deep religious reconversion. He returned to Cairo two years later a fervent adherent of the Muslim Brotherhood and was soon arrested as its leading propagandist. After eleven years in prison, he was hung in 1966 on trumped-up charges of conspiring to overthrow Nasser. Qutb is universally acclaimed as the major philosopher of radical Islamism, if not literally, as the *New York Times* proposes, the "intellectual grandfather to Osama bin Laden and his fellow terrorists." His masterpiece, *Milestones* (1964), is routinely described as the Islamist version of Lenin's *What Is to Be Done?*[32]

Why did Qutb become the Anti-Whitman, recoiling in disgust from the legendary excitement of Manhattan? Understanding his hostility to the self-proclaimed "capital of the twentieth century" might shed some light on the genealogy of anti-Americanism in the milieux that have applauded the destruction of US capitalism's most monumental symbol. Pop analysis, of course, fits the person into the prefabricated stereotype. Thus for Robert Worth and Judith Shulevitz (writing separately in the *Times*), the 42-year-old Egyptian literary critic and poet was, like all Muslim fanatics, a prude scandalized by big city "decadence," by the *Kinsey Report*, by dancing and sexual promiscuity. Indeed, Qutb did complain about the "pornographic" content of much American popular culture, just as he criticized the national obsession with tending lawns to the neglect of family life and the crass materialism that smothered charity. But the great scandal of New York—and his reaction was the same as Garcia Lorca's twenty years before—was *"evil and fanatic racial discrimination."* No doubt Qutb, a Black man from Upper Egypt, had wounding encounters with Jim Crow.[33]

Qutb's tourist experiences today might be even more traumatic. He might be in solitary confinement, without access to relatives or a lawyer, for the "terrorist" crime of having overstayed his visa or having simply aroused the suspicion of his neighbors. The real burden of the new urban fear—the part that is not hallucinatory or hyperbolized—is borne by those who fit the racial profile of white anxiety: Arab and Muslim Americans, but also anyone with an unusual head-covering, Middle Eastern passport, or unpopular beliefs about Israel. For those caught squarely in the middle of this paranoid *gestalt*—say a Pakistani cabdriver in New York or a Sikh electronics engineer in California—there is the threat of violence but, even more, the certainty of surveillance by powers "vast and cool and unsympathetic."[34] "Otherness"—in the form of Arabs, Korans, and spores—has become the central obsession of that interminable Pentagon briefing and George W. Bush celebration that passes for American television. Indeed, the "Threat to America" (another network branding) is depicted as essentially extraterrestrial: the Middle East is the Angry Red Planet sending its monsters to live among us and murder us.

Tous Martiens
Very little of the violent domestic backlash has been reported in the mainstream media. The big city dailies and news networks have shown patriotic concern for the US image abroad by downplaying what otherwise might have been recognized as the good ole boy equivalent of *Kristallnacht*. Yet even the fragmentary statistics are chilling. In the six weeks after 11 September, civil rights groups estimate that there were at least six murders and a thousand serious assaults committed against people perceived as "Arab" or "Moslem," including several hundred attacks on Sikhs.[35] *The Texas Observer,* a progressive weekly that has refused to low-profile domestic terror, reported in early October on the violence that had "ricocheted" through Dallas suburbs in the immediate aftermath of the New York and D.C. attacks. In addition to the hate murder of an immigrant Pakistani grocery proprietor, three mosques were bombed or shot at, a Romanian jogger was beaten because he looked "Middle Eastern," and two Ethiopians were stabbed while touring the Fort Worth botanical gardens. Local Moslem leaders blamed the news media, particularly the *Dallas Morning News,* for helping instigate violence with inflammatory headlines like "Soldiers of Terror Living Next Door!"[36]

If such incidents recall the "Arab hunts" in metropolitan France during the Algerian War that Franz Fanon described ("even a South American was riddled with bullets because he looked like a North African"),[37] then the Justice Department's frenzied search for al Qaeda "sleepers" stirs memories of that other great "terrorist manhunt," the notorious Palmer raids of 1919–20, when thousands of immigrant radicals were arrested without warrant or cause, and then hundreds deported, after a series of package bomb explosions in Washington, D.C. (The bombing of Wall Street was assumed to have been anarchist revenge for the deportations.) This time more than eleven hundred "terrorist suspects" have disappeared into a secretive federal maze where many have been denied lawyers, beaten by guards and inmates, blindfolded and subject to sensory deprivation, and forced to take lie-detector tests. At least one detainee has died, and scores, against whom no criminal charges have been filed, are being held under the indefinite detention permitted by immigration law. Only four are rumored to have any direct connection to bin Laden. Most simply have overstayed visas or used false

IDs: a not uncommon status in a nation where an estimated five to seven million undocumented immigrants provide indispensable cheap labor.

Fanon probably would not be surprised that frustrated FBI investigators, like the French Sûreté before them, are lobbying to take recalcitrant suspects down to the scream-proof basement where the batteries and electrodes are kept. For the first time in American history there is a serious public campaign to justify torture. With the op-ed support of leading liberals like Jonathan Alter in *Newsweek*, the FBI wants access to methods that the *Washington Post* euphemistically characterized "as employed occasionally by Israeli interrogators." If US courts balk at such rough work, the alternative is to export the task to overseas professionals like the Mossad. "Another idea," the *Post* explains, "is extraditing the suspects to allied countries where security services sometimes employ threats to family members or resort to torture."[38] In the same vein, the *Los Angeles Times* interviewed a CIA veteran about current debates inside the Agency. "A lot of people are saying we need someone at the Agency who can pull fingernails out…. Others are saying, 'Let others use interrogation methods that we don't use.' The only question then is do you want to have CIA people in the room?"[39]

Short of electrodes, however, Congress (minus an opposition party) has recently given the Justice Department a cornucopia of vaguely worded and sinister powers. The "Proved Appropriate Tools Required to Intercept and Obstruct Terrorism" (PATRIOT) Act cages noncitizens, including millions of Latino and Asian immigrants, within ruthless new categories of surveillance, prosecution, and liability to deportation. But it is only a cornerstone for the full-fledged Homeland Security State envisioned by the junior Bush administration. At a Halloween press conference, Colin Powell, sounding as if he had just finished reading *Neuromancer*, gloated over plans for a vast centralized data warehouse that would store "every derogatory piece of information" on visitors and would-be immigrants. The Justice Department, with strong support from Southern Republican governors, wants to conscript local and state police in a massive campaign to round up and deport illegal immigrants. Federal law enforcement is being restructured so that the FBI can permanently focus on the War against Terrorism—meaning that it will largely become an elite immigration police—while a mysterious new

Pentagon entity, the Homeland Defense Command, will presumably adopt the Mexican border as a principal battlefield. Both Mexico and Canada are under tremendous pressure to tighten their immigration policies to Washington's standards. Indeed, to the howling delight of nativists and neofascists everywhere, the entire OECD bloc seems to be raising drawbridges and bolting doors against the rest of humanity.

The globalization of fear thus becomes a self-fulfilling prophecy. Automatically, NATO endorsed the blank check that Congress issued the White House to "rid the world of evil," leaving American fighter pilots to drop cluster bombs chalked with the names of dead Manhattan firefighters on the ruins of Kabul—a city infinitely more tragic than New York. Terror has become the steroid of Empire. And Imperialism is again politically correct. However nervously, the established order everywhere has rallied around the Stars and Stripes. As a gloating and still undead Henry Kissinger has pointed out, it is the best thing since Metternich last dined with the Czar.

2002

Notes

1. H. G. Wells, *The War in the Air*, pp. 181–82.
2. Ibid, pp. 182–83 and 186.
3. Ibid, p. 211.
4. José Clemente Orozco, *An Autobiography*, Mineola, N.Y. 1962, p. 132. An excellent color reproduction of *The Dead* appears in Renato Gonzalez Mello and Diane Miliotes, eds., *José Clemente Orozco in the United States, 1927–1934*, New York 2002, p. 58. Many of his New York prints and paintings show the obvious influence of German Expressionism: Beckmann, Dix, Meidner, and Grosz.
5. Lorca, "Lecture: A Poet in New York, " ibid., pp. 192–93
6. "This storm is what we call progress" (Walter Benjamin, "Theses on the Philosophy of History," *Illuminations*, New York 1969, pp. 257ff).
7. Some representative studies: Marina Warner, *No Go the Bogeyman: Scaring, Lulling and Making Mock*, New York 1998; Jane Franklin, ed., *The Politics of the Risk Society*, Oxford 1998; Nancy Schultz, ed., *Fear Itself: Enemies Real and Imagined in American Culture*, West

Lafayette, Ind. 1999; Paul Newman, *A History of Terror: Fear and Dread Through the Ages*, New York 2000; and Robert Goldberg, *Enemies Within: The Culture of Conspiracy in Modern America*, New Haven, Conn. 2001.

8. Barry Glassner, *The Culture of Fear: Why Americans Are Afraid of the Wrong Things*, New York 1999, p. 203.

9. *Los Angeles Times*, 21 September 2001; and Reuters, 8 October 2001.

10. "The Uncanny" (1919) in *Volume 14: Art and Literature*, The Penguin Freud Library, London 1985, p. 367.

11. Yolanda Gampel, "The Interminable Uncanniness," in Leo Rangell and Rena Moses-Hrushovski, eds., *Psychoanalysis at the Political Border*, Madison, Conn. 1996, pp. 85–86.

12. Ibid.

13. "The uncanny [*unheimlich*] is something which is secretly familiar [*heimlich-heimisch*], which has undergone repression and then returned from it..." (Freud, pp. 358 and 368).

14. "These two stanzas, these eight lines [of Van Hoddis's poem] seem to have transformed us into different beings, to have carried us up out of a world of apathetic bourgeoisie which we despised...." Becher would call it the "Marseillaise of the Expressionist Revolution" (Becher, p. 44).

15. Quoted in Carol Eliel, *The Apocalyptic Landscapes of Ludwig Meidner*, Los Angeles 1989, pp. 65 and 72.

16. Ernst Bloch, "The Anxiety of the Engineer," in *Literary Essays*, Stanford, Calif. 1998, pp. 306–8 and 312.

17. Ibid.

18. Ibid.

19. Ernst Bloch, *The Principle of Hope, Volume One*, Cambridge, Mass. 1986, pp. 434–35.

20. Paul Avrich, *Sacco and Vanzetti: The Anarchist Background*, Princeton, N.J. 1991, p. 205.

21. Dos Passos, *Manhattan Transfer*, pp. 366 and 404.

22. Wayne Barrett, *Rudy! An Investigative Biography of Rudolph Giuliani*, New York 2000, p. 2.

23. Ibid, p. 6.

24. Testimony paraphrased by Bill Keller ("Nuclear Nightmares," *New York Times Magazine*, 26 May 2002, p. 51).

25. Robert Fitch points out that the clearances for the WTC displaced 30,000 jobs and, through leveraging the development of adjacent Battery Park City, eliminated the critical lower Manhattan docks as well. "Something had gone seriously wrong with the priorities and politics of a city where 30,000 people can be made to disappear from their jobs and stores for a state office building [the WTC is owned by the Port Authority]" (Robert Fitch, *The Assassination of New York*, New York 1993, pp. 140–41.)

26. David Rothkopf, "Business Versus Terror," *Foreign Policy* (May/June 2002), p. 58.

27. Ibid., p. 56.

28. For a sobering discussion of the costs of protecting Times Square, Grand Central Station, St. Patrick's Cathedral, and the Statue of Liberty, see David Barstow, "Envisioning

an Expensive Future in the Brave New World of Fortress New York," *New York Times,* 16 September 2001.

29. Felix Rohatyn "Fiscal Disaster the City Can't Face Alone," *New York Times,* 9 October 2001.

30. *Business Week,* 19 November 2001, p. 45.

31. "In the new ethic of shared sacrifice, what sacrifice will he [the next mayor] ask of the business community? Neither Bloomberg nor Green could give me a satisfactory answer" (James Traub, "No-Fun City," *New York Times Magazine,* 4 November 2001, p. 41).

32. Robert Worth, "The Deep Intellectual Roots of Islamic Terror," *New York Times,* 13 October 2001; and Anthony Shadid, *Legacy of the Prophet: Despots, Democrats, and the New Politics of Islam,* Boulder, Colo. 2001, p. 58. For a balanced assessment of Qutb's thought—a fascinating combination of anarcho-humanism and Koranic chiliasm—see Ahmad Moussalli, *Moderate and Radical Islamic Fundamentalism,* Tallahassee, Fla. 1999, chapt. five.

33. Worth, ibid.; Judith Shulevitz, "The Close Reader: At War with the World," *New York Times Book Review,* 21 October 2001; and Shadid, p. 57. See also John Calvert, "The World Is an Undutiful Boy: Sayyid Qutb's American Experience," *Islam and Christian–Muslim Relations* 11, no. 1 (2000).

34. H. G. Wells, *The War of the Worlds,* London 1898, p. 1.

35. Tally of hate crimes from Council on American-Islamic Relations, 22 October; for the number of murders, see *Washington Post,* 26 October 2001.

36. Karen Olsson, "Letter from Dallas," *Texas Observer,* 12 October 2001.

37. Fanon, "Racist Fury in France," p. 163.

38. Walter Pincus, "Silence of 4 Terror Probe Suspects Poses Dilemma," *Washington Post,* 21 October 2001.

39. Bob Drogin and Greg Miller, as syndicated in the *Miami Herald,* 28 October 2001.

PART I

NEON WEST

James Mooney (1893)

1

'White People Are Only a Bad Dream …'

The whirlwind! The Whirlwind!
The whirlwind! The Whirlwind!
Ghost Dance chant

The Searcher

Mason Valley, Nevada: 1 January 1892.

It is a bitterly cold, moonless evening. A recent blizzard has mantled the sage-brush steppe in knee-deep snow. A small party of men, nervous and half-frozen, are riding in file, followed by a wagon. Their solitary lantern bathes the ground ahead in pale yellow light. The faint marking of a trail eventually dissolves into a criss-crossing maze of cattle paths in the snow. They have no idea which direction to go. The riders and their horses are tired and disoriented. The wind is beginning to howl ominously, and each man fights a small knot of panic in his stomach.

It is a bad night to be lost. In such conditions, cowboys stay indoors, shepherds abandon their flocks, posses give up pursuit, and outlaws freeze to death along lonely trails. But this party—three Paiutes and two white men—has a mission of unusual urgency. They are searching for a Messiah named Wovoka.

"After vainly following a dozen false trails and shouting repeatedly in the hope of hearing an answering cry," they try a desperate expedient. Using the frost-covered buckboard as a stationary point of reference, each man rides a short distance in a different direction. When they fail to find the trail, they move the wagon and start over again. Finally, the wagon-driver hears sounds. A few hundred yards farther on, the party finds four small tule wikiups. In one of them, the Messiah is quietly waiting by his fire.

The leader of the party is James Mooney, a self-taught linguist and anthropologist. He works for the Bureau of Ethnology, which has recently been moved from the US Geological Survey to the Smithsonian Institution. His boss is the legendary John Wesley Powell, one-armed Civil War hero and explorer of the Grand Canyon. The Bureau's somber, eleventh-hour mission is to bear scientific witness to the extinction of Native America. The nation's most eminent savants have written off the survival of most Indian cultures. Now Mooney and his colleagues are attempting to document as much as possible of these doomed lifeways before they are obliterated by white settlement and industrial progress.

Mooney has the perfect melancholy temperament for his job. An Irish nationalist from Ohio, he finds a profound parallel between the decline of the American Indians and the tragedy of the Celts. Like the Kiowa and Cherokee, whose languages he speaks fluently, Mooney's own tribe, the Gaelic-speaking Irish, are human anachronisms in an age of steel cities, commodity exchange, and Hotchkiss guns. His Indian informants and friends—with stirring names like Standing Bear, Fire Thunder, American Horse, George Sword, Black Coyote, and Sitting Bull—detect a sensibility in Mooney deeper than his whiteness. Perhaps he has quoted to them, in their own language, the bitter epitaph of Irish patriot John Mitchell: "The very nation that I knew as Ireland is broken and destroyed; and the place that knew it shall know it no more." At any event, his empathy is compelling and he has won admission to a world otherwise sealed to whites forever by massacres and broken promises. Vouchsafed by the honor of his famous friends, Mooney passes from tribe to tribe in a quest for information about a strange new religion.

Some weeks earlier, Mooney had been with the Lakota in South Dakota. Every doomed people dreams of magical rebirth. In the misery of their reservations, the Lakota, who only a decade earlier had been the most powerful horsemen on the continent, embraced a ceremony of renewal known as the "Ghost Dance." With the speed of hope, this prophecy of a world restored had spread from the wikiup in Mason Valley to every corner of indigenous America. Like a great bellows, it awoke the dying fires of Indian spiritual self-confidence and resistance.

Mooney would later see a resonance of Indian millenarianism in a Hibernian context. The Irish version of the Ghost Dance—no less fabulous and utopian—was the Celtic Revival of the 1890s, which, following the successive defeats of the Fenians, Davitt and then Parnell, sought to rebuild an Irish nation on the foundation of its forgotten culture. Its prophets and medicine men were named Yeats, Synge, Hyde, and Pearse. (And for those willing to make the conceptual journey, there really was an earlier point of contact between the Lakota and the Irish in the astonishing prairie rebellion of the Métis mystic and utopian socialist Louis Riel. He counted both Sitting Bull and the Irish Republican Brotherhood as allies in his struggle against English Canada.)

The Irish awakening eventually produced a small republic on a divided island. The Lakota, however, immediately paid a terrible price for their dreaming: 146 refugees from Pine Ridge, including 44 women and 18 children, were blown apart on the frozen banks of Wounded Knee Creek by the big explosive rounds of the Army's new Hotchkiss guns. (Some of the survivors, grotesquely, were then paraded across Europe by Buffalo Bill's Wild West Show.) Mooney, in between ethnographic errands for the World's Columbian Exposition in Chicago, is assembling the first comprehensive account of the massacre, which the US government still falsely characterizes as a "uprising."

Later, in the preface to his *The Ghost Dance Religion and the Sioux Outbreak of 1890*, published in 1896, he will note that he has traveled an astounding 32,000 miles in twenty-two months (1890–92) and spent time with twenty tribes. His painstakingly detailed account of the new religion's avowal of love and nonviolence, contrasted with photographs of slain Lakota women and children lying in heaps on the prairie, will morally impeach his own employer, the federal government, for dishonesty and murder. It will also ensure his own ostracism in official circles.

Wounded Knee happened on 29 December 1890. Almost exactly a year later, Mooney is sitting by a sagebrush fire with Wovoka.

The Messiah

Wovoka—the name means "the Man with the Axe"—is thirty-five years old. He is, ironically, the son of Tavibo, or "White Man," and when he was four he witnessed the famous battle of Pyramid Lake. White silver miners had kidnapped several Paiute women, and when their husbands rescued them, it was considered "an Indian outrage," although none of the miners was harmed. A large posse of whites was sent to destroy the Indian camp, but they were ambushed by the Paiutes in a narrow pass. With bows and arrows alone, Wovoka's band killed nearly fifty miners and forced the rest to flee in terror. Undefeated on the battlefield, the Paiute nonetheless lost their freedom to the inexorable expansion of the white mining and cattle frontiers. Thus, following the death of his father, Wovoka was indentured to a local rancher named David Wilson. Although he refuses to learn English or move into a house, he is now called "Jack Wilson" by the whites and is well-regarded as a reliable and hard-working ranchhand. He will spend his entire life tending cattle and sheep in Mason Valley.

Like other great prophets from Moses to Joseph Smith, his revelation occurred on a mountain. One day in late 1888 or early 1889, he was cutting wood for David Wilson when the sky began to darken. Looking up he saw the "sun dying" (an eclipse), followed by a great clamor in the trees. Laying down his axe, he ran in the direction of the tumult. He instantly "died" or passed into unconsciousness. Then, as he explains to Mooney, he "was taken up to the other world."

> Here he saw God, with all the people who had died long ago engaged in their old-time sports and occupations, all happy and forever young. It was a pleasant land and full of game. After showing him all, God told him he must go back and tell his people they must be good and love one another, have no quarreling, and live in peace with the whites.... they must put away all the old practices that savored of war; that if they faithfully obeyed his instructions they would at last be reunited with their friends in this other world, where there would be no death or sickness or old age [or, by implication, whites] ... (p. 772)

As a gift to the Indian people, God gave him a sacred dance. Performed at inter-vals, for five days at a time, the dance would beatify the performers and hasten the advent of the new time. Wovoka was also given powers to predict and control the weather. Mooney would later interview a former Indian agent at Walker Lake who claimed that Wovoka "had once requested him to draw up and forward to the President [Grover Cleveland] a statement of his supernatural claims, with a proposition that if he could receive a small regular stipend he would take up his residence on the reservation and agree to keep Nevada people informed of all the latest news from heaven and to furnish rain whenever wanted." The agent, chuckling to himself while pretending to take Wovoka seriously, never forwarded the letter as promised.

Local whites, including those who knew and liked "Jack Wilson," patroniz-ingly interpreted the revelation as little more than a hysterical reaction to the unexpected solar eclipse. To the Paiutes, on the other hand, it was a long-awaited sign. The first Ghost Dance took place on the Walker Lake reservation in Janu-ary 1889. It produced a spiritual earthquake half a continent in diameter. Soon, every native people of the Great Basin—the Washoo, Ute, Shoshone, Bannock, and Gosiute—had heard that "Christ" himself was at Walker Lake, teaching the Paiute a sacred dance that would restore the world of their fathers. These tribes spread the good news to their neighbors, who in turn passed it on to their neighbors. Delegations quickly arrived at Walker Lake from reservations in Cali-fornia, Utah, Idaho, Montana, and eventually Oklahoma, Nebraska, Iowa, and South Dakota. Within two years, nearly forty nations had entered Wovoka's holy circle. Only the Navajo, whose religion proscribes all mention of the dead, were immune to the resurrectionist fervor that swept the native West from the Pacific to the Mississippi.

Against scores of conflicting oral or second-hand accounts of Wovoka's teach-ing, there is an actual letter—published by Mooney—which the prophet dictated to an Arapahoe delegate in the party of Black Short Nose in 1891. The young Arapahoe wrote down Wovoka in "Carlisle English," which was then transcribed into proper English by Black Short Nose's schoolgirl daughter back in Oklahoma. As Mooney emphasizes, "it is the genuine official statement of the Ghost-dance doctrine as given by the messiah himself to his disciples." It is also a model of spiritual concision.

> Do not tell the white people about this. Jesus is now upon the earth. He appears like a cloud. The dead are all alive again. I do not know when they will be here; maybe this fall or in the spring. When the time comes there will be no more sickness and everyone will be young again.
>
> Do not refuse to work for the whites and do not make any trouble with them until you leave them. When the earth shakes—at the coming of the new world—do not be afraid. It will not hurt you.
>
> I want you to dance every six weeks. Make a feast at the dance and have food that everybody may eat. Then bathe in the water. That is all. You will receive good words again from me some time. Do not tell lies.

As Mooney reminded the readers of his report, the moral code preached by Wovoka is "as pure and comprehensive in its simplicity as anything found in religious systems from the days of Gautama Buddha to the time of Jesus Christ." Indeed, Mooney argues, the Ghost Dance is the New Testament, or at least its spiritual core, in distinctively Indian garb. To a divided and defeated people, Wovoka preaches unity, love, and the hope of renewal. "Only those who have known the deadly hatred that once animated Ute, Cheyenne, and Pawnee, one toward another, and are able to contrast it with their present spirit of mutual brotherly love can know what the Ghost-dance religion has accomplished. … It is such a revolution as comes but once in the life of a race."

To further illustrate the human meaning of the Ghost Dance, Mooney cites the example of an Arapahoe friend who had just lost his small son. "I shall not shoot any ponies, and my wife will not gash her arms. We used to do this when our friends died, because we thought we would never see them again, and it made us feel bad. But now we know we shall all be united again." As to the ceremony itself, especially the hypnotic role of dance, Mooney cautions his white readers to consider how some of their own religious practices might look to strangers. "In a country which produces magnetic healers, shakers, trance mediums, and the like, all these things may very easily be paralleled without going far from home."

The Mirage

On the centenary of Mooney's conversation with Wovoka, January 1991, I visited Walker Lake, Nevada. I had little in mind other than a simple desire to view the prophet's grave and the living conditions of his great-great-grandchildren. Initially I had no luck finding the headstone, so I sought advice from a young Paiute man in his early twenties. He was sitting in his red Ford Ranger pickup, drinking a cup of coffee and listening to Ice-T rapping, "Fuck the police...." He turned down the volume on his cassette player and gave me directions with a smile. Then he turned up Ice-T again.

On Wovoka's modest grave I found an abalone shell containing an eagle feather and several 30-30 shells. There were also some flowers, recently left. It was obvious that this was more than commemoration. The Ghost Dance religion did not die at Wounded Knee and Wovoka remains a living presence to many Native Americans: his spiritual legacy is dynamic and still undergoing elaboration.

Mooney had expressed wonderment at the diverse ways in which Wovoka's message had been assimilated in its fundamentals while being reworked in its details to fit the individual visions and specific histories of each Indian culture. The Arapahoe, for example, believed that the restored world would advance behind a wall of fire that would drive the whites back to Europe, while the Lakota, more embittered, had a vision of white civilization buried alive by earthquakes and landslides, with the survivors transformed into small fishes in the river. The Cheyenne preferred a native version of the Rapture, with Indian peoples ascending through the clouds to a beautiful hunting-ground, while the Shoshone foresaw a paradise where Indians and whites would dwell together in peace. In almost every version, however, the actual spiritual passageway to the new earth was the deep sleep induced by the Ghost Dance and lasting for four days. Upon awakening, the First People would realize that whites had simply been a bad dream.

The year after visiting Wovoka's grave, I had an opportunity to discuss his legacy with some of the Paiute and Shoshone activists who were sponsoring the "Global Healing" demonstrations at the Nevada Nuclear Test Site. With the nuclearized desert as a dramatic backdrop, they emphasized that while the immediate millenarian expectations of their great-grandparents (like the early Christians) had been frustrated, Wovoka's vision of a unified Indian people resuming stewardship of the West after white civilization has been destroyed by a cataclysm of its own making is more compelling than ever. Sweeping the horizon with his finger, one of them challenged me: "Do you really think all this can last?" His gesture was meant to encompass not only the Test Site, but all the chief monuments to the past century's work of conquest: the dams, casinos, instant suburbs, bombing ranges, prisons, theme parks, toxic dumps, trophy homes, and trailer parks.

The latterday Paiute and Shoshone, of course, themselves live in electrified homes, drive pickup trucks, send their kids to college, and lobby members of the Congress—but they do so with a keen awareness of the radical instability of this artificial world and its neon landscapes. It is a catastrophe to which they have painfully adjusted, which has transformed the outward trappings of their lives, but which they continue to resist inwardly. The essence of the Ghost Dance is, perhaps, precisely the moral stamina to outlast this great mirage.

The Prophecy

It is interesting to speculate what Frederick Jackson Turner was doing the day that Mooney met with Wovoka. As he sat down to New Year's dinner with his family in front of a warm hearth was he already ruminating over the famous speech that he would give two years later at the World's Columbian Exposition in Chicago? What had been his reaction to the lurid newspaper accounts of the recent Sioux uprising in South Dakota? Did he know who Wovoka was? Or Mooney?

I have never visited Turner's grave, so I don't know if anyone has placed fresh flowers there recently, much less seashells, eagle feathers, and Winchester cartridges. Yet his somber cult continues to endure. Historians have marched by his tomb for four generations now, and rare is the article or monograph, even in this postmodernist era, that doesn't doff its cap to the boss. True, there are other major traditions in the historiography of the West, most notably Bolton's school of comparative frontier studies and Innes's historical geographies of the staple trades. Enough intermarriage, moreover, has occurred within the field to produce revisionists who are still American exceptionalists and Turnerians who study cities and commodity circuits. But virtually everyone working in the field, whether their theoretical genealogy begins in Berkeley, Toronto, Madison, or even Paris, and whether they mourn with the victims or gloat with the victors, accepts the evolutionary pathway from frontier to region, from colonial periphery to Sunbelt. Virtually by definition, they acknowledge a certain stable core of regional identity and historical continuity.

Everyone, that is, except the heirs to Wovoka. They reject the *telos* of the finished product, the conquered landscape, the linear historical narrative, the managed ecosystem. They see a chaos more ontological than the boom-and-bust cycle. They know the supposedly "permanent" structures of tradition and meaning in the white West seldom endure more than a single generation before they are overthrown and replaced. Like a certain German philosopher, they are all too aware that "all that is solid melts into air," including our most dearly held conceptions of the West as a region.

Wovoka, in other words, sustains his great-great grandchildren with an apocalyptic vision of the history of the American West. Since "apocalypse" is such an over-used and cheapened term, it is important to recall its precise meaning in the Abrahamic religions. An apocalypse is literally the revelation of the Secret History of the world as becomes possible under the terrible clarity of the Last Days. It is the alternate, despised history of the subaltern classes, the defeated peoples, the extinct cultures. I am claiming, in other words, that Wovoka offers us a neo-catastrophist epistemology for reinterpreting Western history from the standpoint of certain terminal features of the approaching millennial landscape. He invites us to reopen that history from the vantage-point of an already visible future when sprawl, garbage, addiction, violence, and simulation will have overwhelmed every vital life-space west of the Rockies. This is Turnerian history, if you will, stripped down to its ultimate paranoia: the West become Los Angeles.

For those who retain the Ghost Dance tradition, this end point is also paradoxically the point of renewal and restoration. It is through this black hole that the West will disappear into the singularity of catastrophe, only to reemerge, on the other side, with streams full of salmon and plains black with bison.

1999

Richard Misrach, *Dead Animals #327*

2

Ecocide in Marlboro Country

Was the Cold War the Earth's worst eco-disaster in the last ten thousand years? The time has come to weigh the environmental costs of the great "twilight struggle" and its attendant nuclear arms race. Until recently, most ecologists have underestimated the impact of warfare and arms production on natural history.[1] Yet there is implacable evidence that huge areas of Eurasia and North America, particularly the militarized deserts of Central Asia and the Great Basin, have become unfit for human habitation, perhaps for thousands of years, as a direct result of weapons testing (conventional, nuclear, and biological) by the Soviet Union, China, and the United States.

Part One: Portraits of Hell

These "national sacrifice zones,"[2] now barely recognizable as parts of the biosphere, are also the homelands of indigenous cultures (Kazakh, Paiute, Shoshone, among others) whose peoples may have suffered irreparable genetic damage. Millions of others—soldiers, armament workers, and "downwind" civilians—have become the silent casualties of atomic plagues. If, at the end of the old superpower era, a global nuclear apocalypse was finally averted, it was only at the cost of these secret holocausts.[3]

This hidden history has come unraveled most dramatically in the ex-Soviet Union where environmental and anti-nuclear activism, first stimulated by Chernobyl in 1986, emerged massively during the crisis of 1990–91. Grassroots protests by miners, schoolchildren, health-care workers, and indigenous peoples forced official disclosures that confirmed the sensational accusations by earlier *samizdat* writers like Zhores Medvedev and Boris Komarov (Ze'ev Wolfson). *Izvestiya* finally printed chilling accounts of the 1957 nuclear catastrophe in the secret military city of Chelyabinsk-40, as well as the poisoning of Lake Baikal by a military factory complex. Even the glacial wall of silence around radiation accidents at the Semipalatinsk "Polygon," the chief Soviet nuclear test range in Kazakhstan, began to melt.[4]

As a result, the ex-Soviet public now has a more ample and honest view than their American or British counterparts of the ecological and human costs of the Cold War. Indeed, the Russian Academy of Sciences has compiled an extraordinary map that shows environmental degradation of "irreparable, catastrophic proportions" in forty-five different areas, comprising no less than 3.3 percent of the surface area of the former USSR. Not surprisingly, much of the devastation is concentrated in those parts of the southern Urals and Central Asia that were the geographical core of the USSR's nuclear military-industrial complex.[5]

Veteran kremlinologists, in slightly uncomfortable green disguises, have fastened on these revelations to write scathing epitaphs for the USSR. According to Radio Liberty and Rand researcher D.J. Peterson, "the destruction of nature had come to serve as a solemn metaphor for the decline of a nation."[6] For Lord Carrington's ex-advisor Murray Feshbach, and his literary sidekick Al Friendly (ex-*Newsweek* bureau chief in Moscow), on the other hand, the relationship between ecological cataclysm and the disintegration of the USSR is more than metaphor: "When historians finally conduct an autopsy on the Soviet Union and Soviet Communism, they may reach the verdict of death by ecocide."[7]

Peterson's *Troubled Lands* and especially Feshbach and Friendly's *Ecocide in the USSR* have received spectacular publicity in the American media. Exploiting the new, uncensored wealth of Russian-language sources, they describe an environmental crisis of biblical proportions. The former Land of the Soviets is as a dystopia of polluted lakes, poisoned crops, toxic cities, and sick children. What Stalinist heavy industry and mindless cotton monoculture have not ruined, the Soviet military has managed to bomb or irradiate. For Peterson, this "ecological terrorism" is conclusive proof of the irrationality of a society lacking a market mechanism to properly "value" nature. Weighing the chances of any environmental cleanup, he holds out only the grim hope that economic collapse and radical de-industrialization may rid Russia and the Ukraine of their worst polluters.[8]

Pentagon eco-freaks Feshbach and Friendly are even more unsparing. Bolshevism, it seems, has been a deliberate conspiracy against Gaia, as well as against humanity. "Ecocide in the USSR stems from the force, not the failure, of utopian ambitions." It is the "ultimate expression of the Revolution's physical and spiritual brutality." With Old Testament righteousness, they repeat the opinion that "there is no worse ecological situation on the planet."[9]

Obviously Feshbach and Friendly have never been to Nevada or western Utah.[10] The environmental horrors of Chelyabinsk-40 and the Semipalatinsk Polygon have their eerie counterparts in the poisoned, terminal landscapes of Marlboro Country.

Misrach's Inferno

A horse head extrudes from a haphazardly bulldozed mass grave. A dead colt—its forelegs raised gracefully as in a gallop—lies in the embrace of its mother. Albino tumbleweed are strewn randomly atop a tangled pyramid of rotting cattle, sheep, horses, and wild mustangs. Bloated by decay, the whole cadaverous mass seems to be struggling to rise. A Minoan bull pokes its eyeless head from the sand. A weird, almost Jurassic skeleton—except for a hoof, it might be the remains of a pterosaur—is sprawled next to a rusty pool of unspeakable vileness. The desert reeks of putrefaction.

Photographer Richard Misrach shot this sequence of 8 x 10 color photographs in 1985–87 at various dead-animal disposal sites located near reputed plutonium "hot spots" and military toxic dumps in Nevada. As a short text explains, it is commonplace for local livestock to die mysteriously, or give birth to monstrous offspring. Ranchers are officially encouraged to dump the cadavers, no questions offspring. Ranchers are officially encouraged to dump the cadavers, no questions asked, in unmarked county-run pits. Misrach originally heard of this "Boschlike" landscape from a Paiute poet. When he asked for directions, he was advised to drive into the desert and watch for flocks of crows. The carrion birds feast on the eyes of dead livestock.[11]

"The Pit" has been compared to Picasso's *Guernica*. It is certainly a nightmare reconfiguration of traditional cowboy clichés. The lush photographs are repellent, elegiac, and hypnotic at the same time. Misrach may have produced the single most disturbing image of the American West since ethnologist James Mooney countered Frederic Remington's popular paintings of heroic cavalry charges with stark photographs of the frozen corpses of Indian women and children slaughtered by the Seventh Cavalry's Hotchkiss guns at Wounded Knee in 1890.[12]

But this holocaust of beasts is only one installment ("Canto VI") in a huge mural of forbidden visions called *Desert Cantos*. Misrach is a connoisseur of trespass who since the late 1970s has penetrated some of the most secretive spaces of the Pentagon Desert in California, Nevada, and Utah. Each of his fourteen completed cantos (the work is still in progress) builds drama around a "found metaphor" that dissolves the boundary between documentary and allegory. Invariably there is an unsettling tension between the violence of the images and the elegance of their composition.

The earliest cantos (his "desert noir" period?) were formal aesthetic experiments influenced by readings in various cabalistic sources. They are mysterious phantasmagorias detached from any explicit sociopolitical context: the desert on fire, a drowned gazebo in the Salton Sea, a palm being swallowed by a sinister sand dune, and so on.[13] By the mid 1980s, however, Misrach put aside Blake and Castaneda, and began to produce politically engaged exposés of the Cold War's impact upon the American West. Focusing on Nevada, where the military controls 4 million acres of land and 70 percent of the airspace, he was fascinated by the strange stories told by angry ranchers: "night raids by Navy helicopters, laser-burned cows, the bombing of historic towns, and unbearable supersonic flights." With the help of two improbable anti-Pentagon activists, a small-town physician named Doc Bargen and a gritty bush pilot named Dick Holmes, Misrach spent eighteen months photographing a huge tract of public land in central Nevada that had been bombed, illegally and continuously, for almost forty years. To the Navy this landscape of almost incomprehensible devastation, sown with live ammo and unexploded warheads, is simply "Bravo 20." To Misrach, on the other hand, it is "the epicenter … the heart of the apocalypse":

> It was the most graphically ravaged environment I had ever seen…. I wandered for hours amongst the craters. There were thousands of them. Some were small, shallow pits the size of a bathtub, others were gargantuan excavations as large as a suburban two-car garage. Some were bone dry, with walls of "traumatized earth" splatterings, others were eerie pools of blood-red or emerald-green water. Some had crystallized into strange salt formations. Some were decorated with the remains of blown-up jeeps, tanks, and trucks.[14]

Although Misrach's photographs of the pulverized public domain, published in 1990, riveted national attention on the bombing of the West, it was a bitter-sweet achievement. His pilot friend Dick Holmes, whom he had photographed raising the American flag over a lunarized hill in a delicious parody on the Apollo astronauts, was killed in an inexplicable plane crash. The Bush administration, meanwhile, accelerated the modernization of bombing ranges in Nevada, Utah, and Idaho. Huge swathes of the remote West, including Bravo 20, have been updated into electronically scored, multitarget grids, which from space must now look like a colossal Pentagon video gameboard.

In his most recent collection of cantos, *Violent Legacies* (which includes "The Pit"), Misrach offers a haunting, visual archeology of "Project W-47," the super-secret final assembly and flight testing of the bombs dropped on Hiroshima and Nagasaki. The hangar that housed the *Enola Gay* still stands (a sign warns: "Use of deadly force authorized") amid the ruins of Wendover Air Base in the Great Salt Desert of Utah. In the context of incipient genocide, the fossil flight-crew humor of 1945 is unnerving. Thus a fading slogan over the A-bomb assembly building reads BLOOD, SWEAT AND BEERS, while graffiti on the administrative head-quarters commands EAT MY FALLOUT. The rest of the base complex, including the atomic bomb storage bunkers and loading pits, has eroded into megalithic abstractions that evoke the ground-zero helter-skelter of J. G. Ballard's famous short story "The Terminal Beach." Outlined against ochre desert mountains (the Newfoundland Range, I believe), the forgotten architecture and casual detritus of the first nuclear war are almost beautiful.[15]

In cultivating a neo-pictorialist style, Misrach plays subtle tricks on the sublime. He can look Kurtz's Horror straight in the face and make a picture postcard of it. This attention to the aesthetics of murder infuriates some partisans of traditional black-and-white political documentarism, but it also explains Misrach's extraordinary popularity. He reveals the terrible, hypnotizing beauty of Nature in its death throes, of Landscape as Inferno. We have no choice but to look.

If there is little precedent for this in previous photography of the American West, it has a rich resonance in contemporary—especially Latin American—political fiction. Discussing the role of folk apocalypticism in the novels of García Márquez and Carlos Fuentes, Lois Zamora inadvertently supplies an apt characterization of *Desert Cantos:*

> The literary devices of biblical apocalypse and magical realism coincide in their hyperbolic narration and in their *surreal images of utter chaos and unutterable perfection.* And in both cases, [this] surrealism is not principally conceived for psychological effect, as in earlier European examples of the mode, but is instead grounded in social and political realities and is designed to communicate the writers' objections to those realities.[16]

Resurveying the West

Just as Marquez and Fuentes have led us through the hallucinatory labyrinth of modern Latin American history, so Misrach has become an indefatigable tour guide to the Apocalyptic Kingdom that the Department of Defense has built in the desert West. His vision is singular, yet, at the same time, *Desert Cantos* claims charter membership in a broader movement of politicized western landscape photography that has made the destruction of nature its dominant theme.

Its separate detachments over the last fifteen years have included, first, the so-called New Topographics in the mid 1970s (Lewis Baltz, Robert Adams, and Joel

Deal),[17] closely followed by the Rephotographic Survey Project (Mark Klett and colleagues),[18] and, then, in 1987, by the explicitly activist Atomic Photographers Guild (Robert Del Tredici, Carole Gallagher, Peter Goin, Patrick Nagatani, and twelve others).[19] If each of these moments has had its own artistic virtue (and pretension), they share a common framework of revisionist principles.

In the first place, they have mounted a frontal attack on the hegemony of Ansel Adams, the dead pope of the "Sierra Club school" of Nature-as-God photography. Adams, if necessary, doctored his negatives to remove any evidence of human presence from his apotheosized wilderness vistas.[20] The new generation has rudely deconstructed this myth of a virginal, if imperilled nature. They have rejected Adams's Manichean division between "sacred" and "profane" landscape, which "leaves the already altered and inhabited parts of our environment dangerously open to uncontrolled exploitation."[21] Their West, by contrast, is an irrevocably social landscape, transformed by militarism, urbanization, the interstate highway, epidemic vandalism, mass tourism, and the extractive industries' boom-and-bust cycles. Even in the "last wild places," the remote ranges and lost box canyons, the Pentagon's jets are always overhead.

Secondly, the new generation has created an alternative iconography around such characteristic but previously spurned or "unphotographable" objects as industrial debris, rock graffiti, mutilated saguaros, bulldozer tracks, discarded girlie magazines, military shrapnel, and dead animals.[22] Like the surrealists, they have recognized the oracular and critical potencies of the commonplace, the discarded, and the ugly."[23] But as environmentalists, they also understand the fate of the rural West as the national dumping ground.

Finally, their projects derive historical authority from a shared benchmark: the photographic archive of the great nineteenth-century scientific and topographic surveys of the intermontane West. Indeed, most of them have acknowledged the centrality of "resurvey" as strategy or metaphor. The New Topographers, by their very name, declared an allegiance to the scientific detachment and geological clarity of Timothy O'Sullivan (famed photographer for Clarence King's 1870s survey of the Great Basin), as they turned their cameras on the suburban wastelands of the New West. The Rephotographers "animated" the dislocations from past to present by painstakingly assuming the exact camera stances of their predecessors and producing the same scene a hundred years later. Meanwhile, the Atomic Photographers, in emulation of the old scientific surveys, have produced increasingly precise studies of the landscape tectonics of nuclear testing.

Resurvey, of course, presumes a crisis of definition, and it is interesting to speculate why the new photography, in its struggle to capture the meaning of the postmodern West, has been so obsessed with nineteenth-century images and canons. It is not because, as might otherwise be imagined, Timothy O'Sullivan and his colleagues were able to see the West pristine and unspoiled. As Klett's "rephotographs" startlingly demonstrate, the grubby hands of Manifest Destiny were already all over the landscape by 1870. What was more important was the exceptional scientific and artistic integrity with which the surveys confronted landscapes that, as Jan Zita Grover suggests, were culturally "unreadable."[24]

The regions that today constitute the Pentagon's "national sacrifice zone" (the Great Basin of eastern California, Nevada, and western Utah) and its "plutonium periphery" (the Columbia-Snake Plateau, the Wyoming Basin and the Colorado Plateau) have few landscape analogues anywhere else on earth."[25] Early accounts of the intermontane West in the 1840s and 1850s (John Frémont, Sir Richard Burton, the Pacific railroad surveys) chipped away eclectically, with little success, at the towering popular abstraction of "the Great American Desert." Nevada and Utah, for instance, were variously compared to Arabia, Turkestan, the Takla Makan, Timbuktu, Australia, and so on, but in reality, Victorian minds were traveling through an essentially extraterrestrial terrain, far outside their cultural experience.[26] (Perhaps literally so, since planetary geologists now study lunar and Martian land-forms by analogy with strikingly similar landscapes in the Colorado and Columbia-Snake River plateaus.)[27]

The bold stance of the survey geologists, their artists and photographers, was to face this radical "Otherness" on its own terms.[28] Like Darwin in the Galapagos, John Wesley Powell and his colleagues (especially Clarence Dutton and the great Carl Grove Gilbert) eventually cast aside a trunkful of Victorian preconceptions in order to recognize novel forms and processes in nature. Thus Powell and Gilbert had to invent a new science, geomorphology, to explain the amazing landscape system of the Colorado Plateau where rivers were often "antecedent" to high-lands and the "laccolithic" mountains were really impotent volcanoes. (Similarly, decades later, another quiet revolutionary in the survey tradition, Harlen Bretz, would jettison uniformitarian geological orthodoxy in order to show that cata-clysmic Ice Age floods were responsible for the strange "channeled scablands" carved into the lava of the Columbia Plateau.)[29]

If the surveys "brought the strange spires, majestic cliff facades, and fabulous canyons into the realm of scientific explanation," then (notes Gilbert's biogra-pher), they "also gave them a critical aesthetic meaning" through the stunning photographs, drawings, and narratives that accompanied and expanded the technical reports.[30] Thus Timothy O'Sullivan (who with Mathew Brady had pho-tographed the ranks of death at Gettysburg) abandoned the Ruskinian paradigms of nature representation to concentrate on naked, essential form in a way that presaged modernism. His "stark planes, the seemingly two-dimensional curtain walls, [had] no immediate parallel in the history of art and photography. No one before had seen the wilderness in such abstract and architectural forms."[31] Simi-larly Clarence Dutton, "the *genius loci* of the Grand Canyon," created a new land-scape language—also largely architectural, but sometimes phantasmagorical—to describe an unprecedented dialectics of rock, color, and light. (Wallace Stegner says he "aestheticized geology"; perhaps, more accurately, he eroticized it.)[32]

But this convergence of science and sensibility (which has no real twentieth-century counterpart) also compelled a moral view of the environment as it was laid bare for exploitation. Setting a precedent which few of his modern descen-dants have had the guts to follow, Powell, the one-armed Civil War hero, laid out the political implications of the western surveys with exacting honesty in his famous 1877 *Report on the Lands of the Arid Region.* His message, which Stegner has called "revolutionary" (and others "socialistic"), was that the intermontane region's only salvation was Cooperativism based on the communal management and conservation of scarce pasture and water resources. Capitalism pure and simple, Powell implied, would destroy the West.[33]

The surveys, then, were not just another episode in measuring the West for conquest and pillage; they were, rather, an autonomous moment in the history of

American science when radical new perceptions temporarily created a pathway for a utopian alternative to the future that became Project W-47 and The Pit. That vantage point is now extinct. In reclaiming this tradition, contemporary photographers have elected to fashion their own clarity without the aid of the Victorian optimism that led Powell into the chasms of the Colorado. But "Resurvey," if a resonant slogan, is a diffuse mandate. For some it has meant little more than checking to see if the boulders have moved after a hundred years. For others, however, it has entailed perilous moral journeys deep into the interior landscapes of the Bomb.

Jellyfish Babies

If Richard Misrach has seen "the heart of the apocalypse" at Bravo 20, Carole Gallagher has spent a decade at "'American Ground Zero'" (the title of her new book) in Nevada and southwestern Utah photographing and collecting the stories of its victims.[34] She is one of the founders of the Atomic Photographers Guild, arguably the most important social-documentary collaboration since the 1930s, when Roy Stryker's Farm Security Administration Photography Unit brought together the famous lenses of Walker Evans, Dorothea Lange, Ben Shahn, Russell Lee, and Arthur Rothstein. Just as the FSA photographers dramatized the plight of the rural poor during the Depression, so the Guild has endeavored to document the human and ecological costs of the nuclear arms race. Its accomplishments include Peter Goin's revelatory *Nuclear Landscapes* (photographed at test sites in the American West and the Marshall Islands) and Robert Del Tredici's biting exposé of nuclear manufacture, *At Work in the Fields of the Bomb*.[35]

But it is Gallagher's work that proclaims the most explicit continuity with the FSA tradition, particularly with Dorothea Lange's classical black-and-white portraiture. Indeed, she prefaces her book with a meditation on a Lange motto and incorporates some haunting Lange photographs of St. George, Utah, in 1953. There is no doubt that *American Ground Zero* is intended to stand on the same shelf with such New Deal–era classics as *An American Exodus, Let Us Now Praise Famous Men,* and *You Have Seen Their Faces*.[36] Hers, however, is a more painful book.

In the early 1980s, Gallagher moved from New York City to St. George to work full-time on her oral history of the casualties of the American nuclear test program. Beginning with its first nuclear detonation in 1951, this small Mormon city, due east of the Nevada Test Site, has been shrouded in radiation debris from scores of atmospheric and accidentally "ventilated" underground blasts. Each lethal cloud was the equivalent of billions of x-rays and contained more radiation than was released at Chernobyl in 1988. Moreover, the Atomic Energy Commission (AEC) in the 1950s had deliberately planned for fallout to blow over the St. George region in order to avoid Las Vegas and Los Angeles. In the icy, Himmlerian jargon of a secret AEC memo unearthed by Gallagher, the targeted communities were "a low-use segment of the population."[37]

As a direct result, this downwind population (exposed to the fallout equivalent of perhaps fifty Hiroshimas) is being eaten away by cumulative cancers, neurological disorders, and genetic defects. Gallagher, for instance, talks about her quiet dread of going into the local K-Mart and "seeing four- and five-year-old children wearing wigs, deathly pale and obviously in chemotherapy."[38] But such horror has become routinized in a region where cancer is so densely clustered that virtually any resident can matter-of-factly rattle off long lists of tumorous or deceased friends and family. The eighty-some voices—former Nevada Test Site workers and "atomic GIs" as well as downwinders—that comprise *American Ground Zero* are weary with the minutiae of pain and death.

In most of these individual's stories, a single moment of recognition distills the terror and awe of the catastrophe that has enveloped their lives. For example, two military veterans of shot Hood (a 74-kiloton hydrogen bomb detonated in July 1957) recall the vision of hell they encountered in the Nevada desert:

> We'd only gone a short way when one of my men said, "Jesus Christ, look at that!" I looked where he was pointing, and what I saw horrified me. There were people in a stockade—a chain-link fence with barbed wire on top of it. Their hair was falling out and their skin seemed to be peeling off. They were wearing blue denim trousers but no shirts....
>
> I was happy, full of life before I saw that bomb, but then I understood evil and was never the same. ... I seen how the world can end.[39]

For sheep ranchers it was the unsettling spectacle they watched season after season in their lambing sheds as irradiated ewes attempted to give birth: "Have you ever seen a five-legged lamb?" For one husband, on the other hand, it was simply watching his wife wash her hair.

> Four weeks after that [the atomic test] I was sittin' in the front room reading the paper and she'd gone into the bathroom to wash her hair. All at once she let out the most ungodly scream, and I run in there and there's about half her hair layin' in the washbasin! You can imagine a woman with beautiful, raven-black hair, so black it would glint green in the sunlight just like a raven's wing, and it was long hair down onto her shoulders. There was half of it in the basin and she was as bald as old Yul Brynner.[41]

Perhaps most bone-chilling, even more than the anguished accounts of small children dying from leukemia, are the stories about the "jellyfish babies": irradiated fetuses that developed into grotesque hydatidiform moles.

> I remember being worried because they said the cows would eat the hay and all this fallout had covered it and through the milk they would get radioactive iodine. ... From four to about six months I kept a-wondering because I hadn't felt any kicks. ... I hadn't progressed to the size of a normal pregnancy and the doctor gave me a sonogram. He couldn't see any form of a baby.... He did a D and C. My husband was there and he showed him what he had taken out of my uterus. There were little grapelike cysts. My husband said it looked like a bunch of peeled grapes.[42]

The ordinary Americans who lived, and still live, these nightmares are rendered in great dignity in Gallagher's photographs. But she cannot suppress her frustration with the passivity of so many of the Mormon downwinders. Their unquestioning submission to a Cold War government in Washington and an authoritarian church hierarchy in Salt Lake City disabled effective protest through the long decades of contamination. To the cynical atomocrats in the AEC, they were just gullible hicks in the sticks, suckers for soapy reassurances and idiot "the atom is your friend" propaganda films. As one subject recalled his Utah childhood: "I remember in school they showed a film once called *A is for Atom, B is for the Bomb*. I think most of us who grew up in that period ... [have now] added C *is for Cancer*.

D is for Death."[43]

Indeed, most of the people interviewed by Gallagher seem to have had a harder time coming to grips with government deception than with cancer. Ironically, Washington waged its secret nuclear war against the most patriotic cross-section of the population imaginable, a virtual Norman Rockwell tapestry of Americana: gung-ho Marines, ultraloyal Test Site workers, Nevada cowboys and tungsten miners, Mormon farmers, and freckle-faced Utah schoolchildren. For forty years the Atomic Energy Commission and its successor, the Department of Energy, have lied about exposure levels, covered up Chernobyl-sized accidents, suppressed research on the contamination of the milk supply, ruined the reputation of dissident scientists, abducted hundreds of body parts from victims, and conducted a ruthless legal battle to deny compensation to the downwinders. A 1980 congressional study accused the agencies of "fraud upon the court," but Gallagher uses a stronger word—"genocide"—and reminds us that "lack of vigilance and control of the weaponeers" has morally and economically "played a large role in bankrupting ... not just one superpower but two."[45]

And what has been the ultimate cost? For decades the AEC cover-up prevented the accumulation of statistics or the initiation of research that might provide some minimal parameters. However, an unpublished report by a Carter administration taskforce (quoted by Philip Fradkin) determined that 170,000 people had been exposed to contamination within a 250-mile radius of the Nevada Test Site. In addition, roughly 250,000 servicemen, some of them cowering in trenches a few thousand yards from ground zero, took part in atomic war games in Nevada and the Marshall Islands during the 1950s and early 1960s. Together with the Test Site workforce, then, it is reasonable to estimate that at least 500,000 people were exposed to intense, short-range effects of nuclear detonation. (For comparison, this is the *maximum* figure quoted by students of the fallout effects from tests at the Semipalatinsk Polygon.)[46]

But these figures are barely suggestive of the real scale of nuclear toxicity. Another million Americans have worked in nuclear weapons plants since 1945, and some of these plants, especially the giant Hanford complex in Washington, have contaminated their environments with secret, deadly emissions, including radioactive iodine.[47] Most of the urban Midwest and Northeast, moreover, was downwind of the 1950s atmospheric tests, and storm fronts frequently dumped carcinogenic, radioisotope "hot spots" as far east as New York City. As the commander of the elite Air Force squadron responsible for monitoring the nuclear test clouds during the 1950s told Gallagher (he was suffering from cancer): "There isn't anybody in the United States who isn't a downwinder. ... When we followed the clouds, we went all over the United States from east to west. ... Where are you going to draw the line?"[48]

Part Two: Healing Global Wounds

Yet over the last decade Native Americans, ranchers, peace activists, downwinders, and even members of the normally conservative Mormon establishment have attempted to draw a firm line against further weapons testing, radiation poisoning, and ecocide in the deserts of Nevada and Utah. The three short field reports that follow (written in 1992, 1993, and 1996–97) are snapshots of the most extraordinary social movement to emerge in the postwar West.

Humbling "Mighty Uncle"

Flash back to fall 1992. The (private) Wackenhut guards at the main gate of the Nevada Nuclear Test Site (NTS) nervously adjust the visors on their riot helmets and fidget with their batons. One block away, just beyond the permanent traffic sign that warns "Watch for Demonstrators!" a thousand antinuclear protestors, tie-dyed banners unfurled, are approaching at a funeral pace to the sombre beat of a drum.

The unlikely leader of this youthful army is a rugged-looking rancher from the Ruby Mountains named Raymond Yowell. With a barrel chest that strains against his pearl-button shirt, and calloused hands that have roped a thousand mustangs, he makes the Marlboro Man seem wimpy. But if you look closely, you will notice a sacred eagle feather in his Stetson. Mr. Yowell is chief of the Western Shoshone National Council.

When an official warns protestors that they will be arrested if they cross the cattleguard that demarcates the boundary of the Test Site, Chief Yowell scowls that it is the Department of Energy who is trespassing on sacred Shoshone land. "We would be obliged," he says firmly, "if *you* would leave. And please take your damn nuclear waste and rent-a-cops with you."

While Chief Yowell is being handcuffed at the main gate, scores of protestors are breaking through the perimeter fence and fanning out across the desert. They are chased like rabbits by armed Wackenhuts in fast, low-slung dunebuggies. Some try to hide behind Joshua trees, but all will be eventually caught and returned to the concrete and razor-wire compound that serves as the Test Site's hoosegow. It is 11 October, the day before the quincentenary of Columbus's crash landing in the New World.

The US nuclear test program has been under almost constant siege since the Las Vegas–based American Peace Test (a direct-action offshoot of the old Moratorium) first encamped outside the NTS's Mercury gate in 1987. Since then more than ten thousand people have been arrested at APT mass demonstrations or in smaller actions ranging from Quaker prayer vigils to Greenpeace commando raids on ground zero itself. (In *Violent Legacies* Misrach includes a wonderful photograph of the "Princesses Against Plutonium," attired in radiation suits and death masks, illegally camped inside the NTS perimeter.) Dodging the Wackenhuts in the Nevada desert has become the rite of passage for a new generation of peace activists.

The fall 1992 Test Site mobilization—"Healing Global Wounds"—was a watershed in the history of antinuclear protest. In the first place, the action coincided with Congress's nine-month moratorium on nuclear testing (postponing until this September a test blast codenamed "Mighty Uncle"). At long last, the movement's strategic goal, a comprehensive test ban treaty, seemed tantalizingly within grasp. Secondly, the leadership within the movement has begun to be assumed by the indigenous peoples whose lands have been poisoned by nearly a half century of nuclear testing.

These two developments have a fascinating international connection. Washington's moratorium was a grudging response to Moscow's earlier, unilateral cessation of testing, while the Russian initiative was coerced from Yeltsin by unprecedented popular pressure. The revelation of a major nuclear accident at the Polygon in February 1989 provoked a nonviolent uprising in Kazakhstan. The famed writer Olzhas Suleimenov used a televised poetry reading to urge Kazakhs to emulate the example of the Nevada demonstrations. Tens of thousands of protestors, some brandishing photographs of family members killed by cancer, flooded the streets of Semipalatinsk and Alma-Ata, and within a year the "Nevada-Semipalatinsk Movement" had become "the largest and most influential public organization in Kazakhstan, drawing its support from a broad range of people—from the intelligentsia to the working class."[49] Two years later, the Kazakh Supreme Soviet, as part of its declaration of independence, banned nuclear testing forever.

It was the world's first successful antinuclear revolution, and its organizers tried to spread its spirit with the formation of the Global Anti-Nuclear Alliance (GANA). They specifically hoped to reach out to other indigenous nations and communities victimized by nuclear colonialism. The Western Shoshones were among the first to respond. Unlike many other western tribes, Chief Yowell's people have never conceded US sovereignty in the Great Basin of Nevada and Utah, and even insist on carrying their own national passport when traveling abroad. In conversations with the Kazakhs and activists from the Pacific test sites, they discovered a poignant kinship that eventually led to the joint GANA-Shoshone sponsorship of "Healing Global Wounds" with its twin demands to end nuclear testing and restore native land rights.

In the past some participants had criticized the American Peace Test encampments for their overwhelmingly countercultural character. Indeed, last October as usual, the bulletin board at the camp's entrance gave directions to affinity groups, massage tables, brown rice, and karmic enhancements. But the Grateful Dead ambience was leavened by the presence of an authentic Great Basin united front that included Mormon and Paiute Indian downwinders from the St. George area, former GIs exposed to the 1950s atmospheric tests, Nevada ranchers struggling to demilitarize public land (Citizen Alert), a representative of workers poisoned by plutonium at the giant Hanford nuclear plant, and the Reese River Valley Rosses, a Shoshone country-western band. In addition there were friends from Kazakhstan and Mururoa, as well as a footsore regiment of European cross-continent peace marchers.

The defeat of George Bush a month after "Healing Global Wounds" solidified optimism in the peace movement that the congressional moratorium would become a permanent test ban. The days of the Nevada Test Site seemed numbered. Yet to the dismay of the Western Shoshones, the downwinders and the rest of the peace community, the new Democratic administration evinced immodest enthusiasm for the ardent wooing of the powerful nuclear-industrial complex. Cheered on by the Tory regime in London, which was eager to test the nuclear warhead for the RAF's new "TASM" missile in the Nevada desert, the Pentagon and the three giant atomic labs (Livermore, Los Alamos, and Sandia) came within a hairsbreadth of convincing Clinton to resume "Mighty Uncle." Only a last-minute revolt by twenty-three senators—alarmed that further testing might undermine the US-led crusade against incipient nuclear powers like Iraq and North Korea—forced the White House to extend the moratorium.

Although the ban has remained in force, there is some evidence that the Pentagon has participated in tests by proxy in French Polynesia. In 1995 the White House, breaking with the policy of the Bush years, allowed the French military to airlift H-bomb components to Mururoa through US airspace, using LAX as a stopover point. The British Labour Party, echoing accounts in the French press, charged that Washington and London were silent partners in the internationally denounced Mururoa test series, sharing French data while providing Paris with logistical and diplomatic support.[50]

More recently, the spotlight has shifted back to Nevada where peace activists in spring 1997 were preparing for protests against NTS's new program of "zero yield" tests. The Department of Energy is planning to use high explosives to compress "old" plutonium to the brink of chain reaction, an open violation of the Comprehensive Test Ban agreement, in order to generate data for a computer study of "the effects of age on nuclear weapons." This is part of the Clinton administration's science-based Stockpile Stewardship program, which, critics allege, has merely shifted the nuclear arms race into high-tech labs like Livermore's $1 billion National Ignition Facility where superlasers will produce miniature nuclear explosions that will in turn be studied by the next generation of "teraflop" (1 trillion calculations per second) supercomputers. Great Basin peace activists, like their Global Healing counterparts fear that such "virtual atomic tests," combined with data from "zero yield" blasts in Nevada, will encourage not only the maintenance, but the further development of strategic nuclear weapons. In the meantime, motorists will still have to "Stop for Demonstrators" at the Mercury exit.[51]

The Death Lab

January 1993. It has been one of the coldest winters in memory in the Great Basin. Truckers freeze in their stalled rigs on ice-bound Interstate 80 while flocks of sheep are swallowed whole by huge snow drifts. It is easy to miss the exit to Skull Valley.

An hour's drive west of Salt Lake City, Skull Valley is typical of the basin and range landscape that characterizes so much of the intermontane West. Ten thousand years ago it was an azure-blue fjord-arm of prehistoric Lake Bonneville (mother of the present Great Salt Lake), whose ancient shorelines are still etched across the face of the snow-capped Stansbury Mountains. Today the valley floor (when not snowed in) is mostly given over to sagebrush, alkali dust, and the relics of the area's incomparably strange history.

A half-dozen abandoned ranch houses, now choked with tumbleweed, are all that remain of the immigrant British cottonmill workers—Engels's classic Lancashire proletariat—who were the Valley's first Mormon settlers in the late 1850s. The nearby ghost town of Iosepa testifies to the ordeal of several hundred native Hawaiian converts, arriving a generation later, who fought drought, homesickness, and leprosy. Their cemetery, with beautiful Polynesian names etched in Stansbury quartzite, is one of the most unexpected and poignant sites in the American West.[52]

Further south, a few surviving families of the Gosiute tribe—people of Utah's Dreamtime and first cousins to the Western Shoshone—operate the "Last Pony Express Station" (actually a convenience store) and lease the rest of their reservation to the Hercules Corporation for testing rockets and explosives. In 1918, after refusing to register for the draft, the Skull Valley and Deep Creek bands of the Gosiutes were rounded up by the Army in what Salt Lake City papers termed "the last Indian uprising."[53]

Finally, at the Valley's southern end, across from an incongruously large and solitary Mormon temple, a sign warns spies away from Dugway Proving Ground: since 1942, the primary test-site for US chemical, biological, and incendiary weapons. Napalm was invented here and tried out on block-long replicas of German and Japanese workers' housing (parts of this eerie "doom city" still stand). Also tested here was the supersecret Anglo-American anthrax bomb (Project N) that Churchill, exasperated by the 1945 V-2 attacks on London, wanted to use to kill 12 million Germans. Project W-47—which did incinerate Hiroshima and Nagasaki—was based nearby, just on the other side of Granite Mountain.[54]

In the postwar years, the Pentagon carried out a nightmarish sequence of live-subject experiments at Dugway. In 1955, for example, a cloud generator was used to saturate thirty volunteers—all Seventh-Day Adventist conscientious objectors—with potentially deadly Q fever. Then, between August and October 1959, the Air Force deliberately let nuclear reactors melt down on eight occasions and "used forced air to ensure that the resulting radiation would spread to the wind. Sensors were set up over a 210-mile area to track the radiation clouds. When last detected they were headed toward the old US 40 (now Interstate 80)."[55]

Most notoriously, the Army conducted 1635 field trials of nerve gas, involving at least 500,000 pounds of the deadly agent, over Dugway between 1951 and 1969. Open-air nerve-gas releases were finally halted after a haywire 1968 experiment asphyxiated six thousand sheep on the neighboring Gosiute Reservation. Although the Army paid $1 million in damages, it refused to acknowledge any responsibility. Shrouded in secrecy and financed by a huge black budget, Dugway continued to operate without public scrutiny.[56]

Then in 1985 Senator Jeremy Stasser and writer Jeremy Rifkin teamed up to expose Pentagon plans to use recombinant genetic engineering to create "Andromeda strains" of killer microorganisms. Despite the American signature on the 1972 Biological Weapons Convention that banned their development, the army proposed to build a high-containment laboratory at Dugway to "defensively" test its new designer bugs.[57]

Opposition to the Death Lab was led by Downwinders, Inc., a Salt Lake City–based group that grew out of solidarity with the radiation victims in the St. George area. In addition to local ranchers and college students, the Downwinders were able to rally support from doctors at the Latter Day Saints (Mormon) Hospital and, eventually, from the entire Utah Medical Association. Local unease with Dugway was further aggravated by the Army's admission that ultratoxic organisms were regularly shipped through the US mail.

The Pentagon, accustomed to red-carpet treatment in superpatriotic Utah, was stunned by the ensuing storm of public hearings and protests, as well as the breadth of the opposition. In September 1988 the Army reluctantly cancelled plans for its new "BL-4" lab. In a recent interview, Downwinders' organizer Steve Erickson pointed out that "this was the first grassroots victory anywhere, ever, over germ or chemical warfare testing." In 1990, however, the Dugway authorities unexpectedly resurrected their biowar lab scheme, although now restricting the range of proposed tests to "natural" lethal organisms rather than biotech mutants.[58]

A year later, while Downwinders and their allies were still skirmishing with the Army over the possible environmental impact of the new lab, Desert Shield suddenly turned into Desert Storm. Washington worried openly about Iraq's terrifying arsenal of biological and chemical agents, and Dugway launched a crash program of experiments with anthrax, botulism, bubonic plague, and other micro-toxins in a renovated 1950s facility called Baker Lab. Simulations of these organisms were also tested in the atmosphere.

The Downwinders, together with the Utah Medical Association (dominated by Mormon doctors), went to the US District Court to challenge the resumption of tests at the veteran Baker Lab as well as the plan for a new "life sciences test facility." Their case was built around the Army's noncompliance with federal environmental regulations as well as its scandalous failure to provide local hospitals with the training and serums to cope with a major biowar accident at Dugway. The fantastically toxic botulism virus, for example, has been tested at Dugway for decades, but not a single dose of the anti-toxin was available in Utah (indeed in 1993, there were only twelve doses on the entire West Coast).[59]

In filing suit, the Downwinders also wanted to clarify the role of chemical and biological weapons in the Gulf War. In the first place, they hoped to force the Army to reveal why it vaccinated tens of thousands of its troops with an experimental, and possibly dangerous, antibotulism serum. Were GIs once again being used as Pentagon guinea pigs? Was there any connection between the vaccinations and the strange sickness—the so-called "Gulf War Syndrome"—brought home by so many veterans?

Secondly, the Downwinders hoped to shed more light on why the Bush administration allowed the sales of potential biological agents in the months before the invasion of Kuwait. "If the Army's justification for resuming tests at Dugway was the imminent Iraqi biowar threat," said Erickson, "then why did the Commerce Department previously allow $20 million of dangerous 'dual-use' biological materials to be sold to Iraq's Atomic Energy Commission? Were we trying to defend our troops against our own renegade bugs?"[60]

In the event, the Pentagon refused to answer these questions and the Downwinders lost their lawsuit, although they remain convinced that bio-agents are prime suspects in Gulf War Syndrome. Meanwhile, the Army completed the controversial Life Sciences Test Facility, and rumors began to fly of research on superlethal fibroviruses. Then, in 1994, Lee Davidson, a reporter for the Mormon *Deseret News,* used the Freedom of Information Act to excavate the details of human-subject experiments in Dugway during the 1950s and 1960s. Two years later, former Dugway employees complained publicly for the first time about cancers and other disabilities they believe were caused by chemical and biological testing. The Defense Department finally admitted that cleanup of Dugway's 143 major toxic sites may cost billions and take generations, if ever, to complete.[61]

The Great "Waste" Basin?

Grassroots protest in the intermontane states has repeatedly upset the Pentagon's best-made plans. Echoing sentiments frequently expressed at "Healing Global Wounds," Steve Erickson of the Downwinders boasts of the peace movement's dramatic breakthrough in the West over the past decade. "We have managed to defeat the MX and Midgetman missile systems, scuttle the proposed Canyon-lands Nuclear Waste Facility, stop construction of Dugway's BL-4, and impose a temporary nuclear test ban. That's not a bad record for a bunch of cowboys and Indians in Nevada and Utah: two supposedly bedrock pro-military states!"[62] Yet the struggle continues. The Downwinders and other groups, including the Western Shoshones and Citizen Alert, see an ominous new environmental and public health menace under the apparently benign slogan of "demilitarization." With the abrupt ending of the Cold War, millions of aging strategic and tactical weapons, as well as six tons of military plutonium (the most poisonous substance that has ever existed in the geological history of the earth), must somehow be dis-posed of. As Seth Schulman warns, "the nationwide military toxic waste problem is monumental—a nightmare of almost overwhelming proportion."[63]

The Department of Defense's reaction, not surprisingly, has been to dump most of its obsolete missiles, chemical weaponry, and nuclear waste into the thinly populated triangle between Reno, Salt Lake City, and Las Vegas: an area that already contains perhaps one thousand "highly contaminated" sites (the exact number is a secret) on sixteen military bases and Department of Energy facilities.[64] The Great Basin, as in 1942 and 1950, has again been nominated for sacrifice. The Pentagon's apocalyptic detritus, however, is a new regional cor-nucopia—the equivalent of postmodern Comstock—for a handful of powerful defense contractors and waste-treatment firms. As environmental journalist Triana Silton warned a few years ago, "a full-fledged corporate war is shaping up as part of the old military-industrial complex transforms itself into a new toxic waste-disposal complex."[65]

There are huge profits to be made disposing of old ordnance, rocket engines, chemical weapons, uranium trailings, radioactive soil, and the like. And com-pany bottom lines look even better when military recycling is combined with the processing of imported urban solid waste, medical debris, industrial toxins, and nonmilitary radioactive waste. The big problem has been to find compliant local governments willing to accept the poisoning of their natural and human landscapes.

No locality has been more eager to embrace the new political economy of toxic waste than Tooele County, just west of Salt Lake City. As one prominent activist complained to me, "The county commissioners have turned Tooele into the West's biggest economic red light district."[66] In addition to Dugway Proving Ground and the old Wendover and Deseret bombing ranges, the county is also home to the sprawling Tooele Army Depot, where nearly half of the Pentagon's chemical-weapons stockpile is awaiting incineration. Its nonmilitary toxic assets include Magnesium Corporation of America's local smelter (the nation's leading producer of chlorine gas pollution) and the West Desert Hazardous Industry Area (WDHIA), which imports hazardous and radioactive waste from all over the country for burning in its two towering incinerators or burial in its three huge landfills.[67]

Most of these facilities have been embroiled in recent corruption or health-and-safety scandals. Utah's former state radiation-control director, for example, was accused in late December 1996 of extorting $600,000 ("in everything from cash to coins to condominiums") from Khosrow Semnani, the owner of Environcare, the low-level radioactive waste dump in the WDHIA. Semnani has contributed heavily to local legislators in a successful attempt to keep state taxes and fees on Environcare as low as possible. Whereas the two other states that license commercial sites for low-level waste, South Carolina and Washington, receive $235.00 and $13.75 per cubic foot, respectively, Utah charges a negligible $.10 per cubic foot. As a result, radioactive waste has poured into the WDHIA from all over the country.[68]

Meanwhile, anxiety has soared over safety conditions at the half-billion-dollar chemical-weapons incinerator managed by EG&G Corporation at Tooele Army Depot. As the only operational incinerator in the continental United States (another, accident-ridden incinerator is located on isolated Johnson Atoll in the Pacific), the Depot is the key to the Pentagon's $31 billion chemical demilitarization program. Vehement public opposition has blocked incinerators originally planned for sites in seven other states. Only in job- and tax-hungry Tooele County, which has been promised $13 million "combat pay" over seven years, did the Army find a welcome mat.[69]

Yet as far back as 1989, reporters had obtained an internal report indicating that in a "worst case scenario" an accident at the plant could kill more than two thousand Tooele residents and spread nerve gas over the entire urbanized Wasatch Front. (The National Gulf War Resource Center in Washington, D.C. later warned that "if there is a leak of sarin from the Tooele incinerator or one of the chemical warfare agent bunkers, residents of Salt Lake City may end up with Gulf War illness coming to a neighborhood near you.") Federal courts nonetheless rejected a last-minute suit by the Sierra Club and the Vietnam Veterans of America Foundation to prevent the opening of the incinerator in August 1996.[70]

Within seventy-two hours of ignition, however, a nerve gas leak forced operators to shut down the facility. Another serious leak occurred a few months later. Then, in November 1996, the plant's former director corroborated the testimony of earlier whistleblowers when he publicly warned EG&G officials that "300 safety, quality and operational deficiencies" still plagued the operation. He also complained that the "plan is run by former Army officers who disregard safety risks and are too focused on ambitious incineration schedules." Environmental groups, meanwhile, have raised fears that even "successful" operation of the incinerator might release dangerous quantities of carcinogenic dioxins into the local ecosystem.[71]

Indeed, there is disturbing evidence that a sinister synergy of toxic environments may already be creating a slow holocaust comparable to the ordeal of the fallout-poisoned communities chronicled by Carole Gallagher. At the northeastern end of Tooele Valley, for example, Grantsville (pop. 5000) is currently under the overlapping shadows of the chlorine plume from Magnesium Corporation, the emissions from the hazmat incinerators in the WDHIA, and whatever is escaping from the Chemical Demilitarization Facility. In the past it has also been downwind of nuclear tests in Nevada and nerve gas releases from Dugway, as well as clearly from the open detonation of old ordnance at the nearby but now closed North Area of the Army Depot.

For years Grantsville had lived under a growing sense of dread as cancer cases multiplied and the cemetery filled up with premature deaths, especially women in their thirtie. As in a Stephen King novel, there was heavy gossip that something was radically wrong. Finally in January 1996, a group of residents, organized into the West Desert Healthy Environment Coalition (HEAL) by county librarian Chip Ward and councilwoman Janet Cook, conducted a survey of 650 local households, containing more than half of Grantsville's population.

To their horror, they discovered 201 cancer cases, 181 serious respiratory cases (not including bronchitis, allergies, or pneumonia), and 12 cases of multiple sclerosis. Although HEAL volunteers believe that majority Mormon residents reported only a fraction of their actual reproductive problems, they recorded 29 serious birth defects and 38 instances of major reproductive impairment. Two-thirds of the surveyed households, in other words, had cancer or a major disability in the family: many times the state and national averages. As Janet Cook told one journalist, "Southern Utah's got nothing on Grantsville."[72]

"Most remarkable," Ward observed, "was the way that cancer seemed to be concentrated among longtime residents." One source of historical exposure that Ward and others now think has been underestimated were the Dugway tests. "One respondent, for example, reported that she gave birth to seriously deformed twins several months after the infamous sheep kill in Skull Valley in 1968. Her doctor told her that he'd never seen so many birth defects as he did that year."[73]

But more than one incident is undoubtedly indicted in Grantsville's terrifying morbidity cluster. Environmental health experts have told HEAL that "cumulative, multiple exposures, with 'synergistic virulence' [the whole is greater than the sum of the parts]" best explain the local prevalence of cancers and lung diseases. In a downwind town where a majority of people work in hazardous occupations, including the Army Depot, Dugway, WDHIA, and the Magnesium Corporation, environmental exposure has been redoubled by occupational exposure, and vice versa.[74]

Consequently, West Desert HEAL, supported by Utah's Progressive Alliance of labor, environmental, and women's groups, has been demanding increased environmental monitoring, a moratorium on emissions and open detonation, complete documentation of past military testing, and a baseline regional health study with meaningful citizen participation. By early 1997 it had won legislative approval for the health survey and a radical reduction in Army detonations. The Magnesium Corporation, however, was still spewing chlorine and the Pentagon was still playing chicken with 13,616 tons of nerve gas.[75]

From a bar stool in the Dead Dog Saloon, Grantsville still seems like a living relic of that Old West of disenfranchised miners, cowboys, and Indians. Yet just a few miles down the road is the advance guard of approaching suburbia. Since 1995 metropolitan Salt Lake City has expanded, or, rather, exploded into northern Tooele Valley. The county seat, Tooele, has been featured in the *New York Times* as "one of the fastest-growing cities in the West," and a vast, billion-dollar planned suburb, Overlake, has been platted on its fringe.[76]

Although local developers deride the popular appeal of "tree-huggers" and antipollution groups, significant segments of the urban population will soon be in the toxic shadow of Tooele's nightmare industries. "When the new suburbanites wake up one morning and realize that they too are downwinders," Chip Ward predicts, "then environmental politics in Utah will really get interesting."[77]

1992/1997

Postscript

Interesting, indeed. In 1997 the tiny Gosiute band in Skull Valley (most of whom actually live in Grantsville) stunned the rest of Utah by signing a contract to open their reservation to 40,000 tons of high-level nuclear waste imported from out-of-state utilities. Although most Native Americans in the West have rejected nuclear storage as the highest form of "enivornmental racism," tribal leader Leon Bear (a nontraditionalist who doesn't speak Gosiute) persuaded a majority of his members that the proposed $3 billion facility would not only make the tribe wealthy but would ensure the preservation of its endangered language. Against the bitter opposition of tribal traditionalists, who claimed that the storage facility would destroy a sacred landscape, Bear argued that it would only be sweet revenge against a white society that had expelled the Gosiute from Tooele Valley and driven their culture to the edge of exintction. Environmental critics, like Downwinders Inc., respond that while the Skull Valley Indians have a strong case for reparations, they are best not paid in plutonium.

In any event, the Gosiute mouse continues to roar in 2002 as vexed state officials struggle to block construction in Skull Valley. The proposed Private Fuel
Storage complex is envisioned as a sinister environmental art park: four thousand
18-foot-tall stainless steel canisters over some 820 acres. (The reflection would
be visible from space.) The utilities envision shipping thousands of fuel assem-
blies, each containing ten times the long-term radioactivity released by the Hiroshima bomb, over a thirty-year period. Although the spent fuel is designated to
be ultimately stored for eons in Yucca Mountain's underground crypt, it is unclear
whether the Nevada site—under siege from scientists and local opponents—will
ever open. Thus Skull Valley by default would become the permanent graveyard
of the nuclear age, and in the nightmares of nearby Salt Lake City residents, the
nation's most inviting terrorist target.

Notes

1. Although whale-hunting and sewerage are considered at length, the environmental
impact of twentieth-century militarism is an inexplicably missing topic among the forty-
two studies that comprise the landmark global audit: B. L. Turner et al., eds., *The Earth as
Transformed by Human Action: Global and Regional Changes in the Biosphere over the Past 100
Years*, Cambridge 1990.
2. This is the term used by Michael Carricato, the Pentagon's former top environmental
official. See Seth Shulman, *The Threat at Home: Confronting the Toxic Legacy of the US
Military*, Boston 1992, p. 8.
3. Nuclear landscapes, of course, also include parts of the Arctic (Novaya Zemlya and
the Aleutians), Western Australia, and the Pacific (the Marshall Islands and Mururoa).
4. Zhores Medvedev, *Nuclear Disaster in the Urals*, New York 1979; and Boris Komarov, *The
Destruction of Nature in the Soviet Union*, White Plains, N.Y. 1980.
5. See D. J. Peterson, *Troubled Lands: The Legacy of Soviet Environmental Destruction*, a Rand
research study, Boulder 1993, pp. 7–10.
6. Ibid., p. 23.
7. Murray Feshbach and Alfred Friendly Jr., *Ecocide in the USSR*, New York 1992, p. 1.
8. Peterson, p. 248. He also quotes Russian fears that Western joint ventures and
multinational investment may increase environmental destruction and accelerate the
conversion of the ex-USSR, especially Siberia, into an "ecological colony" (pp. 254–57).

9. Feshbach and Friendly, pp. 11, 28, and 39.

10. Indeed, their sole citation of environmental degradation in the United States concerns the oyster beds of Chesapeake Bay (ibid., p. 49).

11. Richard Misrach, *Violent Legacies: Three Cantos*, New York 1992, pp. 38–59, 86. Misrach's interpretation of the pits is controversial. Officially, they are burial sites for animals infected with brucellosis and other stock diseases. Paiute ranchers whom I interviewed, however, corroborated the prevalence of mystery deaths and grotesque births.

12. An Irish nationalist who sympathized with the struggle of the Plains Indians, Mooney risked professional ruin by including the Ogalala account of the massacre in his classic *The Ghost-Dance Religion and the Sioux Outbreak of 1890*, Fourteenth Annual Report of the Bureau of Ethnology, Washington 1896, pp. 843–86. The actual photographer was George Trager. See Richard Jensen et al., *Eyewitness at Wounded Knee*, Lincoln, Neb. 1991.

13. Richard Misrach, *A Photographic Book*, San Francisco 1979; and *Desert Cantos*, Albuquerque, N.M. 1987.

14. Richard Misrach (with Myriam Weisang Misrach), *Bravo 20: The Bombing of the American West*, Baltimore 1990, p. xiv.

15. Misrach, *Violent Legacies*, pp. 14–37; 83–86.

16. Lois Parkinson Zamora, *Writing the Apocalypse: Historical Vision in Contemporary US and Latin American Fiction*, Cambridge 1989, p. 189 (my emphasis).

17. William Jenkins, *New Topographics: Photographs of a Man-Altered Landscape*, International Museum of Photography, Rochester, N.Y. 1975.

18. Mark Klett et al., *Second View: The Rephotographic Survey Project*, Albuquerque, N.M. 1984.

19. San Francisco Camerawork, *Nuclear Matters*, San Francisco 1991.

20. See Adams's own account of how he retouched a famous photograph of Mount Whitney to eliminate a town name from a foreground hill: *Examples: The Making of Forty Photographs*, Boston 1983, p. 165.

21. Barry Lopez, paraphrased by Thomas Southall, "I Wonder What He Saw," from Klett et al., *Second View*, p. 150.

22. Aside from Misrach, see especially Mark Klett, *Traces of Eden: Travels in the Desert Southwest*, Boston 1986; and *Revealing Territory*, Albuquerque, N.M. 1992.

23. Revealingly, a decisive influence on the New Topographics was the surrealist photographer Frederick Sommer. See the essay by Mark Haworth-Booth in Lewis Baltz, *San Quentin Point*, New York 1986.

24. Jan Zita Cover, "Landscapes Ordinary and Extraordinary," *Afterimage*, December 1983, pp. 7–8.

25. The cold deserts and sagebrush *(Artemisia)* steppes of the Great Basin and the high plateaux are floristic colonies of Central Asia (see Neil West, ed., *Ecosystems of the World 5: Temperate Deserts and Semi-Deserts*, Amsterdam 1983), but the physical landscapes are virtually unique (see W. L. Graf, ed., *Geomorphic Systems of North America*, Boulder, Colo. 1987).

26. It is important to recall that the initial exploration of much of this "last West" occurred only 125 years ago. Cf. Gloria Cline, *Exploring the Great Basin*, Reno, Nev. 1963; William Goetzmann, *Army Exploration in the American West, 1803–1863*, New Haven, Conn. 1959; and *New Lands, New Men*, New York 1986.

27. The aeolian processes of the Colorado Plateau have provided valuable in sights into the origin of certain Martian landscapes (Julie Laity, "The Colorado Plateau in Planetary Geology Studies," in Graf, pp. 288–97), while the Channeled Scablands of Washington are the closest terrestrial equivalent to the great flood channels discovered on Mars in 1972. (See Baker et al., "Columbia and Snake River Plains," in Graf, pp. 403–68.) Finally, the basalt plains and calderas of the Snake River in Idaho are considered the best analogues to the lunar *mare* (ibid.).

28. There were four topographical and geological surveys afoot in the West between 1867 and 1879. The Survey of the Fortieth Parallel was led by Clarence King, the Survey West of the One Hundredth Meridian was under the command of Lieutenant George Wheeler, the Survey of the Territories was directed by Ferdinand Vandeveer Hayden, and the Survey of the Rocky Mountain Region was led by John Wesley Powell, They produced 116 scientific publications, including such masterpieces as Clarence Dutton, *Tertiary History of the Grand Canyon*, Washington 1873; Grove Karl Gilbert, *Report on the Geology of the Henry Mountains*, Washington 1877; and John Wesley Powell, *Exploration of the Colorado River of the West*, Washington 1873. John McPhee has recently repeated King's survey of the fortieth parallel (now Interstate 80) in his four-volume "cross-section of human and geological time": *Annals of the Former World*, New York, 1980–1993.

29. Cf. R. J. Chorley, A. J. Dunn and R. P. Beckinsale, *The History of the Study of Landforms, Volume 1: Geomorphology before Davis*, London 1964, pp. 469–621; and Baker et al.

30. Stephen Pyne, *Grove Karl Gilbert*, Austin, Tex. 1980, p. 81.

31. Ann-Sargent Wooster, "Reading the American Landscape," *Afterimage*, March 1982, pp. 6–8.

32. Consider "relapsing chasms," "wilted, drooping faces," "waving cones of the Uinkaret," and so on. See Wallace Stegner, *Beyond the Hundredth Meridian: John Wesley Powell and the Second Opening of the West*, Boston 1954, chapter 2.

33. Ibid., chapter 3. The ironic legacy of Powell's *Report* was the eventual formation of a federal Reclamation Agency that became the handmaiden of a Western powerstructure commanded by the utility monopolies and corporate agriculture.

34. Carole Gallagher, *American Ground Zero: The Secret Nuclear War*, Boston 1993.

35. Peter Goin, *Nuclear Landscapes*, Baltimore 1991; and Robert Del Tredici, *Work in the Fields of the Bomb*, New York 1987. See also Patrick Nagatani, *Nuclear Enchantment*, Albuquerque 1990; John Hooton, *Nuclear Heartlands*, 1988; and Jim Leager, *In the Shadow of the Cloud*, 1988. Work by independent filmmakers includes John Else, *The Day after Trinity* (1981); Dennis O'Rouke, *Half Life* (1985), and Robert Stone, *Radio Bikini* (1988).

36. Dorothea Lange and Paul Taylor, *An American Exodus*, New York 1938; James Agee and Walker Evans, *Let Us Now Praise Famous Men*, Boston 1941; Erskine Caldwell and Margaret

Bourke-White, *You Have Seen Their Faces*, New York 1937.

37. Gallagher, p. xxiii.

38. Ibid., p. xxxii.

39. Israel Torres and Robert Carter, quoted in ibid., pp. 61–62. Gallagher encountered the story about the charred human guinea pigs (prisoners?) "again and again from men who participated in shot Hood" (p. 62).

40. Delayne Evans, quoted in ibid., p. 275.

41. Isaac Nelson, quoted in ibid., p. 134.

42. Ina Iverson, quoted in ibid., pp. 141–43. Gallagher points out that molar pregnancies are also "an all too common experience for the native women of the Marshall Islands in the Pacific Testing Range after being exposed to the fallout from the detonations of hydrogen bombs" (p. 141).

43. Jay Truman, quoted in ibid., p. 308.

44. The literature is overwhelming. See House Subcommittee on Oversight and Investigations, *The Forgotten Guinea Pigs*, 96th Congress, 2nd session, August 1980; Thomas Saffer and Orville Kelly, *Countdown Zero*, New York 1982; John Fuller, *The Day We Bombed Utah: America's Most Lethal Secret*, New York 1984; Richard Miller, *Under the Cloud: The Decades of Nuclear Testing*, New York 1986; Howard Ball, *Justice Downwind: America's Atomic Testing Program in the 1950s*, New York 1986; A. Costandina Titus, *Bombs in the Backyard: Atomic Testing and Atomic Politics*, Reno, Nev. 1986; and Philip Fradkin, *Fallout: An American Nuclear Tragedy*, Tucson, Ariz. 1989.

45. Gallagher, pp. xxxi–xxxii.

46. Fradkin, p. 57; Peterson, pp. 203 and 230 (fn 49).

47. See "From the Editors," *The Bulletin of the Atomic Scientists*, September 1990, p. 2.

48. Colonel Langdon Harrison, quoted in Gallagher, p. 97.

49. Peterson, p. 204; see also Feshbach and Friendly, pp. 238–39.

50. See my "French Kisses and Virtual Nukes," in *Capitalism, Nature, Socialism* 7, no. 1 (March 1996).

51. Cf. Kealy Davidson, "The Virtual Bomb," *Mother Jones*, March/April 1995; Jacqueline Cabasso and John Burroughs, "End Run Around the NPT," *Bulletin of the Atomic Scientists*, September–October 1995; and Jonathan Weissman, "New Mission for the National Labs," *Science*, 6 October 1995. On protest plans: interview with Las Vegas Catholic Workers, May 1997.

52. See Tracey Panek, "Life at Iosepa, Utah's Polynesian Colony," *Utah Historical Quarterly*; and Donald Rosenberg, "Iosepa," talk given on centennial, Salt Lake City, 27 August 1989 (special collections, University of Utah library).

53. Ronald Bateman, "Goshute Uprising of 1918," *Deep Creek Reflections*, pp. 367–70.

54. Barton Bernstein, "Churchill's Secret Biological Weapons," *Bulletin of the Atomic Scientists*, January–February 1987.

55. See Ann LoLordo, "Germ Warfare Test Subjects," (first appeared in *Baltimore Sun*), reprinted in *Las Vegas Review-Journal*, 29 August 1994; and Lee Davidson, "Cold War

Weapons Testing," *Deseret News*, 22 December 1994.

56. Lee Davidson, "Lethal Breeze," *Deseret News*, 5 June 1994.

57. For fuller accounts, see Jeanne McDermott, *The Killing Winds*, New York 1987; and Charles Piller and Keith Yamamoto, *Gene Wars: Military Control over the New Genetic Technologies*, New York 1988.

58. Steve Erickson, Downwinders, Inc., interviewed September, November 1992 and January 1993.

59. *Downwinders, Inc. v. Cheney and Stone*, Civil No. 91-C-681j, United States Court, District of Utah, Central Division.

60. Erickson refers to information revealed in December 1990 by Ted Jacobs, chief counsel to the House Subcommittee on Commerce, Consumer and Monetary Affairs.

61. Interviews with Steve Erickson and Cindy King, Salt Lake City, October 1996.

62. Ibid.

63. Schulman, p. 7.

64. The estimate is from figures in Schulman, appendix B.

65. Interview with Triana Silton, September 1992.

66. Interview with Chip Ward, Grantsville, Utah, October 1996.

67. For a description of the WDHIA and its natural setting, see Barry Wolomon, "Geologic Hazards and Land-use Planning for Tooele Valley and the Western Desert Hazardous Industrial Area," Utah Geological Survey, *Survey Notes*, November 1994.

68. Jim Wolf, "Does N-Waste Firm Pay Enough to Utah?" *Salt Lake Tribune*, 10 January 1997.

69. Ralph Vartabedian, "Startup of Incinerator Is Assailed," *Los Angeles Times*, 4 March 1996.

70. Lee Davidson, "An Accident at TAD Could Be Lethal," *Deseret News*, 23 May 1989; and Lee Siegel, "Burn Foes Fear Outbreak of Gulf War Ills," *Salt Lake Tribune*, 12 January 1997.

71. Joseph Bauman, "Former Tooele Manager Calls Plant Unsafe," *Deseret News*, 26 November 1996.

72. West Desert Healthy Environment Alliance, *The Grantsville Community's Health: A Citizen Survey*, Grantsville, Utah 1996; Diane Rutter, "Healing Their Wounds," *Catalyst*, April 1996. See also "Listen to Cancer Concerns," editorial, *Salt Lake Tribune*, 6 April 1996.

73. Interview with Chip Ward, Grantsville, Utah, October 1996.

74. Ibid.

75. Interview with Chip Ward, January 1997.

76. James Brooke, "Next Door to Danger, a Booming City," *New York Times*, 6 October 1996.

77. Interview with Chip Ward, January 1997. His landmark *Canaries on the Rim: Living Downwind in the West*, a comprehensive account of the Grantsville nightmare, was published by Verso in 1999.

German Village, Dugway Proving Ground (1998)

3

Berlin's Skeleton in Utah's Closet

All scattered lies Berlin.
Günter Grass

Berlin's most far-flung, secret, and orphan suburb sits in the saltbrush desert about ninety miles southwest of Salt Lake City. "German Village," as it is officially labeled on declassified maps of the US Army's Dugway Proving Ground, is the remnant of a much larger, composite German/Japanese "doomtown" constructed by Standard Oil in 1943. It played a crucial role in the New Deal's last great public works project: the incineration of the cities of eastern Germany and Japan.

In 1997, the Army allowed me to briefly tour German Village with a dozen of my students from the Southern California Institute of Architecture. Dugway, it should be pointed out, is slightly bigger than Rhode Island and more toxically contaminated than the Nuclear Test Site in nearby Nevada. As the devil's own laboratory for three generations of US chemical, incendiary, and biological weapons, it has always been shrouded in official secrecy and Cold War myth. The threat of base restructuring, however, has prompted the Army to mount a small public relations campaign on Dugway's behalf. Since napalm, botulism, and binary nerve gas are not conventional tourist attractions, Dugway Proving Ground instead extolls its preservation of an original section of the Lincoln Highway.[1] Most visitors are pioneer-motoring enthusiasts who come to admire the decrepit, one-lane bridge that fords a swampy patch in Baker Area, not far from the controversial bio-warfare lab, guarded by a double perimeter of razor wire, where the Army tinkers with Andromeda strains.

German Village is a dozen or so miles farther west, in a sprawling maze of mysterious test sites and target areas which Dugway's commander is not eager to add to the visitor itinerary. He relented only when we convinced his press office that the Village had an important aura that might enhance "base heritage": It was designed by one of Modernism's gods, the German-Jewish architect Eric Mendelsohn.

Bombing Brecht

In 1943, the Chemical Warfare Corps secretly recruited Mendelsohn to work with Standard Oil engineers and RKO set designers to create a miniature Hohenzollern slum in the Utah desert. Nothing in the appearance of the surviving structure—the double tenement block known as Building 8100—gives any hint that it is the product of the same hand that designed such landmarks of Weimar Berlin as the offices of the *Berliner Tageblatt*, the Columbushaus, the Sternefeld villa in Charlottenburg, or the Woga Complex on the Kurfurstendamm. Absolute "typicality" in all aspects of layout and construction was what the Chemical Warfare Corps wanted.[2]

They were in a hurry. Despite the horrifying successes of their thousand-bomber fire raids against Cologne and Hamburg, their British allies were increasingly frustrated by their inability to ignite a firestorm in the Reich's capital. The top Allied science advisors urged a crash program of incendiary experimentation on exact replicas of working-class housing. Only the United States—or, rather, the combined forces of Hollywood and the oil industry—had the resources to complete the assignment in a few months. The design and construction processes were dovetailed with parallel secret research on the fire characteristics of Japanese homes coordinated by the architect Antonin Raymond, who had worked in

Japan before the war.[3] The eventual test complex was five square miles in area.

Mendelsohn's achievement was the anonymity of his result: six iterations of the steeply gabled brick tenements—*Mietskasernen* or "rent barracks"—that made the Red districts of Berlin the densest slums in Europe. Three of the apartment blocks had tile-on-batten roofs, characteristic of Berlin construction, while the other three had slate-over-sheathing roofs, more commonly found in the factory cities of the Rhine. Although not as tall as their seven-story counterparts in Wedding or Kreuzberg, the test structures were otherwise astonishingly precise replicas, far surpassing in every specification what the British had achieved at their own German target complex at Harmondsworth.

Before drawing any blueprints, Mendelsohn exhaustively researched the roof area coverage—a critical incendiary parameter—of target neighborhoods in Berlin and other industrial cities. His data were "extended and confirmed," reported the Standard Oil Development Company, "by a member of the Harvard Architecture School, an expert on German wooden frame building construction." (Could it have been Walter Gropius?) The builders, working with fire protection engineers, then gave extraordinary attention to ensuring that the framing (authentic woods imported from as far away as Murmansk) duplicated the aging and specific gravity of older German construction. When the fire experts objected that Dugway's climate was too arid, their Standard Oil counterparts contrived to keep the wood moist by having GI's regularly "water" the targets in simulation of Prussian rain.

The interior furnishing, meanwhile, was subcontracted to RKO's Authenticity Division, the wizards behind *Citizen Kane*. Using German-trained craftsmen, they duplicated the cheap but heavy furniture that was the dowry of Berlin's proletarian households. German linen was carefully studied to ensure the typicality of bed coverings and drapes. While the authenticators debated details with Mendelsohn and the fire engineers, the construction process was secretly accelerated by the wholesale conscription of inmates from the Utah State Prison. It took them only forty-four days to complete German Village and its Japanese counterpart (twelve double apartments fully furnished in *hinoki* and *tatami*). The entire complex was fire-bombed with both thermite and napalm, and completely reconstructed at least three times between May and September 1943. The tests demonstrated conclusively the superiority of the newly invented M-69 napalm munition.[4] It was a splendid example of the characteristic American "approach [to] war as a vast engineering project whose essential processes are as precisely calculated as the tensile strength requirements of a dam or bridge."[5]

Mendelsohn's secret signature on German Village is also rich in irony. Like all of his progressive Weimar contemporaries, he had a deep interest in housing reform and the creation of a *neue Wohnkultur* (new culture of living). Yet, as all of his biographers have noted, he never participated in the big social housing competitions organized by the Social Democrats in the later 1920s, which were such crucial showcases for the urbanist ideas of the emergent Modern movement. His absence was most dramatic (and mysterious) in the case of the 1927 Weissenhof Siedlung—the model housing project coordinated by Mies van der Rohe and sponsored by Stuttgart's leftwing government—which Philip Johnson has called "the most important group of buildings in the history of modern architecture." In his biography, Bruno Zevi says that Mendelsohn was *"excluded* from the large works of the Siedlung." (Is he implying anti-Semitism?)[6]

If so, Dugway's German Village was his revenge. Here was workers' housing perversely designed to accelerate the campaign "to dehouse the German industrial worker," as the British bluntly put it. The Weissenhof masterpieces of Gropius and the Taut brothers were included in the 45 percent of the 1939 German housing stock that Bomber Command and the Eighth Air Force managed to destroy or damage by the spring of 1945.[7] Indeed, Allied bombers pounded into rubble more 1920s socialist and modernist utopias than Nazi villas. (Ninety-five percent of the Nazi Party membership is estimated to have survived the Second World War.)[8]

Did Mendelsohn and the other anti-Nazi refugees who worked on German Village have any qualms about incendiary experimentation that involved only plebian housing? Did they apprehend the agony that the Chemical Warfare Corps was meticulously planning to inflict upon the Berlin proletariat? (Standing in front of Building 8100, I couldn't help but think: "This is like bombing Brecht.") No memoir or correspondence—Mendelsohn was notoriously tight-lipped—authorizes any surmise. Historians of the US Army Air Force, on the other hand, have excavated a complex, sometimes tortured debate (one that never occurred in the racial inferno of the Pacific Theater) over the ethics of firebombing Berlin.

The Zoroastrian Society

During the early days of the Second World War, tens of millions of American voters of German and Italian ancestry were reassured that the Army Air Force would never deliberately make a target out of "the ordinary man in the street." Americans were officially committed to the clean, high-tech destruction of strictly military or military-industrial targets. The Eighth Air Force sent its crews in daylight "precision" raids against visually identified targets, in contrast to its Blitz-embittered British allies, who saturation-bombed German cities at night by radar, hoping to terrorize their populations into flight or rebellion. The extraordinary technologies of the B-17 and the Norden bombsight allowed the United States to bomb "with democratic values." (Then, as now, "collateral damage" was smugly swept under the rug of national conscience.)

But, as the construction of German Village dramatizes, the uncensored story is considerably more sinister. While staff doctrine, aircraft technology, and domestic public opinion preserved a huge investment in precision bombing, counter-civilian or "morale" bombing had never been excluded from US war planning against Germany. As Ronald Schaffer and other historians have shown, AWPD-1—the secret strategy for an air war against Germany that was adopted months before Pearl Harbor—specifically envisioned that it might be "highly profitable to deliver a large-scale, all-out attack on the civil population of Berlin" after precision bombing had disrupted the Ruhr's industries. As preparation for attacking an industrial metropolis of Berlin's scale, the Air Corps Tactical School had already "bombed" the critical infrastructures of New York City in a 1939 targeting exercise.[9]

The British, moreover, fiercely pressured the American Eighth Air Force to join their "area bombing" crusade. Even before the Battle of Britain, Churchill had advocated an "absolutely devastating, exterminating attack by very heavy bombers from this country upon the Nazi homeland."[10] The Blitz quickly generated a vengeful public opinion that supported this strategy of bombing enemy civilians. But neither Churchill nor his chief science advisor and Dr. Strangelove, Lord Cherwell, were primarily interested in revenge per se. As they unleashed the fury of Bomber Command in March 1942, they were testing the hypothesis long advanced by Lord Trenchard, Britain's pioneer theorist of strategic bombing, that domestic morale (as in 1918) was Germany's achilles heel. It soon became the *idée fixe* around which all British air policy revolved.[11]

Of course there were different ways to terrorize Germans from the sky. For example, a case might have been made for singling out the mansions of the Nazi political and industrial elites for aerial punishment. But this risked retaliation against Burke's Peerage and was excluded by Cherwell from the outset. "The bombing must be directed essentially against working-class houses. Middle-class houses have too much space around them, and so are bound to waste bombs." Thus the squalid Mietskasernen were the bullseye, and "area bombing" was adopted as the official euphemism for Churchill's earlier "extermination."[12] "It has been decided," read the official order to air crews in February 1942, "that the primary objective of your operations should now be focused on the morale of the enemy civil population and in particular of the industrial workers."[13] By November 1942, when thousand-bomber night raids had become common over western Germany, Churchill was able to boast to FDR about the heroic quotas that the RAF had pledged to produce: nine hundred thousand civilians dead, one million seriously injured, and twenty-five million homeless.[14]

A. J. P. Taylor would later write of "the readiness, by the British, of all people, to stop at nothing when waging war. Civilised constraints, all considerations of morality, were abandoned." At the time, the only significant public dissent was British writer Vera Brittain's powerful protest, *Massacre by Bombing*, which was published in the United States by the Fellowship of Reconciliation. Socialist leader Norman Thomas then defended Brittain in a famous radio debate with Norman Cousins, the bellicose editor of *The Saturday Evening Post*. Although Brittain and Thomas were generally excoriated in the press, some of the US air chiefs, like General George McDonald, the director of Air Force intelligence, privately shared their revulsion against "indiscriminate homicide and destruction."[15] General Cabell, another "precisionist," complained about the "the same old baby killing plan of the get-rich-quick psychological boys."[16] Secretary of War Henry Stimson and Chief of Staff George Marshall also quietly struggled to maintain a moral distinction between the Nazi leadership and the German working class. (Stimson, not wanting "the United States to get the reputation of outdoing Hitler in atrocities," equally opposed the fire-bombing of Japan.)[17] Meanwhile, reports to FDR complained that Eighth Air Force crews, harboring "no particular hatred of the Germans," lacked the vengeful racial motivation of their brothers in the Pacific.[18] But the Commander in Chief, influenced by his own Strangelovian advisors and his friendship with Churchill, was more broadminded about massacring enemy civilians. When RAF's Operation Gomorrah in July and August 1943 succeeded in kindling tornadic firestorms in the heart of Hamburg (seven thousand children were amongst the carbonized victims), Roosevelt was reported to be greatly impressed.[19]

Gomorrah also strengthened the hand of the fire war advocates within the Army Air Force and the National Defense Research Committee. Six months before Pearl Harbor, the Chemical Warfare Service had secretly dispatched Enrique Zanetti, a Columbia University chemist, to study incendiary warfare in London. He became a fervent and influential lobbyist for the Churchillian method of brimstone and pitch. After the arrival of the Eighth Air Force, the ambitious head of its Chemical Section, Colonel Crawford Kellogg, also sought out British expertise. The RAF accordingly organized a discussion group, the so-called Zoroastrian Society, to share technical information and promote the city-burning strategy.[20] It soon became an intellectual home for aggressive young commanders like Curtis Le May who were infected with the British enthusiasm for incendiary weapons and wanted to see their deployment greatly expanded in every theater. Their views were endorsed by Assistant Secretary of War Robert Lovett. In a meeting to discuss the adoption of a nightmare anti-personnel bomb loaded with napalm and white phosphorus, he argued: "If we are going to have a total war we might as well make it as horrible as possible."[21]

On the home front, civilians were often more avid advocates of total warfare than their military counterparts. Walt Disney, for instance, popularized the chilling ideas of Russian émigré Alexander P. de Seversky—a fanatical advocate of bombing cities—in the film, *Victory Through Airpower*.[22] After the fall of Bataan, *Harper's* published a widely discussed article that extolled fire-bomb attacks on Kyoto and Kobe: "The suffering that an incendiary attack would cause is terrible to contemplate. But the fact remains that this is the cheapest possible way to cripple Japan."[23] In addition, incendiary warfare enjoyed powerful support from influential Harvard scientists (led by the "father of napalm," Louis Feiser), the oil companies, psychologists (who studied Axis morale),[24] and the fire protection industry. The fire insurance experts, one historian emphasizes, "did not simply advise the Army Air Force. They pushed it as hard as they could to make it wage incendiary warfare against factories and homes." They loved to point out to airmen the overlooked fire potentials of structures like churches, which were "quite vulnerable to small incendiaries."[25] Top operations analyst William B. Shockley (the future inventor of the transistor and a notorious advocate of the intellectual inferiority of people of color) buttressed the case for fire bombs with a clever accounting of their higher destructive "profitability."[26]

German Village was constructed in May 1943, on the eve of Churchill's burnt offering at Hamburg, to address opportunities and problems that were beyond the moral perimeter of precision bombing. It was a trade show for the burgeoning fire-war lobby hungry for "profits." Those planning the coming air offensive against urban Japan were eager to see how newly invented incendiaries, including napalm and an incredible "bat bomb" (Project X-Ray) that released hundreds of live bats booby-trapped with tiny incendiaries, performed against Dugway's Japanese houses.[27] Meanwhile, the Zoroastrian Society was looking for clues on how to set ablaze Berlin's massive masonry shell.

Churchill's "Marxism"

In his authoritative postwar report "The Fire Attacks on German Cities," Horatio Bond, the National Defense Research Committee's chief incendiary expert, underscored Allied frustration. "Berlin was harder to burn than most of the other German cities. There was better construction and better 'compartmentation.' In other words, residential buildings did not present as large fire divisions or fire areas. Approximately twice as many incendiaries had to be dropped to assure a fire in each fire division." As the German Village tests demonstrated, "little [could] be expected in the way of the free spread of fire from building to build-ing." Buildings were lost "because they were hit by bombs rather than because fire spread from other buildings."[28]

Yet until Zhukov was literally spitting in the Spree, the British clung to the belief (or dementia, as many Americans saw it) that Berlin could be bombed out of the war. What the Mietskasernen refused to oblige in terms of combustibility, RAF planners argued, could be compensated for by more bombers and greater incendiary density. They assumed that intolerable civilian suffering would inevi-tably produce a proletarian revolt in the heart of the Third Reich. "The British," explains Robert Pape, had distinctively married "the Air scare to the Red scare of the 1920s. Air power, according to this logic, would bomb industrial centers, creating mass unemployment and panic, especially among the working classes, who in turn would overthrow the government. In short, air attack against popula-tions would cause workers to rise up against the ruling classes."[29] Churchill, who thought enough Lancaster bombers could turn Berlin's workers back into anti-fascists, remained a more orthodox Marxist than Stalin, who alone seems to have understood the enormity of Hitlerism's moral hold on the Reich's capital.

Promising the British people that "Berlin will be bombed until the heart of Nazi Germany ceases to beat," Sir Arthur Harris (whose enthusiasm for bombing civilians dated back to the Third Afghan War in 1919)[30] unleashed the RAF's heavy bombers on 18 November. In a new strategy that the Germans called *Bombentep-pich* or "carpet bombing," the Lancasters, flying in dangerously tight formations, concentrated their bomb loads on small, densely populated areas. Mission per-formance was measured simply by urban acreage destroyed. Incendiary attacks were followed up by explosives with the deliberate aim of killing firefighters, rescue workers, and refugees. In line with the Churchillian doctrine of targeting Weimar's Red belts to maximize discontent, the famous KPD stronghold of Wed-ding was thoroughly pulverized and set afire.[31]

The Zoo was also a major target, which inadvertently increased the meat ration of the city's poorer residents. "Berliners discovered to their surprise that some unusual dishes were extremely tasty. Crocodile tail, for instance, cooked slowly in large containers, was not unlike fat chicken, while bear ham and sausages proved a particular delicacy." Although Harris was unable to fuel a Hamburg-style firestorm over the Tiergarten, the Lancasters did flatten almost a quarter of the metropolitan core. The BBC boasted that as many as a million Berliners had been killed or injured.[32]

Yet as Harris himself had to acknowledge to Churchill, the RAF's all-out effort "did not appear to be an overwhelming success." For one thing, Goebbels, the city's real ruler, mounted a brilliant defense with his flak towers, squadrons of deadly nightfighters, and fire brigades conscripted from all over Germany. Five percent of Harris's air crews were shot out of the sky every night, an unsustainable sacrifice for Bomber Command. Moreover, despite terrible damage to the slums, the real machinery of power and production in Berlin remained remarkably undamaged. The Americans, who had broken the Japanese codes, found no reports of crippling damage in intercepted wires from Japan's embassy in Berlin. Strategic bombing analysts, for their part, marveled at the ability of the city's industries "to produce war material in scarcely diminished quantities almost up to the end."[33]

As for the calculus of suffering that firebombing was supposed to instruct, Goebbels cunningly shifted the parameters. "Issue no denials of the English claim to have killed a million in Berlin," he ordered his propagandists. "The sooner the English believe there's no life in Berlin, the better for us."[34] Meanwhile, he evacuated more than one million nonessential civilians—especially children—into the countryside. Conversely, he moved hundreds of thousands of Russian and Polish prisoners of war directly under Allied bombsights. As Alexander Richie has described their plight: "They had almost no protection from air-raids, were kept in concentration camp conditions, received low rations and were inevitably given the most difficult, filthy and dangerous jobs. ... [O]f the 720 people killed in a typical raid on 16 December 1943, 249 were slave labourers...."[35]

While Hitler was throwing tantrums in his bunker, Goebbels was holding stirring rallies in the ruins of the Red Belt, harvesting the populist anger against the Allies that carpet bombing had aroused in working-class neighborhoods. At the same time, he massively reinforced his incomparable network of surveillance and terror, ensuring that any seed of discontent would be promptly destroyed before it could germinate into a larger conspiracy. If the British were dumbly oblivious to the possibility that "morale bombing" actually strengthened the Nazi state, Goebbels's own internal enemies had no doubts:

> The terror of the bombings forged men together. In rescue work there was no time for men to ask one another who was for and who against the Nazis. In the general hopelessness people clung to the single fanatical will they could see, and unfortunately Goebbels was the personification of that will. It was disgusting to see it, but whenever that spiteful dwarf appeared, people still thronged to see him and felt beatified to receive an autograph or a handshake from him.[36]

The RAF clung with fanaticism to its flawed paradigm. Harris convinced Churchill—whose own penchant was for massive, first-use poison gas attacks—that "we can [still] wreck Berlin from end to end if the US air force will come in on it. It will cost us between 400–500 aircraft. It will cost Germany the war."[37] In late winter and spring 1944, as the sensational new American long-range fighters began to give B-17s unprecedented protection over eastern Germany, the Eighth Air Force, while still theoretically selecting only precision targets, became partners with British area bombers in a series of thousand-plane raids on what the crews always called "the Big City." The offensive culminated in April with a second carpet-bombing of bolshevik Wedding and its red sister, Pankow. One and a half million Berliners were made homeless, but industrial output, once again, quickly rebounded.[38]

Operation Thunderclap

Roosevelt had thus far in the war reconciled the divergent philosophies of strategic bombing by accepting at the 1943 Casablanca Conference the British concept of a Combined Bomber Offensive "to undermine the morale of the German people," but at the same time preserving the Army Air Force's tactical option for daylight, precision targets. After Hitler retaliated for D-Day with his V-1 and then V-2 attacks on London, this compromise became untenable. Indeed, Churchill's initial reaction to Germany's secret weapons was to demand poison gas attacks or worse on Berlin: "It is absurd to consider morality on this topic," he hectored RAF planners in early July, "I want the matter studied in cold blood by sensible people, and not by psalm-singing uniformed defeatists."[39]

As Barton Bernstein has shown, Churchill asked Roosevelt to speed up the delivery of 500,000 top-secret "N-bombs" containing deadly anthrax, which had been developed at Dugway's Granite Peak complex.[40] The RAF, writes Bernstein, "was putting together a bombing plan for the use of anthrax against six German cities: Berlin, Hamburg, Stuttgart, Frankfurt, Aachen, and Wilhelmshafen. The expectation was that 40,000 of the 500-pound projectiles, containing about 4.25 million four-pound bombs, could kill at least half the population 'by inhalation,' and many more would die later through skin absorption."[41]

Poison gas and anthrax were too much for the White House, but Roosevelt passionately wanted to offer a gift to the British. In August 1944, he complained angrily to his Secretary of Treasury, Henry Morgenthau Jr.: "We have got to be tough with Germany and I mean the German people not just the Nazis. We either have to castrate the German people or you have got to treat them in such a manner so they can't just go on reproducing people who want to continue the way they have in the past."[42] Churchill the same month proposed to FDR "Operation Thunderclap," an RAF plan that would guarantee to "castrate" 275,000 Berliners (dead and injured) with a single 2000-bomber super-raid against the city center. Roosevelt, following Chief of Staff George Marshall's advice, accepted the plan in principle.[43]

Key Air Force leaders were disturbed by the unsavory character of Thunderclap. Major General Laurence Kuter protested to colleagues that "it is contrary to our national ideals to wage war against civilians." Intelligence chief McDonald railed against a plan that "repudiates our past purposes and practices ... [and] places us before our allies, the neutrals, our enemies and history in conspicuous contrast to the Russians whose preoccupation with wholly military objectives has been as notable as has been our own up to this time."[44] Lt. General Carl Spaatz, the commander of the US bombers in Europe, had "no doubt ... that the RAF want very much to have the US Air Forces tarred with the morale bombing aftermath which we feel will be terrific." (Spatz was already smarting from international criticism of the hideous civilian casualties, more than 12,000 dead, caused by an errant American "precision" raid on Bucharest in September.)[45] War hero Jimmy Doolittle, the Eighth Air Force's commander, remonstrated bitterly after being ordered by Eisenhower to be ready to drop bombs "indiscriminately" on Berlin.[46]

Nor did Air Force commanders in Europe easily buy the argument of planners in Washington who thought that Stalin had grown too potent on the battlefield and needed a dramatic demonstration of the destructive power of Allied bombers. The RAF Air Staff had added that frosting to Thunderclap's cake in an August 1944 briefing: "A spectacular and final object lesson to the German people on the consequences of universal aggression would be of continuing value in the postwar period. Again, the total devastation of the centre of a vast city such as Berlin would offer incontrovertible proof to all peoples of the power of a modern air force. ... [I]t would convince our Russian allies and the Neutrals of the effectiveness of Anglo-American air power."[47]

In the end, Thunderclap (which now included Dresden and Leipzig in its menu) was unleashed for competing and contradictory reasons, having as much to do with starting the Cold War as with ending the Second World War. Meanwhile, the murderous potential of what American planners called "promiscuous bombardment" had been dramatically increased by the influx of hundreds of thousands of panicked refugees fleeing the advancing Red Army in early 1945. When the leaden winter skies finally cleared over Berlin on 3 February, Doolittle stubbornly withheld his more vulnerable B-24s, but sent in 900 B-17s and hundreds of fighter escorts. It was not the *Gotterdammerung* that the British had envisioned, but 25,000 Berliners nonetheless perished while deep under the burning Reich Chancellery Hitler listened to Wagner.[48]

Dresden, a month later, was closer to the original apocalyptic conception of Thunderclap. Although the last unscathed city on Harris's bombing menu, the approaching Red Army had not requested its targeting. Crowded with desperate refugees, slave laborers, and Allied prisoners, the cultural center's only strategic role was as a temporary transport junction on the imploding Eastern Front. "The impetus within British circles to attack Dresden itself came more from Churchill," whose objective, as always, was "increasing the terror." Thus American bombers concentrated on the railyards, while the British went after the residential areas. "Dresden's marginal war industries, though sometimes cited as justification for the attacks, were not even targeted."[49]

It was the biggest firestorm since Hamburg: "complete burnout" in the jargon of ecstatic British planners. The death toll, given the huge number of refugees, is unknowable, although estimates range from 35,000 to 300,000. After reducing it to cinder, Harris savagely bombed the city again with high explosives to kill off the survivors in the cellars. An official history called it Bomber Command's "crowning achievement."[50] The RAF then infuriated Spaatz and Doolittle with a gloating press conference that implied that the US Army Air Force now fully embraced Churchillian strategy. (The AP wire read: "Allied air bosses have made the long-awaited decision to adopt deliberate terror bombing of the great German population centers as a ruthless expedient to hasten Hitler's doom.")[51]

Back in Berlin, Hitler, who had always hated the city and its bolshevik-infected working class, issued his infamous "Nero" order. Every civic installation and structure of potential value to the Russians was to be systematically destroyed in advance of their arrival. When Speer protested that "such demolitions would mean the death of Berlin," the Fuhrer responded that this was exactly his intention. "If the war is lost, the nation will also perish. Besides, those who remain after the battle are only the inferior ones, for the good ones will have been killed." The end of the Reich would be a vast exercise in terminal eugenics.[52]

Roosevelt's endorsement of Thunderclap, which paved the way for US complicity in Dresden, was a moral watershed in the American conduct of the war. The city burners had finally triumphed over the precision bombers. By committing the Air Force to British doctrine in Germany, Thunderclap also opened the door to the Zoroastrian Society alumni who wanted an unrestricted incendiary campaign against Japan. The hundred thousand or so civilians whom the Eight Air Force burnt to death in the cities of eastern Germany during the winter of 1945 were but a prelude to the one million Japanese consumed in the B-29 autos-da-fé later that spring.

The secret napalm tests at Dugway's "Japanese Village" and later at Eglin

Field's "Little Tokyo" in Florida, together with Curtis Le May's experimental "incendiary only" raid on the Chinese city of Hankow in December 1944, gave American planners the confidence that they could achieve bombing pioneer Billy Mitchell's old dream of incinerating Japan's "paper cities" ("the greatest aerial targets the world has ever seen").[53] The Committee of Operations Analysts—whose Brahmin membership included Thomas Lamont of J. P. Morgan, W. Barton Leach of Harvard Law, and Edward Mead Earle of Princeton's Institute of Advanced Study—was convinced it had cracked the scientific puzzle of how to generate holocausts whose "optimum result" would be "complete chaos in six [Japanese] cities killing 584,00 people." In the event, the Twenty-First Bomber Command's attack on Tokyo on 10 March 1945 exceeded all expectations: General Norstad described it as "nothing short of wonderful."[54]

The target of "Operation Meetinghouse"—the most devastating air raid in world history—was Tokyo's counterpart to Wedding or the Lower East Side, the congested working-class district of Asakusa. The Fifth Air Force's commander, Curtis Le May, regarded the Japanese in the same way that a Heydrich or an Eichmann regarded Jews and Communists: "We knew we were going to kill a lot of women and kids when we burned that town. Had to be done. ... For us, there are no civilians in Japan."[55] Since Japan had hardly any nightfighters, Le May stripped his B-29 Superfortresses of armaments in order to make way for maximum bombloads. Two thousand tons of napalm and magnesium incendiaries were dropped in the dense pattern that Dugway tests had shown to maximize both temperature and fire spread. The resulting inferno (*Akakaze* or "red wind" in Japanese) was deadlier than Hiroshima, killing an estimated 100,000 people. American "know-how" manufactured the fires of hell.

Most died horribly as intense heat from the firestorm consumed the oxygen, boiled water in canals, and sent liquid glass rolling down the streets. Thousands suffocated in shelters and parks; panicked crowds crushed victims who had fallen in the streets as they surged toward waterways to escape the flames. Perhaps the most terrible incident came when one B-29 dropped seven tons of incendiaries on and around the crowded Kokotoi Bridge. Hundreds of people turned into fiery torches and "splashed into the river below in sizzling hisses." One writer described the falling bodies as resembling "tent caterpillars that had been burned out of a tree." Tail gunners were sickened by the sight of the hundreds of people burning to death in flaming napalm on the surface of the Sumida River. ... B-29 crews fought super-heated updrafts that destroyed at least ten aircraft and wore oxygen masks to avoid vomiting from the stench of burning flesh.[56]

The macabre "success" of the raid, which made Le May the most "profitable" air commander of the war, was kept secret from the US public for nearly three months. Then, on 30 May, the *New York Times* shrieked with proud hyperbole: "1,000,000 Japanese Are Believed to Have Perished." As Air Force historian Thomas Searle dryly notes, "few Americans complained."[57] The horrors of Hiroshima and Nagasaki a few months later were mere anticlimax to the million deaths in Tokyo that most Americans believed had already been inflicted in revenge for Pearl Harbor. The mass extermination of Japanese civilians had passed the muster of public opinion long before the *Enola Gay* locked Hiroshima's city hall into its bombsight.

These ghosts of the Good War's darkest side—perhaps two million Axis civilians—still haunt the lifeless waste around German Village. The ghastly history of modern incendiary warfare is archived here. Now that Potsdamer Platz and the other open wounds of Berlin's history have been healed into showpieces of reunified prosperity, Mendelsohn's forlorn Mietskasernen suddenly seems monumental: reproof to the self-righteousness of punishing "bad places" by bombing them. German Village is Berlin's secret heartache, whispering in the contaminated silence of the Utah desert.

2002

Notes

1. Department of the Interior, *Historic Properties Report: Dugway Proving Ground*, Washington, D.C. 1984.

2. What follows is taken from Standard Oil Development Company, "Design and Construction of Typical German and Japanese Test Structures at Dugway Proving Grounds, Utah," 27 May 1943 (copy provided by Dugway public relations office).

3. Antonin Raymond, *An Autobiography*, Rutland, Vt. 1973. "As I was then very busy

at Fort Dix, New Jersey, we constructed a prefabrication factory near Fort Dix, and established a line of trucks from Dix in Jersey to the Utah Proving Grounds, thousands of miles away, to transport the prefabricated parts. The parts were then assembled at the Proving Grounds and were subjected to bombarding. As soon as they were destroyed, new ones were erected, until the result was satisfactory. The buildings were fully furnished with *futon, zabuton* and everything that one finds usually in a Japanese house. They even had *amado* (sliding shutters), and bombarding was tried at night and in the daytime, with the *amado* closed or open" (p. 189).

4. Louis Fieser, *The Scientific Method: A Personal Account of Unusual Projects in War and Peace*, New York 1964, pp. 129–30; and Kenneth Werrell, *Blankets of Fire: US Bombers over Japan During World War II*, Washington, D.C. 1996, p. 49.

5. Barry Watts, *The Foundations of US Air Doctrine*, Maxwell Air Force Base 1984, p. 106.

6. Bruno Zevi, *Erich Mendelsohn*, London 1985, p. 140 (my emphasis).

7. Richard Pommer and Christian Otto, *Weissenhof 1927 and the Modern Movement in Architecture*, Chicago 1991, pp. 156–57.

8. Alexander Richie, *Faust's Metropolis: A History of Berlin*, London 1999, p. 533.

9. Robert Pape, *Bombing to Win: Air Power and Coercion in War*, Ithaca, N.Y. 1996, p. 64.

10. John Terraine, *The Right of the Line: The Royal Air Force in the European War, 1939–1945*, London 1985, p. 259.

11. Ibid., p. 263. Advocates of terror bombing, like Cherwell, Trenchard, Sir Charles Portal, and, of course, Bomber Command's Arthur Harris, argued that a sustained campaign would bring complete defeat of the Reich by 1944 with the help of only "a relatively small land force" (p. 504).

12. "Air policy, Bomber Command policy, the whole course of the strategic offensive, were now drawn inexorably towards the method which not even Churchill now called 'extermination,' although 'morale' would still be widely used, and even more generally, 'area bombing'": ibid., p. 262.

13. Lee Kennett, *A History of Strategic Bombing*, New York 1982, p. 129.

14. Terraine, p. 507. "This was a prescription for massacre; nothing more or less."

15. Ronald Schaffer, *Wings of Judgment: American Bombing in World War II*, New York 1985, p. 102.

16. Ibid., p. 92.

17. Conrad Crane, *Bombs, Cities and Civilians: American Airpower Strategy in World War II*, Lawrence, Kans. 1993, pp. 29–30 and 34–37. Unfortunately, the secretary of war found no significant constituency that echoed his scruples. "Robert Oppenheimer recalled that Stimson thought it was 'appalling' that no one protested the heavy loss of life caused by the air raids against Japan" (p. 37).

18. Ibid., p. 58.

19. Kenneth Hewitt, "Place Annihilation: Area Bombing and the Fate of Urban Places," *AAAG* 73, no. 2 (1983), p. 272; Werrell, p. 41; Sherry, p. 156; and Crane, pp. 32–33; 1.3 million incendiary bombs were dropped on Hamburg. The death toll has been estimated

at 45,000, but "exact figures could not be obtained out of a layer of human ashes" (Brooks Kleber and Dale Birdsell, *The Chemical Warfare Service: Chemicals in Combat* [United States Army in World War II], Washington, D.C. 1966, p. 619).

20. Crane, p. 91; and Kleber and Birdsell, pp. 617–19. See also Michael Sherry, *The Rise of American Air Power: The Creation of Armageddon*, New Haven, Conn. p. 227

21. Schaffer, p. 93.

22. Crane, p. 24.

23. Charles McNihols and Clayton Carus, "One Way to Cripple Japan: The Inflammable Cities of Osaka Bay," *Harper's*, June 1942.

24. To discover the best method of shattering German morale, one Ohio State psychologist proposed to "make guinea pigs" of civilian Nazi internees to discover what fears or torments would be most demoralizing (Schaffer, p. 91).

25. Schaffer, p. 109

26. Sherry, p. 232.

27. Jack Couffer, *Bat Bomb: World War II's Other Secret Weapon*, Austin, Tex. 1992.

28. Horatio Bond, "The Fire Attacks on German Cities" in National Fire Protection Association, *Fire and the Air War*, Boston 1946, pp. 86 and 243 (see also p. 125).

29. Pape, p. 61. The same idea surfaced later in planning for the fire raids on Japan. "Back in the fall of 1944, when the Joint Target Group was planning the firebombing raids, Professor Crozier had suggested that the air force could intensify class hostility if it destroyed slum areas while leaving wealthier districts intact" (Schaffer, p. 136).

30. Charles Messenger, *"Bomber" Harris and the Strategic Bombing Offensive, 1939–1945*, New York 1984, p. 15.

31. Anthony Read and David Fisher, *The Fall of Berlin*, London 1992, p 130.

32. Ibid., pp. 141–42.

33. Ibid., p. 142; Sherry, p. 156.

34. Kennett, p. 154.

35. Richie, p. 583.

36. Berlin Police Commissioner von Helldorf quoted in Ralf Reuth, *Goebbels*, New York 1993, p. 335.

37. Messenger, p 142; and Stephen Garrett, *Ethics and Airpower in World War II: The British Bombing of German Cities*, New York 1993, p. 17.

38. Crane, pp. 90–91 [poison gas]; and Richie, p. 531.

39. Barton Bernstein, "Churchill's Secret Biological Weapons," *Bulletin of the Atomic Scientists*, Jan./Feb. 1987, p. 49. Churchill had long been an enthusiast of chemical warfare against civilians; as, for example when he notoriously advocated its use against Pushtan villages during the Third Afghan War in 1919 (p. 45).

40. Dugway also conducted extensive testing of phosgene, cyanogen, hydrogen cyanide, and other deadly airborne agents. Some contaminated areas of the base are rumored to be quarantined "for at least 1000 years."

41. Bernstein, p. 50.

42. FDR's surprising metaphor is symptomatic of an elite culture steeped in eugenical values. If, in his past career, he had not been an outspoken public zealot of negative eugenics and forced sterilization like Churchill and Hitler, he certainly shared the mindset: believing, for example, that the Japanese had "less developed skulls" (Crane, p. 120).

43. Ibid., pp. 115–18.

44. Schaffer, p. 102.

45. Crane, pp. 98 and 117.

46. Richard Davis, "Operation 'Thunderclap," *Journal of Strategic Studies*, pp. 94 and 105: "The mission is unique among the approximately 800 Eighth Air Force missions flown under USSTAF's command, for the nature and vehemence of Doolittle's objection to his targets" (p. 105).

47. Davis, p. 96. Likewise, US General David Schlatter: "I feel that our air forces are the blue chips with which we will approach the postwar treaty table, and that [Thunderclap] will add immeasurably to their strength, or rather to the Russian knowledge of their strength" (Schaffer, p. 96).

48. Army Air Forces, p. 726

49. Sherry, p. 260.

50. Crane, p. 115; Garrett, p. 20. When asked by one of Churchill's aides about the effects of the attack, Harris replied: "Dresden? There is no such place as Dresden" (Garrett, p. 42).

51. Schaffer, pp. 98–100.

52. Cf. Michael Burleigh, *The Third Reich*, New York 2000, pp. 789–91; and Robert Payne, *The Life and Death of Adolf Hitler*, New York 1973, p. 541.

53. Kennett, p. 164; and Sherry, p. 58.

54. Schaffer, pp. 111–20 and 138.

55. Ibid., p. 142; and Crane, p. 133.

56. Ibid., p. 132.

57. Thomas Searle, "'It Made a Lot of Sense to Kill Skilled Workers': The Firebombing of Tokyo in March 1945," *Journal of Military History* 66 (January 2002), p. 122. As Searle emphasizes, the incendiary bombing of Japanese cities was luridly reported in the US daily press (albeit with time delays because of military censorship). There can be little doubt that most Americans were aware of the scale and horror of the campaign, including the probable incineration of thousands of small children and their mothers.

Fake cactus (Las Vegas, 1994)

4

Las Vegas Versus Nature

It was advertised as the biggest nonnuclear explosion in Nevada's history. On 27 October 1993, Steve Wynn, the state's official "god of hospitality," flashed his trademark smile and pushed the detonator button. As 200,000 Las Vegans cheered, the Dunes Hotel, former flagship of the Strip, slowly crumbled to the desert floor. The giant dust plume was visible from the California border.

Nobody in Nevada found it the least bit strange that Wynn's gift to the city that he so adores was to blow up an important piece of its past. This was simply urban renewal Vegas-style: one costly facade destroyed to make way for another. Indeed, the destruction of the Dunes merely encouraged other corporate casino owners to blow up their obsolete properties with equal fanfare: the Sands, of Rat Pack fame, came down in November 1996, while the Hacienda was dynamited at the stroke of midnight on New Year's Eve. Extravagant demolitions have become Las Vegas's version of civic festivals.

In place of the old Dunes, Wynn's Mirage Resorts is completing the $1.25 billion Bellagio, a super-resort with lakes large enough for jet-skiing, created using water that came from the allotments of the original Dunes golf course. Wynn's purchase of the Dunes solved his water problem, but not that of other developers of resorts. Impresario Sheldon Adelson, who is building the $2 billion, 6000-room

Venetian Casino Resort on the site of the Sands, with gondolas floating on artificial canals, has not explained where his water will come from; neither has Circus Circus Enterprises, which is transforming the old Hacienda into Project Paradise, "an ancient forbidden city on a lush tropical island with Hawaiian-style waves and a swim-up shark exhibit."[1]

In the five years since Wynn blew up the Dunes, $8 billion has been invested in thirteen major properties along the Strip alone. As a result, the Sphinx now shares an adjacent street address with the Statue of Liberty, the Eiffel Tower, Treasure Island, the Land of Oz, and, soon, the Piazza San Marco. And the boom, still breaking all records in 1997, shows every sign of continuing.[2]

By obscure coincidence, the demolition of the Dunes followed close on the centenary of Frederick Jackson Turner's legendary "end of the frontier" address to the World's Columbian Exposition in Chicago, where the young prairie historian meditated famously on the fate of the American character in a conquered and rapidly urbanizing West. Turner questioned the survival of frontier democracy in the emergent epoch of giant cities and trusts (not to mention Coney Island and movies) and wondered what the West would be like a century hence.

Steve Wynn and the other robber barons of the Strip think they know the answer: Las Vegas is the terminus of western history, the end of the trail. As an overpowering cultural artifact it bestrides the gateway to the twenty-first century in the same way that Burnham's "White City" along the Chicago lakefront was supposed to prefigure the twentieth century. At the edge of the millennium, this strange amalgam of boomtown, world's fair, and highway robbery is the fastest growing metropolitan area in the United States. (It is also, as we shall see, the brightest star in the neon firmament of postmodernism.)

More than 30 million tourists had their pockets picked by its one-armed bandits and crap tables in 1996: a staggering 33 percent increase since 1990. (By the time you read this, Vegas should be hard on the heels of Orlando, Florida, which is, with 35 million visitors to Walt Disney World and Universal-MGM Studios, the world's premiere tourist destination.) While Southern California suffered through its worst recession since the 1930s, Las Vegas has generated tens of thousands of new jobs in construction, gaming, security, and related services. As a consequence, nearly a thousand new residents, half of them Californians, arrive each week.[3]

Some of the immigrants are downwardly mobile blue-collar families—the Californians are called "reverse Okies" by locals—desperately seeking a new start in the Vegas boom. Others are affluent retirees headed straight for a gated suburb in what they imagine is a golden sanctuary from the urban turmoil of Los Angeles. Increasing numbers are young Latinos, the new bone and sinew of the casino-and-hotel economy. In spring 1995, Clark County's population passed the one million mark, and anxious demographers predicted that it will grow by another million before 2010.[4]

Environmental Terrorism

The explosive, and largely unforeseen, growth of southern Nevada has dramatically accelerated the environmental deterioration of the American Southwest. Las Vegas long ago outstripped its own natural-resource infrastructure, and its ecological "footprint" now covers all of southern Nevada and adjacent parts of California and Arizona. The hydro-fetishism of Steve Wynn (he once proposed turning downtown's Fremont Street into a pseudo-Venetian Grand Canal) sets the standard for Las Vegans' prodigal overconsumption of water: 360 gallons daily per capita versus 211 in Los Angeles, 160 in Tucson, and 110 in Oakland. In a desert basin that receives only 7 to 8 inches of annual rainfall, irrigation of lawns and golf courses (60 percent of Las Vegas's total water consumption)—not to mention artificial lakes and lagoons—adds the equivalent of another 20 to 30 inches of rainfall per acre.[5]

Yet southern Nevada has little water capital to squander. As Johnny-come-lately to the Colorado Basin water wars, it has to sip Lake Mead through the smallest straw. At the same time, reckless groundwater overdrafts in the Las Vegas Valley are producing widespread and costly subsidence of the city's foundations. The Strip, for example, is several feet lower today than in 1960, and sections of some subdivisions have had to be abandoned.[6]

Natural aridity dictates a fastidiously conservative water ethic. Tucson, after all, has prospered on a reduced water ration: its residents actually seem to prefer

having cactus instead of bermuda grass in their front yards. But Las Vegas haughtily disdains to live within its means. Instead, it is aggressively turning its profligacy into environmental terrorism against its neighbors. "Give us your water, or
we will die," developers demand of politicians grown fat on campaign contributions from the gaming industry. Las Vegas is currently pursuing two long-term
and fundamentally imperialist strategies for expanding its water resources.

First, the Southern Nevada Water Authority is threatening to divert water
from the Virgin River (a picturesque tributary of the Colorado with headwaters
in Zion National Park) or steal it from ranchers in sparsely populated central
Nevada. In 1989 the Authority (then called the Las Vegas Valley Water District)
stunned rural Nevadans by filing claims on more than 800,000 acre-feet of surface
and groundwater rights in White Pine, Nye, and Lincoln Counties.[7]

This infamous water grab ("cooperative water project" in official parlance)
brought together an unprecedented coalition of rural Nevadans: ranchers,
miners, farmers, the Moapa Band of Paiutes, and environmentalists. Their battle
cry has been "Remember the Owens Valley," in reference to Los Angeles's notorious annexation of water rights in that once-lush valley on the eastern flank of
the Sierra Nevada: an act of environmental piracy immortalized in the film *Chinatown*. Angry residents of the Owens Valley blew up sections of the Los Angeles
Aqueduct during the 1920s, and some central Nevadans have threatened to do the
same to any pipeline highjacking local water to Las Vegas.[8]

Since 1966, the Authority, without abandoning its legal claims to central
Nevada water, has put more emphasis on withdrawing Virgin River water directly
from its terminus in Lake Mead. This conforms to its second, and more important, strategy of increasing Las Vegas's withdrawal of Colorado River water
stored in Lake Mead or in downstream reservoirs. To circumvent the status quo
of the Colorado River Compact, the Authority has teamed up with the powerful Metropolitan Water District of Southern California in what most observers
believe is the first phase of a major water war in the Southwest.

Las Vegas and the Los Angeles area want to restructure the allocation of
Colorado River water away from agriculture and toward their respective metropolitan regions. In most scenarios, this involves a raid on Arizona's allotment,

and Arizona's governor, J. Fife Symington III, retaliated by allying himself with water managers in San Diego and the Imperial Valley. (Other major players in this anti–Metropolitan Water District coalition include the billionaire Bass brothers from Fort Worth, who have bought up tens of thousands of acres of choice agricultural land in the El Centro area in order to sell their federally subsidized water allotments to San Diego.)[8]

Through one or another of these machiavellian gambits, the Authority's general manager, Pat Mulroy, has assured the gaming industry that Clark County will have plenty of water for continued breakneck growth over the next generation. Independent water experts have criticized Mulroy's optimistic projections, however, and one of them, Hal Furman, caused a small sensation in February 1997 with his assertion that "southern Nevada will run dry shortly after the turn of the century." In the event of such a crisis, Las Vegas's last resort probably would be "to help subsidize the costly process of desalting Pacific Ocean water in exchange for some of California's Colorado River share." This, however, would almost certainly double the current artificially low acre-foot cost of water.[10]

Compounding the problem of future water supply is the emergent crisis of water quality in Lake Mead, which operates as both reservoir and wastewater sink for Las Vegas. Federal researchers in 1997 discovered that "female egg protein in blood plasma samples of male carp was causing widespread reproductive deformation in the fish. This catastrophic endocrine disruption, with potential human genetic impact as well, is likely related to the large amounts of toxic waste, especially pesticides and industrial chemicals, that are discharged into the lake."[11] In 1994, moreover, thirty-seven people, most with AIDS, died as a result of a lethal protozoan, *Cryptosporidium parvum*, that experts from the national Centers for Disease Control surmised was carried in tap water drawn from Lake Mead. Public health researchers have become alarmed by the coincidence of these two outbreaks at a time when hypergrowth is overwhelming regional water and waste treatment capabilities. As one biologist recently asked on the op-ed page of the *Las Vegas Review-Journal*, "Will more people die from *Cryptosporidium* contamination of our drinking water when we put more wastewater back in the lake?"[12]

Finally, to return to yet another *Chinatown* parallel, watchdog groups such as the Nevada Seniors Coalition and the Sierra Club are increasingly concerned that the Southern Nevada Water Authority's $1.7 billion water delivery system from Lake Mead, currently under construction, may be irrigating huge speculative real-estate profits along metropolitan Las Vegas's undeveloped edge. For example, one major pipeline (the so-called South Valley Lateral) runs through an area near the suburb of Henderson, where private investors recently acquired huge parcels in a complicated land swap with the Bureau of Land Management, which controls most of Las Vegas's desert periphery. This is the same formula—undervalued land plus publicly subsidized water—that made instant millions for an "inside syndicate" when the Los Angeles Aqueduct was brought to the arid San Fernando Valley in 1913.[13]

Southern Nevada is as thirsty for fossil fuels as it is for water. Most tourists naturally imagine that the world's most famous nocturnal light show is plugged directly into the turbines of nearby Hoover Dam. In fact, most of the dam's output is exported to Southern California. Electricity for the Strip, as well as for the two million lights of downtown Las Vegas's new (and disconcerting) "Fremont Street Experience," is primarily provided by coal-burning and pollution-spewing plants on the Moapa Indian Reservation northeast of the city, and along the Colorado River. Only 4 percent of Las Vegas's current electrical consumption comes from "clean" hydropower. Cheap power for the gaming industry, moreover, is directly subsidized by higher rates for residential consumers.[14]

Automobiles are the other side of the fossil fuel problem. As Clark County's transportation director testified in 1996, the county has the "lowest vehicle occupancy rate in the country" in tandem with the "longest per person, per trip, per day ratio." Consequently, the number of days with unhealthy air quality is dramatically increasing. Like Phoenix and Los Angeles before it, Las Vegas was once a mecca for those seeking the restorative powers of pure desert air. Now, according to the Environmental Protection Agency, Las Vegas has supplanted New York as the city with the fifth highest number of days with "unhealthy air" (as measured from 1991 to 1995). Its smog already contributes to the ochre shroud over the Grand Canyon and is beginning to reduce visibility in California's new East Mojave National Recreation Area as well.[15]

Las Vegas, moreover, is a major base camp for the panzer divisions of motorized toys—dune buggies, dirt bikes, speed boats, jet-skis, and the like—that each weekend make war on the fragile desert environment. Few western landscapes are as a result more degraded than the lower Colorado River Valley, which is under relentless three-pronged attack by the leisure classes of southern Nevada, Phoenix, and Southern California.

In the blast-furnace heat of the Colorado River's Big Bend, Las Vegas's own demon seed, Laughlin, has germinated kudzulike into an important gambling center. Skyscraper casinos and luxury condos share the west bank with the megapolluting Mojave Power Plant, which devours coal slurry pumped with water stolen from Hopi mesas hundreds of miles to the east. Directly across the river, sprawling and violent Mohave County, Arizona—comprising Bullhead City and Kingman—provides trailer-park housing for Laughlin's nonunion, minimum-wage workforce, as well as a breeding ground for antigovernment militias. (It was here that Timothy McVeigh worked as a security guard while incubating his *Turner Diary* fantasies of Aryan vengeance.)

Hyperbolic Growth

The Las Vegas "miracle," in other words, demonstrates the fanatical persistence of an environmentally and socially bankrupt system of human settlement and confirms Edward Abbey's worst nightmares about the emergence of an apocalyptic urbanism in the Southwest. Although postmodern philosophers (who don't have to live there) delight in the Strip's "virtuality" or "hyperreality," most of Clark County is stamped from a monotonously real and familiar mold. Las Vegas, in essence, is a hyperbolic Los Angeles, the Land of Sunshine on fast-forward.

The historical template for all low-density, resource-intensive southwestern cities was the great expansion of the 1920s that brought two million Midwesterners and their automobiles to Los Angeles County. This was the "Ur" boom that defined the Sunbelt. Despite the warnings of an entire generation of planners and environmentalists chastened by the 1920s boom, regional planning and

open-space conservation again fell by the wayside during the post-1945 population explosion in Southern California. In a famous article for *Fortune* magazine
in 1958, sociologist William Whyte described how "flying from Los Angeles to
San Bernardino—an unnerving lesson in man's infinite capacity to mess up his
environment—the traveler can see a legion of bulldozers gnawing into the last
remaining tract of green between two cities." He baptized this insidious growth
form "urban sprawl."[16]

Although Las Vegas's third-generation sprawl incorporates some innovations
(casino-anchored shopping centers, for example), it otherwise recapitulates, with
robotlike fidelity, the seven deadly sins of Los Angeles and its Sunbelt clones such
as Phoenix and Orange County. Las Vegas has (1) abdicated a responsible water
ethic; (2) fragmented local government and subordinated it to private corporate
planning; (3) produced a negligible amount of usable public space; (4) abjured
the use of "hazard zoning" to mitigate natural disaster and conserve landscape;
(5) dispersed land-use over an enormous, unnecessary area; (6) embraced the
resulting dictatorship of the automobile; and (7) tolerated extreme social and,
especially, racial inequality.[17]

In mediterranean California or the desert Southwest, water use is the most
obvious measure of the environmental efficiency of the built environment.
Accepting the constraint of local watersheds and groundwater reservoirs is a
powerful stimulus to good urban design. It focuses social ingenuity on problems
of resource conservation, fosters more compact and efficient settlement patterns,
and generates respect for the native landscape. In a nutshell, it makes for "smart"
urbanism (as seen in modern Israel, or the classical city-states of Andalucia and
the Maghreb), with a bias toward continual economies in resource consumption.[18]

Southern California in the early Citrus era, when water recycling was at a
premium, was a laboratory of environmental innovation, as evinced by such
inventions as the bungalow (with its energy-efficient use of shade and insulation), solar heating systems (widespread until the 1920s), and state-of-the-art
sewage and wastewater recovery technologies. Its departure from the path of
water rectitude, and thus smart urbanism, began with the Owens Valley aqueduct

and culminated in the 1940s with the arrival of cheap, federally subsidized water from the Colorado River. Hoover Dam extended the suburban frontier deep into Southern California's inland basins and in the process underpriced traditional water conservation practices such as sewer-farming and stormwater recovery out of existence.

Unlike Los Angeles, Las Vegas has never practiced water conservation or environmental design on any large scale. It was born dumb. Cheap water has allowed it to exorcise even the most residual semiotic allusion to its actual historical and environmental roots. Visitors to the contemporary Strip, with its tropical islands and Manhattan skylines, will search in vain for any reference to the Wild West (whether dude ranches or raunchy saloons) that themed the first-generation casinos of the Bugsy Siegel era. The desert, moreover, has lost all positive presence as landscape or habitat; it is merely the dark, brooding backdrop for the neon Babel being created by Wynn and his competitors.

Water profligacy likewise dissolves many of the traditional bonds of common citizenship. Los Angeles County is notorious for its profusion of special-interest governments—"phantom cities," "county islands," and geographical tax shelters—all designed to concentrate land use and fiscal powers in the hands of special interests. Clark County, however, manages to exceed even Los Angeles in its radical dilution and dispersal of public authority.

The Las Vegas city limits, for example, encompass barely one-third of the metropolitan population (versus nearly half in the city of Los Angeles). The major regional assets—the Strip, the Convention Center, McCarran International Airport, and the University of Nevada, Las Vegas (UNLV)—are located in an unincorporated township aptly named Paradise, while poverty, unemployment, and homelessness are disproportionately concentrated within the boundaries of the cities of Las Vegas and North Las Vegas.

This is a political geography diabolically conceived to separate tax resources from regional social needs. Huge, sprawling county electoral districts weaken the power of minorities and working-class voters. Unincorporation, conversely, centralizes land-use decision making in the hands of an invisible government of gaming corporations and giant residential and commercial-strip developers.

In particular, the billion-dollar corporate investments along the Strip—with their huge social costs in terms of traffic congestion, water and power consumption, housing, and schools—force the fiscally malnourished public sector to play constant catchup. This structural asymmetry in power between the gaming corporations and local government is most dramatically expressed in the financing of the new public infrastructures to accommodate casino expansion and the growth of tourism. A classic example is the Southern Nevada Water Authority's new water distribution system, whose bonds are guaranteed by sales taxes: the most regressive means available, but the only significant source of undesignated state revenue in Nevada.[19]

Contrary to neoclassical economic dogmas and trendy "public choice" theory, corporate-controlled economic development within a marketplace of weak, competing local governments is inherently inefficient. Consider, for example, the enormous empty squares in the urbanized fabric of Las Vegas, dramatically visible from the air, that epitomize the leapfrog pattern of development that planners have denounced for generations in Southern California because it unnecessarily raises the costs of streets, utilities, and schools. Crucial habitat for humans (in the form of parks), as well as for endangered species such as the desert tortoise, is destroyed for the sake of vacant lots and suburban desolation.

Similarly, both Los Angeles and Las Vegas zealously cultivate the image of infinite opportunity for fun in the sun. In reality, however, free recreation is more accessible in older Eastern and Midwestern cities that cherish their parks and public landscapes. As long ago as 1909, experts were warning Los Angeles's leaders about the region's looming shortage of parks and public beaches. Although the beach crisis was partially ameliorated in the 1950s, Los Angeles remains the most park-poor of major American cities, with only one-third the usable per capita open space of New York City.[20]

Las Vegas, meanwhile, has virtually no commons at all: just a skin-flint 1.4 acres per thousand residents, compared with the recommended national minimum of 10 acres. This park shortage may mean little to the tourist jet-skiing across Lake Mead or lounging by the pool at the Mirage, but it defines an impoverished quality of life for thousands of low-wage service workers who live in the

stucco tenements that line the side streets of the Strip. Boosters' claims about
hundreds of thousands of acres of choice recreational land in Clark County refer
to car-trip destinations, not open space within walking distance of homes and
schools. One is not a substitute for the other.[21]

Some of the most beautiful desert areas near Las Vegas, moreover, are now
imperiled by rampant urbanization. Developers are attempting to raise land
values by privatizing natural amenities as landscape capital. The local chapter of
the Sierra Club, for example, has recently mobilized against the encroachment
of Summerlin West, a segment of the giant Summerlin planned community that
is the chief legacy of Howard Hughes, upon Red Rock Canyon National Conservation Area—native Las Vegans' favorite site for weekend hikes and picnics.
The project, as endorsed by the Las Vegas City Council (which was subsequently
allowed to annex the development), encompasses 20,000 homes, two casinos, five
golf courses, and nearly 6 million square feet of office and commercial space. As
one local paper put it, most environmental activists were "less [than] enthused
about the possibility of lining one end of Red Rock Canyon, one of the valley's
most pristine landmarks, with casinos, businesses and homes."[22]

The recreation crisis in Sunbelt cities is the flip side of the failure to preserve
native ecosystems, another consequence of which is the loss of protection from
natural hazards such as floods and fires. The linkage between these issues is part
of a lost legacy of urban environmentalism espoused by planners and landscape
architects during the City Beautiful era. In 1930, for example, Frederick Law
Olmsted Jr., the greatest city designer of his generation, recommended "hazard
zoning" to Los Angeles County as the best strategy for reducing the social costs
of inevitable floods, wildfires, and earthquakes. In his sadly unrealized vision,
development would have been prohibited in floodplains and fire-prone foothills.
These terrains, he argued, were best suited for preservation as multipurpose
greenbelts and wilderness parks, with the specific goal of increasing outdoor
recreational opportunities for poorer citizens.[23]

Las Vegas is everything Olmsted abhorred. Its artificial deserts of concrete
and asphalt, for example, have greatly exacerbated its summer flash-flood problem (probably the city's best-kept secret, except on occasions, as in 1992, when

unsuspecting tourists drown in casino parking lots). Like Los Angeles, Clark
County has preferred to use federal subsidies to transform its natural hydrology
(the valley literally tilts toward the Colorado River) into an expensive and failure-
prone plumbing system rather than use zoning to exclude development from the
arroyos and washes that should have become desert equivalents to Olmsted's
greenbelts. (The recent declaration of a desert wetlands park in the Las Vegas
Wash riparian corridor is a belated half-measure.)[24]

Off Worlds

Los Angeles was the first world metropolis to be decisively shaped in the era of
its greatest growth by the automobile. One result was the decentralization of
shopping and culture and the steady atrophy of its downtown district. Recently a
group of theorists at the University of California at Irvine have suggested that we
are seeing in Orange County and in other "edge cities" the birth of a "postsuburban metropolis" where traditional central-place functions (culture and sports,
government, high-end shopping, and corporate administration) are radically dispersed among different centers.[25]

Whether or not this is truly a general tendency, contemporary Las Vegas
recapitulates Orange County in an extreme form. The gaming industry has irre-
sistibly displaced other civic activities, with the partial exception of government
and law, from the center to the periphery. Tourism and poverty now occupy the
geographical core of the metropolis. Other traditional downtown features, such
as shopping areas, cultural complexes, and business headquarters, are chaotically
strewn across the Las Vegas Valley with the apparent logic of a plane wreck.

Meanwhile its booming suburbs stubbornly reject physical and social integration with the rest of the city. To use the nomenclature of the futuristic movie
Blade Runner, they are self-contained "off worlds," prizing their security and
social exclusivity above all else. Planning historian William Fulton has recently
described suburban Las Vegas as a "back to the future" version of 1950s Southern
California: "It is no wonder that the Los Angeles homebuilders love Las Vegas.
Not only can they tap into a Los Angeles–style market with Los Angeles–style
products, but they can do things the way they used to do them in the good old

days in L.A." As Fulton points out, while Southern California homebuilders must now pay part of the costs of new schools and water systems, Vegas developers "pay absolutely no fees toward new infrastructure."[26]

The most ambitious of Las Vegas's "off worlds" is Summerlin. Jointly developed by the Summa and Del Webb corporations, and named after one of Howard Hughes's grandmothers, it boasts complete self-sufficiency (it's "a world within itself," according to one billboard slogan) with its own shopping centers, golf courses, hospitals, retirement community, and, of course, casinos. "Our goal is a total community," explains Summerlin president Mark Fine, "with a master plan embracing a unique lifestyle where one can live, work, and play in a safe and aesthetic environment." (Residents rather than the corporations pay for key infrastructural investments, such as the new expressway from Las Vegas, through special assessment districts.) When Summerlin is finally completed in the early twenty-first century, a population of more than 200,000 living in twenty-six income- and age-differentiated "villages" will be hermetically sealed in Las Vegas's own upscale version of Arizona's leaky Biosphere.[27]

The formerly gritty mill town of Henderson, southeast of the Strip, has also become a major growth pole for walled, middle-income subdivisions, and it is becoming Nevada's second-largest city. (For optimal advantage in its utility and tax obligations, Summerlin is divided between the city of Las Vegas and unincorporated Clark County.) On the edge of Henderson is the larval Xanadu of Lake Las Vegas: a Wynnian fantasy created by erecting an eighteen-story dam across Las Vegas Wash. "The largest privately funded development under construction in North America," according to a 1995 brochure, Lake Las Vegas (controlled by the ubiquitous Bass brothers of Fort Worth) is sheer hyperbole, including $2 million lakefront villas in a private gated subdivision *within* a larger guard-gated residential community. The Basses' grand plan envisions the construction of six major resorts, anchored by luxury hotels and casinos, as well as five world-class golf courses, in addition to "restaurants and retail shops that will be the upscale alternative for Las Vegas."[28]

Las Vegas's centrifugal urban structure, with such gravitationally powerful edge cities as Summerlin and Henderson–Lake Las Vegas, reinforces a slavish dependence on the automobile. According to trendy architectural theorists such as Robert Venturi and Denise Scott Brown, whose *Learning from Las Vegas* has been a founding text of postmodernism, Las Vegas Boulevard is supposed to be the apotheosis of car-defined urbanism, the mother of strips. Yet the boom of the last decade has made the Strip itself almost impassable. Las Vegas Boulevard is usually as gridlocked as the San Diego Freeway at rush hour, and its intersection with Tropicana Road is supposedly the busiest street corner in the nation.[29]

As a result, frustrated tourists soon discover that the ride from McCarran Airport (immediately adjacent to the Strip) to their hotel frequently takes longer than the plane flight from Los Angeles. The Brobdingnagian scale of the properties and the savage summer heat, not to mention the constant assault by hawkers of sex-for-sale broadsheets, can turn pedestrian expeditions into ordeals for the elderly and families with children. The absence of coherent planning for the Strip as a whole (the inescapable consequence of giving the gaming corporations total control over the development of their sites) has led to a series of desperate, patchwork solutions, including a few new pedestrian overpasses. In the main, however, the Nevada Resort Association—representing the major gaming corporations—is relying on new freeways and arterials to divert cross-traffic from the Strip and a proposed $1.2 billion monorail to speed customers between the larger casino-hotels.

For most of the 1990s, contemporary Las Vegas has been one vast freeway construction site. Nothing has been learned from the dismal California experience, not even the elementary lesson that freeways increase sprawl and consequently the demand for additional freeways. When completed, the new Las Vegas freeway network will allow most local commuters to bypass the Strip entirely, but it will also centrifuge population growth farther into the desert, with correspondingly higher social costs for infrastructure and schools.

Meanwhile, the Nevada Resort Association has concentrated its overwhelming political clout to ensure that a proposed increase in the 8 percent hotel room tax is spent exclusively on its own Resort Corridor Transportation Master Plan (the monorail). Having engineered the financing of the new water delivery system with regressive sales tax increases, the gaming industry has opposed all

efforts by desperate Clark County School District officials to divert part of the
room tax increase to school construction. As in previous tax fights, school and
welfare advocates are vastly outnumbered by the Resort Association's hired guns.
Nevada is the most notoriously antitax state in the country, and gaming industry
lobbyists, their coffers swollen with the profits of the boom, dominate the legislature in Carson City. The recent flood of retirees to Las Vegas's suburbs has
only reinforced the antitax majority. (Paradoxically, the Clark County electorate is
aging, while the actual median age of the population—thanks especially to young
Latino immigrants and their families—is declining.)[30]

One index of the extraordinary power wielded by the Resort Association
is the fact that the relative contribution of gaming taxes to state revenue actually declined during the *annus mirabilis* of 1995 when hotel-casino construction
broke all records. Yet the industry, shaken by the local "Rodney King" riots in
spring 1992, is not unconscious that eroding education quality and social services
will eventually produce social pathologies that may undermine the city's resort
atmosphere. Their calculated solution, after months of top-level discussion in
the winter of 1996–97, has been to volunteer the room tax increase—which is
directly passed on to tourists and then spent exclusively on the Strip monorail—as
a "heroic" act of social responsibility. This reduced the tax heat on the casino
owners while conveying the clear message, scripted by Resort Association lobbyists, that the time had come for homebuilders and small-business owners to
make a contribution to school finance. As columnist John Smith pointed out,
"By coming out first, they shift the focus away from a potential gaming revenue
tax increase [which would come out of their pockets] and raise the question of
responsibility of Southern Nevada's developers and shopkeepers."[31]

In the meantime, the previous decade's hypergrowth without counterpart
social spending has increased economic inequality throughout Clark County.
Despite the feverish boom, the supply of jobless immigrants has far outpaced
the demand for new workers in the unionized core of the gaming economy. The
difference translates into a growing population of marginal workers trapped in
minimum-wage service jobs, the nonunion gaming sector, the sex industry, and
the drug economy. According to one estimate, Las Vegas's homeless population

increased 750 percent during the superheated boom years of 1990–95. At the same time, a larger percentage of Las Vegans lack health insurance than the inhabitants of any other major city. Likewise, southern Nevada is plagued by soaring rates of violent crime, child abuse, mental illness, lung cancer, epidemic illness, suicide, and—what no one wants to talk about—a compulsive gambling problem that is a major factor in family pathologies.[32]

This obviously provides a poor setting for the assimilation of Las Vegas's new ethnic and racial diversity. Despite consent decrees and strong support for affirmative action from the Culinary Workers Union, the gaming industry remains far from achieving racial or gender equality in hiring and promotions. In the past, Las Vegas more than earned its reputation as "Mississippi West." While African American entertainers such as Sammy Davis Jr. and Nat King Cole were capitalizing the Strip with their talent, Blacks were barred from most hotels and casinos, except as maids, through the 1960s. Indeed, a comparative study during that period of residential discrimination across the United States found that Las Vegas was the "most segregated city in the nation."[33]

More recently, persistent high unemployment rates in the predominantly Black Westside precipitated four violent weekends of rioting following the Rodney King verdict in April 1992. Interethnic tensions, exacerbated by a relatively shrinking public sector, have also increased as Latinos replace African Americans as the valley's largest minority group. Indeed, Black leaders have warned of "creeping Miamization" because some casino owners prefer hiring Latino immigrants instead of local Blacks. Latinos, for their part, point to overcrowded schools (Latinos now constitute 40 percent of the elementary school population in the city of Las Vegas), police brutality, and lack of representation in local government.[34]

Greening the Urban Desert

Let's return, once again, to Las Vegas and the end of western history. In his apocalyptic potboiler, The Stand (1992), Stephen King envisioned Las Vegas as Satan's earthly capital, with the Evil One literally enthroned in the MGM Grand. Many environmentalists, together with the imperiled small-town populations of Las Vegas's desert hinterlands, would probably agree with this characterization of the

Glitterdome's diabolical *zeitgeist*. No other city in the American West seems to be as driven by occult forces or as unresponsive to social or natural constraints. Like Los Angeles, Las Vegas seems headed for some kind of eschatological crackup (in the King novel, Satan ultimately nukes himself).

Confronted with the Devil himself, and his inexorable plan for two-million-plus Las Vegans, what can the environmental community do? The strategic choices are necessarily limited. On one hand, environmentalists can continue to defend natural resources and wilderness areas one at a time against the juggernaut of development: a purely defensive course that may win some individual victories but is guaranteed to lose the larger war. On the other hand, they can oppose development at the source by fighting for a moratorium on further population growth in the arid Southwest. Pursued abstractly, however, this dogmatic option will only pigeonhole Greens as enemies of jobs and labor unions. Indeed, on the margins, some environmentalists may even lose themselves in the Malthusian blind alley of border control, by allying themselves with nativist groups that want to deport hardworking Latino immigrants whose per capita consumption of resources is only a small fraction of that of their native-born employers.

A better approach, even if utopian in the short or medium run, would focus comprehensively on the character of desert urbanization. "Carrying capacity," after all, is not just a linear function of population and the available resource base; it is also determined by the social form of consumption, and that is ultimately a question of urban design. Cities have incredible, if largely untapped, capacities for the efficient use of scarce natural resources. Above all, they have the potential to counterpose public affluence (great libraries, parks, museums, and so on) as a real alternative to privatized consumerism, and thus cut through the apparent contradiction between improving standards of living and accepting the limits imposed by ecosystems and finite natural resources.

In this perspective, the most damning indictment against the Sunbelt city is the atrophy of classical urban (and pro-environmental) qualities such as residential density, pedestrian scale, mass transit, and a wealth of public landscapes. Instead, Los Angeles and its postmodern clones are stupefied by the ready availability of artificially cheap water, power, and land. Bad design, moreover, has unforeseen environmental consequences, as illustrated by southern Nevada's colossal consumption of electric power. Instead of mitigating its desert climate through creative design (for example, proper orientation of buildings, maximum use of shade, minimization of heat-absorbing "hardscape," and so on), Las Vegas, like Phoenix, simply relies on universal air-conditioning. But, thanks to the law of the conservation of energy, the waste heat is merely exported into the general urban environment. As a result, Las Vegas is a scorching "heat island" whose nightly temperatures are frequently 5 to 10 degrees hotter than the surrounding desert.

Fortunately, embattled western environmentalists have some new allies. In their crusade for the New Urbanism, Peter Calthorpe, Andreas Duany, and other young environmentally conscious architects have reestablished a critical dialogue between urban designers and mainstream environmental groups, particularly the Sierra Club. They have sketched, with admirable clarity, a regional planning model that cogently links issues of social equity (economically diverse residential areas, recreational equality, greater housing affordability through elimination of the need for second cars, and a preferential pedestrian landscape for children and seniors) with high-priority environmental concerns (on-site recycling of waste products, greenbelts, integrity of wetland ecosystems, wildlife corridors, and so on). They offer, in effect, elements of a powerful program for uniting otherwise disparate constituencies—inner-city residents, senior citizens, advocates of children, environmentalists—all of whom are fundamentally disadvantaged by the suburban, automobile-dominated city.

The New Urbanism has had many small successes in northern and central California, the Pacific Northwest, and other areas where preservation of environmental quality commands a majority electoral constituency. In the Southwest, by contrast, the Summerlin model—with its extreme segregation of land uses and income groups, as well as its slavish dependence on cheap water and energy—remains the "best practice" standard of the building industry. (Only Tucson, with its self-imposed environmental discipline, constitutes a regional exception.) The West, in other words, is polarizing between housing markets where the New Urbanism has made an impact and those where 1960s Southern California templates remain hegemonic. In the case of Las Vegas, where the contradictions of hypergrowth and inflexible resource demand are most acute, the need for an alternative settlement model has become doubly urgent.

The New Urbanism by itself is a starting point, not a panacea. A Green politics for the urban desert would equally have to assimilate and synthesize decades of international research on human habitats in drylands environments. It would also have to consider the possible alternatives to a regional economy that has become fatally dependent on a casino-themepark monoculture. And it would need to understand that its major ally in the long march toward social and environmental justice must be the same labor movement (particularly the progressive wing represented by unions such as the Las Vegas Culinary Workers) that today regards local environmental activists with barely disguised contempt. These are the new labors of Hercules. Creating a vision of an alternative urbanism, sustainable and democratic, in the Southwest is an extraordinary challenge. But this may be the last generation even given the opportunity to try.

1998

Notes

1. Dave Berns, "Venice in Las Vegas," *Las Vegas Review-Journal*, 27 November 1996; Gary Thompson, "Paradise to Be Part of Vegas Strip," *Las Vegas Sun*, 19 November 1996.

2. Michael Hiltzik, "Stakes Raised Ever Higher in Las Vegas Building Boom," *Los Angeles Times*, 24 December 1996.

3. Darlene Superville, "L.V. Grew Fastest in Nation," *Las Vegas Review-Journal*, 2 October 1995; "Nevada No. 1 in Job Growth," *Las Vegas Review-Journal*, 13 February 1996.

4. Hal K. Rothman, *Devil's Bargains: Tourism and Transformation in the Twentieth-Century American West*, Lawrence, Kans. 1998; Ed Vogel, "Growth Figures Revised," *Las Vegas Review-Journal*, 4 November 1995.

5. Data from telephone interviews with metropolitan water authorities.

6. *Annual Report, 1992–1993*, Las Vegas: Clark County Flood Control District, 1993.

7. Jon Christensen, "Will Las Vegas Drain Rural Nevada?" *High Country News*, 21 May 1990.

8. See Jon Christensen, "Betting on Water," in Mike Davis and Hal Rothman, eds., *The Grit Beneath the Glitter: Tales from the Real Las Vegas*, Berkeley, Calif. 2002.

9. William Kahrl, "Water Wars about to Bubble Over," *San Bernardino Sun*, 4 February 1996; Mike Davis, "Water Pirates," *Los Angeles Weekly*, 23–29 February 1996.

10. Susan Greene, "Water Outlook Revised," *Las Vegas Review-Journal*, 20 February 1997.

11. Frank Clifford, "Lake Mead Carp Deformed," *Las Vegas Review-Journal*, 19 November 1996.

12. Larry Paulson, "Leading the Charge against Growth," *Las Vegas Review-Journal*, 17 January 1997.

13. Ibid.

14. Jay Brigham, "Lighting Las Vegas," in Davis and Rothman, eds., *Grit Beneath the Glitter*.

15. Data provided by the Environmental Protection Agency. See also Keith Rogers, "Scientists Tackle Dirty Air in LV.," *Las Vegas Review-Journal*, 15 January 1996.

16. Mike Davis, "How Eden Lost Its Garden," in Allen Scott and Edward Soja, eds., *The City*, Berkeley, Calif. 1997.

17. Ibid.

18. For a first-rate discussion of the principles of sustainable urban design, with particular application to Southern California case-studies, see John Tillman Lyle, *Design for Human Ecosystems*, New York 1985.

19. For an authoritative discussion of Las Vegas's tax "scissors" (that is, exploding social needs versus artificially constrained tax capacity), see Eugene Moehring, "Growth, Urbanization, and the Political Economy of Gambling, 1970–1996," in Davis and Rothman, eds., *Grit Beneath the Glitter*.

20. Davis, "How Eden Lost Its Garden."

21. *Parks and Recreation Master Plan, 1992–1997*, Las Vegas, Nev.: Clark County Parks and Recreation Department, 1992.

22. Mike Zapler, "Huge Project OK'd Next to Red Rock," *Las Vegas Review-Journal*, 28 January 1997.

23. *Parks, Playgrounds, and Beaches for the Los Angeles Region*, Los Angeles: Olmsted Brothers and Bartholomew, 1930, esp. pp. 97–114.

24. Letters in *Las Vegas Review-Journal*, 1 February 1996.

25. Stuart Olin et al., *Postsuburban California*, Berkeley, Calif. 1992.

26. William Fulton, *The Reluctant Metropolis: The Politics of Urban Growth in Los Angeles*, Point Arena, Calif. 1997, pp. 307–8.

27. Sam Hall Kaplan, "Summerlin," *Urban Land* (September 1994), p. 14.

28. Adam Steinhauer, "Lake Las Vegas Resort Planned," *Las Vegas Review-Journal*, 13 December 1996.

29. Robert Venturi, *Learning from Làs Vegas: The Forgotten Symbolism of Architectural Form*, Cambridge, Mass. 1977.

30. Lisa Bach, "Panel Hears Mass Transit Options," *Las Vegas Review-Journal*, 7 February 1997; Susan Greene, "Schools, Roads Plan Nears," *Las Vegas Review-Journal*, 25 February 1977.

31. John Smith, "Gaming Industry Hits PR Jackpot with Hotel Tax Proposal," *Las Vegas Review-Journal*, 27 February 1997.

32. Robert Parker, "The Social Costs of Rapid Urbanization in Southern Nevada," in Davis and Rothman, eds., *Grit Beneath the Glitter*.

33. Mike Davis, "Racial Cauldron in Las Vegas," *Nation*, 6 July 1992.

34. Ibid.

1946 memorial, Lapahoehoe Point

5

Tsunami Memories

A land which bares itself to tourism veils itself meta-
physically: it sells scenery at the expense of magic.
Gerhard Nebel

Hilo, according to that indefatigable Victorian globetrotter Isabella Bird Bishop
(*Six Months in the Sandwich Islands*), "is the paradise of Hawai'i. What Honolulu
attempts to be, Hilo is without effort." Some of the lotus-eating ambience, to be
sure, has eroded from the capital of the Big Island since portly Isabella was rowed
ashore in 1875. With the construction of the modern breakwater, few "athletes,
like the bronzes of the Naples Museum, ride the waves on their surf-boards,"
and the morning commute, complete with mini-gridlock on Highway 11, has
replaced the "brilliant dressed riders" in their *mu'umu' us* and leis cantering along
the beach.

Yet the fundamental things still apply: a greenness that makes Ireland look
gray encircling a perfect crescent bay with Pele's immense thrones, Mauna Loa
and snow-capped Mauna Kea, piercing the cloudbank in the background. Unlike
balmy but deeply troubled Honolulu, which debates whether it is metamorphos-
ing into Orange County or Las Vegas, soggy Hilo (ten feet of annual rainfall—a

formidable deterrent to gentrification) retains a homespun Hawaiian character. Culturally it is an *ohana* (extended family) not a *malihini* (stranger's) town, and, despite the recent eruption of a regional mall (Prince Kuhio Plaza) and a handful of sub-Waikiki hotel towers, it is still possible to agree with Bishop that this city of 45,000 just below the Tropic of Cancer is "clothed in poetry."

Yet Hilo's broad grin also displays some missing teeth. The city has warily retreated from its waterfront. The scores of cafés, cheap hotels, and warehouses, as well as the famed Hilo Theater, that once lined the bay side of Kamehameha Avenue downtown are now a parking strip. Further south the shoreline neighborhood of Shinmachi ("new town") has become Mooheau Park, while all that remains of boisterous Waiakea Town (or Yashijima) on the Wailoa River is the Social Settlement clock tower, its hands forever frozen at 1:04 a.m. An important part of Hilo past—the very heart of its Japanese working-class tradition—has disappeared.

The Tsunami Museum

Hilo's parks and open spaces gracefully dissimulate a tragic history. No inhabited landscape in the United States (or colonized Polynesia, if you prefer) is so regularly convulsed by extreme natural forces. If Mauna Loa's lavas have repeatedly threatened, but at the last minute, miraculously spared the city (thanks to native Hawaiian prayer or Army Air Corps bombs), the Pacific has been crueler. As the windward flank of the Hawaiian archipelago, Hilo and the adjacent Hamakua Coast are the bullseye for tidal waves originating from earthquakes anywhere in a vast arc from Kamchatka to Tierra del Fuego.

The energy in the earthquakes is bundled in a succession of huge swells that travel across the ocean at nearly 500 miles an hour. Arriving ten to twenty minutes apart, the waves are notoriously unpredictable and the first wave—contrary to popular belief—is not necessarily the largest.

Twice since the Second World War, truly mountainous tsunamis have surged over the breakwater. The "April Fools' Day" 1946 waves—unleashed by a giant earthquake in the Aleutians—devastated Shinmachi and downtown (57 dead), as well as inundating the school at Laupahoehoe Point, twenty-six miles north of

Hilo, and drowning 24 teachers and students. Another 61 people perished in the early morning of 22 May 1960 when an even bigger tsunami hit downtown with such force that all the parking meters along Kamehameha Avenue were bent to the ground. The Waiakea district with its famous fish market and sampan landing, largely spared in 1946, was virtually obliterated.

To mitigate future destruction, the bay front was redeveloped into a greenbelt buffer zone, which, in addition to the famous Waiakea clock, includes a memorial plaza dedicated to the Shinmachi victims of 1946. Beyond this swelling wall of black lava, meant to suggest the lethal beauty of a tsunami, the Army Corps of Engineers constructed a large landfill plateau safely above the 1960 high-water mark. As a gesture of confidence, state and federal offices were relocated here, along with the aptly named Kaiko'o ("violent seas") Mall.

Tsunamis recently have come to vie with orchids and hula (the annual "Merrie Monarch Festival") as Hilo's principal civic symbol. It is a populist phenomenon that defies the rule that cities always repress the memory of disaster. Tourists, after all, are not invited to celebrate "Earthquake Day" in San Francisco or to tour the "Terrorism Museum" in Oklahoma City. Nor did Hilo itself, until the last decade, care much about monumentalizing its unique exposure to the combined wraths—volcanoes, earthquakes, tsunamis, and hurricanes—of nature.

The turning point was the coincidence in 1996 of the fiftieth anniversary of the 1946 tragedy with the "Last Harvest" on the Big Island's sole remaining sugar plantation at Ka'u. For more than a century sugar had been the chief livelihood of Hilo and the neighboring Puna and Hamakua coasts. The region's multi-ethnic popular culture had been shaped by generations of shared toil in the cane fields, as well as by the heroic battle to organize the plantations. (Hawai'i, thanks to local activists of the ILWU, became the unique successful instance of high-wage agricultural unionization in US history.) The death of the island's sugar industry—a victim of nonunion Florida mills and Midwest corn sweeteners—has been the one disaster from which recovery, so far, has proven impossible.

Economic crisis has in turn given urgency to public memory. The dramatic outpouring of tsunami recollections in 1996, which led to the establishment of the Pacific Tsunami Museum two years later, has been a therapeutic catalyst for a larger dialogue about community survival and intergenerational continuity in the face of a decade-long state and local recession that has forced so many young families to emigrate to the mainland. (Hawaiian labor migration to Las Vegas is one of the ten largest interstate population streams, and an estimated 40,000 islanders now live in the Glitterdome.) "Talking tsunami," moreover, potentially opens the way for the recovery of other, more controversial collective memories, including Hilo's militant labor history and the struggle to defend native Hawaiian sovereignty.

Even a casual visitor to the museum, housed in a massive art deco bank building on Kamehameha Avenue that withstood the furies of 1946 and 1960, ends up spellbound by the homeric tales of the survivors who come to "talk story" and visit friends. (The local newspaper now devotes a regular column to these narratives.) Yet the terrain of memory is complex, and the museum is obviously a coalition of different agendas. It is my outsider's impression—and I apologize for any caricature—that at least four kinds of interests participate in the current memorialization of tsunami as "island heritage."

The "boosters"—*qua* the character in John Sayles's *Limbo* who wants to Disneyfy the Alaskan wilderness—perceive tsunami history as a thrilling adventure-theme to attract more visitor dollars from the tourist-rich (and heavily Californianized) Kona side of the island. The "dowagers," like their matronly counterparts in other small towns, are concerned to preserve the hegemony of old ruling *haole* families over anything that smacks of genealogy or local history. The tsunami "buffs," meanwhile, are simply enthralled by the romance of these spectacular cataclysms. But the principal constituency of the museum is the "activists" (although some will disdain such a loaded term) struggling to preserve *ohana* values in the face of the centrifugal forces of deindustrialization, outmigration, and scattered gentrification. Two groups in particular have become exemplary grassroots historians. For both the oldsters of the Waiakea Pirates Athletic Club, custodians of the famous clock, and the youngsters of Laupahoehoe School, curators of an oral history of the 1946 tragedy, tsunamis are a way of evoking solidarity, courage, and community identity.

The Waiakea Pirates

Although the Hilo region offers a diverse menu of tsunami ruins and monuments (including the Shinmachi memorial, the site of the Hakalau Mill, and the
museum itself), it is the Waiakea Clock that tourists take home as an image on
a tee-shirt or even as a ten-inch replica. The clock, saved from the rubble in 1960
and restored to its original location in 1984 by the Waiakea Pirates, was originally
erected in 1939 outside the Social Settlement Gym to honor a prominent haole
planter's wife.

Settlement houses, of course, are usually associated with boweries and back-
of-the-yards neighborhoods. And, indeed, from the perspective of the Hawaiian
Board of Missions, which established the Social Settlement in 1900 "to make
Waiakea a clean, sober, industrious, wholesome and desirable community," this
largely Japanese town of 5000 people across from the Hilo Iron Works and next
to the port was an incipient slum. Certainly it had a tough, rambunctious atmosphere, and Waiakea kids were famous for *Our Gang* monikers that would not
have been out of place in Hell's Kitchen: "Spitoon" Hamano, "Wreck" Matsuno,
"Blackie" Takemura, "Bones" Kondo, "Cowboy Joe" Okahara.

More to the point, however, Waikaea was a source of haole anxiety because it
was the largest concentration of blue-collar workers—mill laborers, sampan fishermen, railroaders, and longshoremen—on the Big Island. Unlike the plantation
villages, moreover, it was not under the constant scrutiny of resident foremen and
managers. A self-contained world, it became a crucible for trade unionism and
the local Communist Party. In 1935 Harry Kamoku, a Chinese-Hawaiian from
Waiakea who had picketed with Harry Bridges during the San Francisco General
Strike, returned to lead a successful revolt of longshoremen at the Hilo Port. The
ILWU hall at 1383 Kamehameha Avenue in Waiakea (now the third hole of the
Country Club Hawai'i golf course) became the regional CIO headquarters.

On 1 August 1938, several hundred longshoremen, unionized laundry women,
and Waiakea residents marched to the wharves in a show of solidarity with the
striking CIO sailors of the Inter-Island Steam Navigation Company. As the
scab freighter S.S. *Waialeale* arrived from Honolulu, police and company goons

armed with shotguns and bayonets savagely attacked the peaceful demonstrators. Nearly a quarter of the crowd, including several children, were wounded in
what became notorious as the "Hilo Massacre." As an outraged Harry Kamoku
reported to the ILWU in Honolulu: "They shot us down like a herd of sheep. We
didn't have a chance. The firing kept up for about five minutes. They just kept on
pumping buckshot and bullets into our bodies. They shot men in the back as they
ran. They shot men who were trying to help wounded comrades and women.
They ripped their bodies with bayonets." (After the war, Kamoku played a leading role in the great 1946 strike—79 days in duration—that organized the sugar
plantations on the Hamakua Coast north of Hilo.)

In this troubled era, the only passion that equaled Waiakea's enthusiasm for
the labor movement was baseball, and the Depression years were the golden age
of Hilo's powerhouse Commercial and Japanese Leagues. Founded in 1924 as
the first club open to all ethnic groups, the Waiakea Pirates quickly became the
Brooklyn Dodgers of island baseball. The big-league caliber of play at Hoolulu
Park was famously demonstrated in 1933 when a haole player from the mainland
went to bat against legendary Pirates' pitcher Taffy Okamura. After repeatedly
striking Babe Ruth out, Okamura was obliged to slow down his fast ball so that
the visitor could hit it. Old-timers still treasure the photograph of Ruth smiling
sheepishly next to the diminutive but triumphant Okamura.

In the postwar years, as the ILWU was winning strikes and organizing sugar
workers into a powerful union, the Pirates maintained their preeminence in the
new 100th Battalion Memorial Baseball League, dedicated to the celebrated Nisei
combat unit in which so many Waiakea men had served. Both the 1960 baseball
season and the sugar harvest were in full swing when the "earthquake of the
century" devastated the southern coast of Chile on 22 May.

It took fifteen hours for the rest of the seismic energy to cross the Pacific.
The warning sirens installed after 1946 began wailing at 8:30 that evening, but
reassuring reports from Tahiti of waves barely three feet high induced a false
sense of security. The Waiakea sampan fleet sought safety in the open ocean, but
dare-devil youngsters gathered at the Suisan Fish Market and along the Wailoa
River Bridge to wait for the tsunami. Some were worried that there would be

nothing to see. At 1:04 a.m., when the Settlement Clock stopped forever, they were suddenly confronted by a nearly vertical wall of black water thirty-five feet high. Roaring ashore, it crushed people, homes, and businesses as far inland as Kekuanaoa Street—almost a mile from the bay.

At the Pirates' seventy-fifth anniversary banquet last year, one of the guests of honor was nonagenarian Mrs. Ito, the most renowned survivor of the destruction of Waiakea. Her home had been uprooted by the wave, then sucked out to sea by its riptide. Unable to swim and less than five feet tall, she clung all night to a flimsy window screen surrounded by hammerhead sharks, before being rescued in the morning by the Coast Guard. Weaker characters might boast of the ordeal, but Mrs. Ito, with superb humility, simply recalled the indelible beauty of the star-jeweled night sky as she floated in the ocean.

Old Waiakea, thanks to the Pirates, also seems indelible. The oral histories and public recollections of the 1960 tragedy, together with the publication last spring of a town and team memoir by Richard Nakamura and Gloria Kobayashi, have rekindled its descendants' abiding affection. Local baseball, even if faded in glory since the homeric era of Lefty Okamura, still replenishes communal identity. Waiakea's militant labor history, on the other hand, has been largely excluded from commemoration. There is no memorial to the victims of the 1938 massacre, and, although the local ILWU hall is named after Harry Kamoku, there is an unsurprising tendency—in the absence of any strong union participation in the making of public memory—to stress "unifying" experiences like natural disasters or sports legends rather than the "divisive" legacies of racism or class struggle.

Laupahoehoe School

At Laupahoehoe, however, five generations of labor in the cane fields and sugar mills remain the pivot of community identity. Frank DeCaires, who was fourteen years old in 1946 when two sisters and a brother were swept away by a tsunami exactly where he is now sitting, at the grassy tip of breathtaking Laupahoehoe Point, is talking about mules. "We used mules in those days, you know, to help us bring the cane down slope. They were good animals and they worked hard, just like the rest of us. Sugar was a difficult way of life, but it made us into a family, close-knit and generous."

The kids sitting at his feet, a typical Hawaiian rainbow of half a dozen inter-mixed ethnicities, smile or nod. Mules are not a comprehensible category, but the tall sugar stalks growing wild in abandoned fields everywhere along the Hamakua Coast are like a distant relation. They are ancestral pride. None of these children will ever pull a graveyard shift in the Pepeekeo Mill or haul tons of cane through a tropical cloudburst in an eighteen-wheeled Peterbilt. But their parents and grand-parents did, just as their parents and grandparents before them cut the treacher-ous cane by hand and hauled it on their backs to wagons or flumes. Cane is family history and a subject deserving respect.

1 April 1999: a chance to see firsthand how history is transmitted on the Big Island. The Laupahoehoe School (or, rather, the "new school," several hundred feet above the ocean) has organized a day-long commemoration of the 1946 trag-edy. Although a weekday, a majority of Laupahoehoe's 152 families have come out to the Point. Since the final sugar harvest here in the fall of 1994, unemploy-ment has skyrocketed and the school has become more than ever the center of community life. As adults prepare lunch or trade gossip, students of all grades gather in "talk story" tents to listen to honored *kupunas* (elders) like DeCaires recollect about work and disaster.

In 1996–97, students transcribed and published a remarkable narrative based on interviews with thirteen tsunami survivors. It is chilling to hear the *kupunas* tell yet again how the sea suddenly retreated, leaving the ocean shelf exposed as far as the reef. It was 6:50 a.m., ten minutes before the first bell, and school buses were discharging kids from nearby plantation camps and hamlets. Some were playing baseball. Unconscious of the danger, they ran to the shore just beyond the teachers' cottages to view the shoals of fish writhing on the seabed. Before adults could recall them to safety, the main swell arrived: not a breaking wave, but an ever-rising, ever-encroaching wall of water.

Leonie Kawaihona Laeha, interviewed by students in 1996, remembers that "'it was very high, coming like, over the coconut trees. All the kids who had been down by the beach were running up the road and through the park.' When she saw that, she started to cry because she 'didn't realize what was happening, but just knew it was bad.'" Nearby, sixteen-year-old Yasu Gusukuma, "'gripped by a horror that would remain forever in her memory,' witnessed 'water coming from all sides of the point and boiling in the center. She saw cottages spinning on top of the water, she saw the grandstand collapse, then she turned and ran up the hill as fast as she could.'"

Some of the children, like DeCaires's adopted sister, Janet Yokoyama, were killed instantly as the tsunami dashed them against rocks and trees. More than a dozen others were swept alive into the shark-infested waters. "Some of the boys tried to swim out, tied to ropes, and grab people floating by, but weren't able to reach any. They helplessly watched many float by." As the screaming kids were sucked out to sea, David Kailimai, superintendent at a neighboring sugar mill, tried to borrow the area's only sailboat from a wealthy haole resident. More concerned about possible damage to his boat than the lost children, the owner stubbornly refused. It was 1 p.m., six hours after the disaster, before he finally relented. Kailimai, a doctor from the plantation clinic, and two others sailed north toward Kohala through an incredible junkyard of floating debris. Before nightfall they were able to rescue two injured boys clinging to a luahala tree and a teacher who had reached one of the rubber rafts dropped by Navy planes. The teacher was the doctor's fiancée. Three more children were rescued the next morning, but others, "seen on makeshift rafts down the coast," were never found.

After an hour or two of such stories, everyone converges on the impressive community center (built for a much larger population in the heyday of Laupahoehoe Sugar Company) for a feast of *luau* pig, rice, and *poi*. A few older people poke around sadly in the vine-covered ruins of the elementary kids' restrooms destroyed in 1946: the tiny, rusted sinks are especially poignant. Then the assembled community, led by the school honor guard, marches down to the modest lava monument engraved with the names of the dead children and their young teachers. There is a moving hymn in Hawaiian, ecumenical prayers from Catholics, Mormons, and Buddhists, and then the principal, Jane Uyehara, reminds her students that the point itself, "this beautiful, special place," is God's grace on the dead. Surrounded on three sides by thundering surf, it is indeed a magnificent monument.

It is also a long way from Littleton, Colorado and other suburban infernos. When asked what careers they imagine for themselves, half the kids reply that they want to be teachers at the school. Teachers, like those who died trying to save children from the monstrous wave in 1946, are still heroes. (Hawaiians, it should be noted, have always cherished education. In the 1880s, when it was still an independent nation, Hawai'i had an astonishing 96 percent literacy rate: higher than that of the United States as whole in the year 2000.) But many of these kids, inexorably, will be forced to leave Laupahoehoe.

Despite tax-funded schemes to turn displaced sugar workers into taro farmers or to reemploy them as low-wage "teleworkers" processing credit-card billings from the mainland, there are no longer enough jobs to sustain the community. Residents are bitterly divided over the jobs-versus-traditional-values proposal to build a maximum-security prison outside Hilo. Thousands of acres of cane fields, meanwhile, have been speculatively recycled into eucalyptus for wood pulp, turning Hamakua into a dangerous fire ecology but generating negligible employment. Sugar, it seems, has no substitute.

Thus, like hundreds—perhaps thousands—of other defunct plantation, fishing, mining, and hunting communities with a soul-stirring view and a heroic proletarian past, Laupahoehoe has little choice but to sell its charms to whomever it can. From Crete and Cornwall to Montana and Hawai'i, the gentrification of wild places (like that of urban centers) is always a theft of tradition, an uprooting of community. Eventually all the world's ruggedly beautiful landscapes of toil and struggle seem destined to be repackaged as "heritage," wrenched from unemployed locals and sold off to scenery-loving burghers fleeing the cities. This is the future of the incomparable Hamakua Coast, once the backbone of the militant Hawaiian labor movement. Inevitably, more aging *malihinis* (like myself) from California will fall under the spell of its beauty and supplant local families who will, in turn, end up in the sprawling, troubled outskirts of Los Angeles or Las Vegas. Their grandchildren will never talk story with the kapunas or learn the lore of the Giant Wave. This is the true "cloud of sadness," as one local writer put it, that hangs over the Big Island.

2001

HOLY GHOSTS

Gospel theater

6

Pentecostal Earthquake

Azusa Street

On the evening of 17 April 1906—just twelve hours before the San Andreas Fault shook San Francisco like a dusty old carpet—the Holy Ghost descended in a "rain of fire" upon the congregation of a small storefront church in downtown Los Angeles, inspiring thirty men and women to pray their hearts out on pine planks stretched across nail kegs. For months these dazed plainfolk Christians under the leadership of William Joseph ("Daddy") Seymour, a Black evangelist from Texas, had prayed nightly for "signs and wonders" like those recently witnessed by revivalist coal miners in Wales and missionaries in India. Suddenly, the worshipers—poor Blacks, whites, Mexicans, and Filipinos—were shaken by a mighty force that some compared to an "electric current," others to a "great wind." The Spirit animated each person in a different way. Some writhed, jerked, shimmied like Jello, and fell on top of each other; others bellowed with holy laughter or slew invisible devils. A woman from the Midwest carried on a lively conversation in fluent Chinese; a man sang hymns in a strange tongue that people later thought might have been Hindi, or perhaps Hungarian. It was difficult for the participants to convey fully what it was like to be "spirit drunk," although one man later volunteered that "each time I would come out from under the power, I would feel so sweet and clean, as though I had been run through a washing machine."

The following morning's terrible leveling of Babylon by the Bay only under-scored the momentous, possibly apocalyptic character of the Pentecost (the descent of the Holy Ghost) at 312 Azusa Street (today an alley in Little Tokyo). Sister Mary Galmond, a Black washerwoman from Pasadena, revealed that the Lord had sent her a vision of the San Francisco earthquake the previous year and had warned her that the destruction of Los Angeles, the "New Jerusalem," would soon follow. "He showed me an eagle flying over Los Angeles and lighting on the highest building, and as it did, the building began to crumble. I asked the Lord what it meant, and He said, 'The eagle means death and the crumbling is an earthquake. There will be a violent shock in the morning and one at night. There will be mangling and tangling with wires, the streetcar rails will bend and twist, and the telegraph wires will come down.'"

For three years the Holy Ghost kept the revival going with high-voltage pulses of libidinal energy that scandalized the corseted Protestant establishment in its gray granite citadels. Los Angeles's Pentecostals were denounced as "howling dervishes," "Sodomites," "demon-possessed seducers," and "the last vomit of Satan." Their visceral surrender to the Spirit was too explicitly sexual, too sub-versive and carnivalesque. Moreover, in emulation of primitive Christianity, they had transgressed the racial and gender boundaries of Teddy Roosevelt's America. "The color line," one of Seymour's white saints boasted, "was washed away in the blood of the lamb."

Equally heretical, the Pentecostals sometimes seemed to confuse Biblical eschatology with Marxist theory, as when Sister Galmond prophesied a great "War of Labor against Capital." "The Lord says, 'The time is coming when the poor will be oppressed and Christians can neither buy nor sell, unless they have the mark of the beast.' Then He says, 'The time will come when the poor man will say that he has nothing to eat, and work will be shut down. And the rich man will go and buy up all the sugar, tea, and coffee, and hold it in his store, and we cannot get it unless we have the mark of the beast.'"

The censure of the big churches only reinforced the conviction of Seymour and his fellow saints that they were the poor, despised people of God's Original

Tabernacle. The "awakening" quickly attracted growing numbers of lower-class people of every race, and soon the Holy Ghost was convulsing huge populist encampments in the oak and sycamore groves of Hermon, an agricultural community between downtown L.A. and Pasadena. "One sister saw a tongue of fire issuing out of the tabernacle. Her daughter also saw it. And a little boy who was in the power of the spirit in the tabernacle saw a ball of fire which broke and filled the whole place with light. Many were the heavenly anthems the Spirit sang through His people. And He gave many beautiful messages in unknown tongues, speaking of His soon coming...."

Daddy Seymour published these and other vivid descriptions of miracles in a newsletter, *The Apostolic Faith*, which was circulated nationally in evangelical circles as well as in overseas missions. For a decade, American Protestants had been roused by the millenarian injunction: "The evangelicalization of the world in our lifetime!" Old-guard evangelicals were beginning to take a serious interest in the Pentecostal earthquake in Los Angeles. Perhaps here was a force to move the heathens. What others dismissed as quackery or mass hysteria, they eventually embraced as the spiritual dynamo of a new global proselytization.

Angelus Temple

Pentecostalism's second wave in Los Angeles was led by Aimee Semple McPherson, who was plucked from Canada by an emissary of the Azusa Street Awakening. Her evangelical comet coincided with the waning boom of the late 1920s and the heartbreak of the Depression. She was God's cheerleader in hard times. "Sister," as everyone called her, introduced the Holy Ghost to Hollywood, and brought vaudeville and radio to the revival stage. She loved pageants, costumes, showers of rose petals, laughter, and spotlights, as well as big cars and handsome hustlers. She filled a vast trophy case in her great Angelus Temple in Echo Park with the discarded crutches, canes, and eye patches of the more than fifteen hundred people she claimed to have successfully touched with God's Healing Power. Even more wondrously, she transformed thousands of lonely and downwardly mobile exiles from the heartland into a crusading army of saints. With all her miracles and scandals, she stayed in the headlines longer than any movie star.

Sister also provided the perfect foil for Bohemian Los Angeles. Long before Edmund Wilson and other literary tourists peeked inside Angelus Temple, Louis Adamic—a brilliant, down-at-the-heels proletarian writer—was churning out ferocious lampoons of Aimee and her followers for the irreverent press of Emanuel Haldeman-Julius:

> The people who come to Aimee's temple are a disappointed, unhappy lot. Old, broken men and women.... They are farmers and petty merchants from Kansas, Illinois, Nebraska, Iowa, Ohio, Tennessee. They are the drudges of the farms and small-town homes, victims of cruel circumstances, victims of life, slaves of their biological deficiencies. They are diseased, neurotic, unattractive, sexually and intellectually starved, warped and repressed. For almost all purposes of this world they are dead.
>
> A stir in the crowd—Aimee appears. A woman in her early thirties, rather good-looking, at least from a distance; with beautiful, abundant hair, well-shaped hands, a pleasant smile; attractive, energetic; a magnetic and radiant personality. If she had not fallen in love with, and been "converted" by the young and beautiful evangelist at Ingersoll, Ontario, Sister Aimee would today be perhaps a stage or movie actress. Her voice must have been pleasant. but now it has grown raspy in the service of the Lord. Her mouth· must have been pretty, but now it has grown big through overexercise. She is full of business, directs everything. First a hymn. "Everybody sing, dear hearts, everybody!" Then: "Now everybody turn around and shake hands with at least four people near you." And I am compelled to take three or four hands—cold, soft, feeble, shaky hands that send shudders through me.

But the theatrical sentimentality that repulsed Adamic was ungirded by a truly sincere and often courageous egalitarianism. Racial bigotry and anti-Semitism were anathema to her Foursquare Gospel. In the official anthology of her sermons and crusades, *This Is That*, there is an astonishing account of how Sister Aimee persuaded a white audience in northern Florida to "please the Holy Ghost" by taking down the rope barrier segregating them from their Black Pentecostal brethren. As Jim Crow was temporarily dismantled, Sister led the applause, "as the Black saints came marching in." Needless to say, almost anyone else attempting to integrate a massive public gathering in the Deep South in the 1920s would have been lynched.

In one of her most famous sermons, she defied the rising power of the Ku Klux Klan in 1920s Los Angeles. When the Klan showed up in force at Angelus Temple, she preached a parable about an elderly Black farmer excluded from a segregated white church. The farmer was sitting dejectedly outside on the stone steps when a hobo Jesus put his hand on the man's shoulder. "Don't feel sad, my brother," said the hobo Jesus, a fellow traveler who seemed neither young nor old, but whose clothes and boots showed the wear of many days and nights on the road. "No, don't feel sad. I too have been trying to get into that church for many, many years. ..." As the Klansmen squirmed in the pews, Sister's voice rose to a crescendo. "You men who pride yourselves on patriotism, you men who have pledged to make America free for white Christianity, listen to me! Ask yourselves, how it is possible to pretend to worship one of the greatest Jews who ever lived, Jesus Christ, and then to despise all living Jews? I say unto you as our Master said, Judge not that ye be not judged!"

In 1936, finding that the spirit at Angelus Temple had grown "cold and dead," she threw the doors open to Black veterans of Daddy Seymour's movement. "A different speaker was scheduled for each hour, beginning with Sister at 6 a.m., and for several months climaxing with the triple whammy of A. Earl Lee, Ora B. (Hurricane) Hurley, and J. D. Long." A typical session included "flashing visions, flaming messages in tongues and interpretation, signs of billowing smoke, sheets of living fire, and floods of prophecies," culminating with the chant "We need the Rain." At this point, the orchestra struck up "Dancing in the Rain," and the congregation, Black and white together, danced in "soul-tingling ecstasy." Sister's Pentecost had become a utopian MGM musical.

The Crystal Cathedral

The epitome of suburbanized evangelicalism in the United States is undoubtedly Garden Grove's Crystal Cathedral, just a few freeway exits south of Disneyland in Orange County. Designed by pagan modernist Philip Johnson in 1980, its 25-story glass steeple and 10,000 silver-colored glass windows have replaced the drive-in movie theater that Robert A. Schuller used to build his ministry in the 1950s and 1960s. Schuller, who preached to his congregation from the top of a drive-in snack bar as they sat in their cars, is one of the great success stories in postwar Protestantism. He and his son control a media empire reaching twenty million viewers in 180 countries, and his reputation now rivals that of Billy Graham. Along with the Reverend Jesse Jackson, he has helped exorcise with prayer the many demons that infest the Clinton White House. In February 1997, the President invited him to sit next to the First Lady at the State of the Union Address (Schuller later remarked that "Jesus specializes in loving sinners").

Sundays in the Crystal Cathedral are to storefront revivals what a Vegas headline act is to unrehearsed burlesque (imagine Siegfried and Roy versus the early Eddie Cantor). Certainly the production values in Garden Grove are as awesome as Schuller's bank account: angels soar seventy feet above the congregation on invisible wires; mega-choirs, often led by country music celebrities, stamp out gospel hits to the accompaniment of the world's largest church organ. Schuller intones the same reassuring sermon of happiness and prosperity that has made his "Hour of Power" television program famous worldwide. There are no seizures, balls of fire, dancing in the aisles, or speaking in tongues. Everything is responsibly staged, without a hint of spontaneity, danger, or shamanism. The largely affluent congregation of six thousand just has a wholesome good time and then adjourns to the gift shop to purchase Cathedral-logo coffee mugs, "sun catchers," refrigerator magnets, and other holy knickknacks for the entire family.

The Light of the World

The glorious La Luz del Mundo is located across from the Pomona Freeway in East Los Angeles. Its neoclassical temple is decorated with a folkloric exuberance of pastel Bible animals and saints that on first sight suggest Santería or Voudoun. In fact, "The Light of the World" is second-generation Pentecostalism from Mexico, as interpreted by an apostle named Eusebio Joaquín González. A soldier in the Revolution, González received the good news of the Holy Ghost Awakening from *santos* who themselves had knelt in prayer with Daddy Seymour at Azusa Street. In 1926, while stationed in Monterrey, González was ordered by the Spirit to return to his native Jalisco and "restore the primitive church established by Jesus Christ." Rechristened Aarón, he became Guadalajara's thundering, charismatic counterpart to Sister Aimee. The extraordinary glass skyscraper zig-gurat that serves as the sect's world headquarters now looks down upon an entire neighborhood of saints (Hermosa Provincia) in that city.

The Los Angeles "Light of the World" is thus the vintage Seymour sacra-ment tropicalized and reimported to slake the spiritual thirst of the Los Angeles area's five million Latinos. Although not an anarchist democracy of revelation like Azusa Street, the First Street Templo is still a first-class emotional "washing machine," with a participatory spectacle that culminates in extraordinary *tableaux vivants* in the basement where members enact scenes from the New Testament. After this cleansing, the congregation leaves—not for brunch, but to conquer worlds in this End Time for the Holy Ghost.

Spanish-language Pentecostalism (of which La Luz del Mundo is merely an unusual instance) is probably the most important social movement in contempo-rary Los Angeles. It is certainly the ungentrified alternative to the suburban God mall with fast food and piped Muzak that passes for mainstream Protestantism. Although some on the Left dismiss these churches and their blue-collar saints as "reactionaries"—because of their alledged association with counterrevolution in Central America—there is no inherent reason why they should not claim the same social role as their older, African American counterparts, as flame-keepers of an uncompromising egalitarianism.

The fire that first ignited in Azusa Street has become a worldwide spiritual conflagration. It burns with a particular intensity wherever the emotional fuel is supplied by poverty and injustice: in Appalachian valleys, big-city ghettoes, migrant labor camps, Black townships in South Africa, and urban *colonias* in Central America. With an estimated fifty million denominational adherents, and perhaps twice that many "charismatic" cousins in more mainstream churches, Pentecostalism is today the fastest-growing religious movement in the world. And after decades of denying their origins in Black-led, racially mixed revivals, today's predominantly white Pentecostal groups now warmly acknowledge the Los Angeles movement led by William Seymour, who died in debt in 1922, as their common hearth.

1999

Bunker Hill (ca 1940)

7

Hollywood's Dark Shadow

Seeding city, saturated with dreams

...

Mysteries everywhere run like sap
Baudelaire

In the beginning, of course, Los Angeles was (in Orson Welles's words) simply a "bright, guilty place" without a murderous shadow or mean street in sight. Hollywood found its own Dark Place belatedly and only through a fortuitous amalgam of older, migrant sensibilities. Once discovered by hardboiled writers and exiled Weimar *auteurs*, however, Los Angeles's Bunker Hill in particular began to exert an occult power of place.[1] Here, overlooking L.A.'s monotonous, Midwestern flatness (Reyner Banham's "plains of Id"), was a hilltop slum whose decaying mansions and sinister rooming houses might have been envisioned by Edgar Allan Poe.[2] Its residents, "women with the faces of stale beer ... men with pulled-down hats" (Chandler), suggested Macheath's minions in Brecht and Weill's *The Threepenny Opera*. The Hill was broodingly urban and mysterious—everything that Los Angeles, suburban and banal from birth, was precisely not. With such star qualities, it is not surprising that it so quickly lodged itself in our nocturnal imagination. Disdaining to be mere location, Bunker Hill was soon stealing the best lines in film noir. Here are some fan notes on a brilliant career.

Five Points

Some cinematic mappings of the metropolis—*Berlin, Symphony of a City* (1927), *Le Million* (1931), *Rome, Open City* (1945)—were truly avant-garde in that they anticipated or paralleled similar conceptions in literature and painting. But studio film has generally preferred to meet the city on the familiar terms of literature (and later, of commercial photography and advertising). Thus 1940s film noir, with its trademark "city of night," was indelibly derivative of classic templates laid down a century before by Dickens, Sue, Poe, and Baudelaire.

Indeed, the antipodal contrast of the city's divided soul—high and low, sunshine and shadow—was a ploy with roots in *The Beggar's Opera* and even ultimately Dante. Christian eschatology made it all the easier to interpret documentary images of urban polarization as spatialized allegories. The urban Lower Depths quickly became the romantic equivalent of Niagara Falls or the Wreck of the Medusa in the competition to satisfy the Victorian public's peculiar need to be simultaneously horrified, edified, and titillated.

Similarly, the slum poor were reimagined as counterparts to the incomparably savage cultures of the forest and desert. The gothic city novel was orientalism brought home. Eugene Sue, for example, famously claimed that his *Les Mysteries de Paris* was nothing less than the urbanization of *The Last of the Mohicans:*

> Everyone has read those admirable pages in which James Fenimore Cooper … traces the ferocious customs of savages, their picturesque, poetic language, the thousand ruses by means of which they flee or pursue their enemies. … The barbarians of whom we speak are in our midst; we can rub elbows with them by venturing into lairs where they congregate to plot murder or theft, and to divvy up the spoils. These men have mores all their own, women different from others, a language incomprehensible to use, a mysterious language thick with baneful images, with metaphors dripping blood.[3]

An American appetite for the underworld and its "savagery" was most memorably stimulated by Dickens's famous expedition to New York's notorious Five Points in the heart of the "Bloody Ould Sixth Ward" in 1842, published in his *American Notes*.[4] Protected by "two heads of the police" (as was his custom in London), the great writer "plunged into" the maze of reeking alleyways, decaying Georgian houses ("debauchery has made the very houses prematurely old"), and fetid cellars that housed New York's poorest people: a cosmopolitan mixture of seafarers, Irish immigrants, German widows, freedmen, and escaped slaves: "What place is this, to which the squalid street conducts us? A kind of square of leprous houses some of which are attainable only by crazy wooden stairs without. What lies beyond this tottering flight of steps, that creak beneath our tread?"

The contemporary reader may have shuddered in suspense, but Dickens knew exactly where he was. The description was already in the can: "hideous tenements, which take their name from robbery and murder: all that is loathsome, drooping, and decayed is here." He had written the same thing about London's Seven Dials and, indeed, makes clear that the Five Points is the generic Victorian slum—"the coarse and bloated faces at the doors have counterparts at home, and all the wide world over"—with a few New World idiosyncrasies like the cheap prints of George Washington on the squalid walls and the "cramped botches full of sleeping Negroes." Dickens was hardly aiming at documentary realism. If anything, he was trying to take his readers somewhere already subliminally familiar to them, a secret city visited in dreams. This archetypal place—slum, casbah, Chinatown—is a museum archiving vices and miseries of potential fascination to the middle class. "Ascend these pitch-dark stairs, heedful of a false footing on the trembling boards," he invited his fellow voyeurs, "and grope your way with me into this wolfish den, where neither ray of light nor breath of air, appears to come...."

With his official bodyguards glowering over residents, the great writer was free to probe, question, and intone judgments. (An English biographer once wrote that his nocturnal slumming "seems indistinguishable from his trips to the London Zoological Garden.") He interrogated a feverish, possibly dying man, stirred exhausted Black women from their sleep, and visited an interracial saloon (or was it a bawdyhouse?) where the proprietor—"with a thick gold ring upon his little finger, and round his neck a gleaming golden watch-guard"—forced two "shy" mulatto girls to dance for their celebrated visitor. Dickens's adventure made the Five Points as infamous as Whitechapel, and spawned an imitative industry of risqué urban travelogues, of which E. Z. C. Judson's *Mysteries and Miseries of New York* ("a perfect daguerreotype from above Bleeker to the horrors of Five Points") and Mathew Hale Smith's *Sunshine and Shadow in New York* became bestsellers.[5] There was now a full-fledged New World franchise for gothic cities and midnight tourism.

Angel's Flight

The early film industry in New York, as in London, Paris, or Berlin, interacted with an urban landscape comprehensively reconnoitered by naturalist writers, modernist poets, muckraking journalists, secessionist painters, and first-generation street photographers. Districts like Greenwich Village, the Lower East Side, Fifth Avenue, and Hell's Kitchen had become famous metonyms for classic aspects of city life (bohemia, immigrant poverty, the swank classes, delinquency) and were instantly recognizable to millions of people who had never actually visited Gotham. The city was too familiar, in an imagined sense, for film to wander very far from literary geography. Movies accordingly entrenched and magnified the hegemonic clichés of O. Henry's generation of writers, including the pathos of the nocturnal slum. Although Five Points was long gone, the Hell's Kitchen waterfront remained potent Dickensian terrain, especially when seen through Josef von Sternberg's monocle in the famous *The Docks of New York* (1928).

Los Angeles, on the other hand, had no compelling image in American letters. When the film industry suburbanized itself in Hollywood, as several historians have pointed out, it acquired a locational freedom from powerful and unavoidable urban referents. While turning a lens on itself and constructing "Hollywood" (the spectacle and pilgrimage site, not the workaday suburb), the industry otherwise had no need to acknowledge the specificity of place. L.A. was all (stage) set, which is to say, it was utopia: literally, no-place (or thus any place). Indeed by the early 1920s, the set had rapidly become the architectural zeitgeist of the region, and movies were, in a very real sense, redesigning the city in their own image. Historicized stucco-façade bungalows—for example, Zorro next door to Robin Hood in adjoining "Spanish Colonial" and "English Tudor" boxes—were the house rage of the 1920s, and the great Griffith and De Mille bible epics produced a small epidemic of "Egyptian revival" apartment houses and masonic temples. A decade later, Westerns gave birth to dude-ranch subdivisions (Rancho Palos Verdes, for example) and an infinity of John Ford–inspired "ranch-style" tracts.

Thanks to Hitler and the Depression, however, the studios were soon crowded with exiled Berliners and New Yorkers whose creative and affective universe was Metropolis not the Frontier. A national obsession with gangsters and a new "hard-boiled" style in popular fiction abetted Hollywood's return to gritty urban locales and classic slums, although this meant Manhattan or, sometimes, San Francisco and its mysterious Chinatown. It was not until the 1940s that breathless location scouts brought word to the writers' huts that the Land of Sunshine itself harbored a dark place of quasi-Weimarian grandeur. With its aging Victorian cliff-dwellings connected by a crazy quilt of stairways, narrow alleys, and two picturesque funic-ular railroads (Angel's Flight and Court Flight), Bunker Hill eventually bedazzled the exiles like an Expressionist mirage. If this neglected understudy for the Evil City seemingly called out for a film career, its big break—Robert Siodmak's *Criss Cross* (1949)—arrived only after it was already a literary icon.

Raymond Chandler, of course, created the noir street map of Los Angeles that Hollywood subsequently took as its guide. He choreographed the class conflict of locales—Pasadena mansions, Bunker Hill tenements, Central Avenue bars, and Malibu beach homes—that made Los Angeles recognizably a city after its long apprenticeship as a back lot. His stroke of genius was the *frisson* created by Marlowe's ceaseless commutes between equally sinister extremes of wealth and immiseration. Yet Chandler worked from prefabricated parts as well as his own perception, and the Bunker Hill ("lost town, shabby town, crook town") which he so famously enshrined in *The High Window* (1943) had already been "invented" by the writer John Fante and the painter Millard Sheets.

Fante was one of the tough-realist "regional" writers like Louis Adamic and James M. Cain whom H. L. Mencken cultivated in the pages of *American Mercury* in the late 1920s and early 1930s. In the early Depression, he worked as a busboy in a downtown restaurant and lived in a cheap hotel room in Bunker Hill, where he pounded out the stories that would become his 1939 novel, *Ask the Dust.* It was the first declension of Bunker Hill's human melodrama. Twenty-year-old Arturo Bandini—like Fante, a first-generation Italian American and desperately aspiring screenwriter—falls into a doomed love triangle with Camilla Lopez, a pretty Chicana waitress, and Sammy, the consumptive who is his rival for her affection. In other short stories, Fante wrote with both poignancy and condescension about the melancholy Filipino bachelors—also busboys and waiters—who were a major ingredient in Bunker Hill's cosmopolitan ethnic mix.

The Bunker Hill that Fante evoked for Mencken's readers was Los Angeles's most crowded and urban neighborhood. According to the 1940 Census, its population increased almost 20 percent during the Depression since it provided the cheapest housing for Downtown's casual workforce as well as for pensioners, disabled war veterans, Mexican and Filipino immigrants, and men whose identities were best kept in shadow. Its nearly two thousand dwellings ranged from oil prospectors' shacks and turn-of-the-century tourist hotels to the decayed but still magnificent Queen Anne and Westlake mansions of the city's circa-1880 elites. Successive Works Progress Administration and city housing commission reports chronicled its dilapidation (60 percent of its structures were considered "dangerous"), arrest rates (eight times the city average), health problems (tuberculosis and syphilis), and drug culture (the epicenter of marijuana and cocaine use).[6] Yet grim social statistics failed to capture the district's *favela*-like community spirit, its multiracial tolerance, or its closed-mouth unity against the police.

As a local historian later conceded, its residents (perhaps like those of the Five Points a century earlier) actually seem to like their aged but colorful perch above the city's bustle:

> The high-ceilinged rooms and low rents suited them. The streets might be steep, but they were remote from traffic rush, with trees along the sidewalks, and a profusion of the more vigorous garden flowers—nasturtiums, hollyhocks, tough-stemmed geraniums—still thriving behind the old carriage houses. At night, there were the rights to watch—office windows in the City Halls, cars lining endlessly...

Millard Sheets's *Angel's Flight*—virtually the only famous Southern California painting from the Depression era (1931)—portrays better than Fante's novel a defiant Bunker Hill attitude. Usually considered a Thomas Hart Benton regionalist, Sheets for a brief moment was a hard-eyed, unpuritanical Otto Dix. Two streetwise but attractive "courtesans of the tenements" in casual conversation look down on the city from the top of the stairway that paralleled the tracks of the little Angel's Flight cable car. Their relaxed insolence, the commanding view, the inference of erotic sanctuary—nothing suggests that they are anywhere other than exactly where they want to be, and probably where young Sheets—doomed to become a gray arts institution in the 1950s—yearned to be as well. It was as close to Montmartre as L.A. would ever get.

But the Hill, at least by the time that Siodmak got around to it, was living on borrowed time. It not only picturesquely overlooked Downtown but brusquely interrupted real-estate values between the new City Hall and the great department stores on Seventh Street. "Bunker Hill," argued a civic leader in 1929, "has been a barrier to progress in the business district, preventing the natural expansion westward. If this Civic Center is to be a success, the removal or regrading of Bunker Hill is practically a necessity." The Second World War only temporarily postponed the crusade of downtown leaders to destroy the "blight" and relocate its 12,000 residents.

Criss Cross

Producer Mark Hellinger, a veteran newspaper reporter, had a clear vision of big cities as actors in their own right. In promoting *Criss Cross* (loosely based on a 1936 pulp by Don Tracy) to the studio, he boasted that it "would do for Los Angeles what [his film] *Naked City* had done [the year before] for New York."[7] Although Hellinger died before production began, director Robert Siodmak—fresh from shooting his Richard Conte noir, *Cry of the City* (1948), in the mean streets of Manhattan's Little Italy—preserved the central idea of shocking the public with a hardcore Los Angeles not usually seen on Grayline tours. However, he considerably "Weimarized" the project, scrapping the original screenplay and (with the help of Daniel Fuchs) importing major elements of his own *Sturme der Leiden-*

schaft—a pathological *Blue Angel*–like tale of male humiliation and jealousy—into the new scenario. The Universum Film AG (Ufa) veteran Franz Planer (Max Ophuls's favorite cinematographer) was brought in to ensure a "Berlin touch."

The story line is industrial-strength sexual obsession unraveling through complex duplicities to the final betrayal of the otherwise-decent protagonist by the *femme fatale*. Steve (Burt Lancaster) is clearly in need of some serious advice from Sam Spade on how to deal with his scheming ex-wife Anna (Yvonne de Carlo) as she plays him off against her new beau, gangster Slim Dundee (Dan Duryea). After a year of cooling his libido elsewhere, Steve is back on the block with the tragic delusion that he has broken Anna's spell. It takes only a few rotations of the famous De Carlo hips, however, to sink Steve as hopelessly back in lust as if he were a clumsy mastodon trapped in the nearby La Brea Tar Pits. While Anna is convincing him that he is still the only one who can really ring her bell, she is also sighing "but Slim gives me diamonds." Police Lieutenant Pete Ramirez, a chum from the old neighborhood (and one of the rare Chicano characters this side of Zorro), accurately warns of what will follow next, but Steve lets his lower anatomy lead him ahead blindly. In a desperate gambit to satisfy Anna's greed, he helps Dundee and his gang plot the robbery of the armored car company where he and his stepfather work. The complicated heist is, of course, also an ambush designed to rid the jealous Dundee of his competition. Steve, although wounded, kills several of Slim's henchmen and is acclaimed as an innocent hero. Dundee tries to have him murdered in the hospital, but Steve bribes his way past the hit man and finally to a rendezvous with Anna, who has betrayed him comprehensively. He is still masticating the full extent of her duplicity when Slim barges in and plugs them both.

It is a fairly grim ending and apparently did not please contemporary audiences. Newspaper critics, with few exceptions, panned *Criss Cross*. Moreover, its release coincided with the beginning of the Cold War and intensified ideological surveillance of film. In England, France, and Germany, *Criss Cross* was heavily censored and even, in some instances, pulled from distribution. Its emphasis on predestination—"it's all in the cards," Steve tells Anna—struck the guardians of public taste as subversively "amoral." Within a decade, however, *Cahiers du Cinéma* was singing its praises, and Borde and Chaumeton, in their influential *Panorama du film noir,* esteemed *Criss Cross* as one of the supreme noirs and "the summit of his [Siodmak's] American career."[7] Subsequent film histories have endorsed this judgment, and it is now generally recognized that *Criss Cross* was also one of Lancaster's finest performances. In my opinion, however, it was Bunker Hill that clearly deserved Parisian adulation.

Siodmak adopted a fascinating strategy for achieving Hellinger's goal of a revelatory, neorealist L.A.: he subsumed the city entirely into Bunker Hill and adjacent downtown streets. Apart from the opening aerial view of L.A. at night and the denouement in a Palos Verdes beach house, *Criss Cross* is entirely located in Bunker Hill and its social space (Union Station and a Terminal Island factory count as the latter). The compression of the city is literally claustrophobic and perfectly matches Steve's jealous self-implosion. It is the first explicitly L.A. film, to my knowledge, that refuses any concession to canonical postcard landscape, except for the implacable, almost sinister sunshine that heightens the emotional tension. Otherwise, Siodmak has annihilated suburbia and drained the Pacific. This is emphatically Mahagonny, not Burbank.

Perhaps the most nostalgic Berlin touch—and undoubtedly Siodmak and Planer were still in crepe over the *Götterdammerung* of their hometown—is the "Rondo Club": the Isherwoodian cabaret at the base of Bunker Hill where the film's most unnerving scenes and violent erotic confrontations take place. Every modern critic of Siodmak has lingered in awe over the notorious rumba scene whatever Steve watches with both growing jealousy and voyeuristic excitement as Anna sways closer to the handsome young gigolo played by (the then still unknown) Tony Curtis. Equally, Planer (who obviously knew Sheets's 1931 painting) must have exulted over such fabulous shots as the visual conjugation through a window of the Angel's Flight cable car ascending the incline and Slim's gang discussing their heist in the drawing room of a ruined Bunker Hill mansion.

But the depiction of the city stops short of pulp cliché. Siodmak (who had inactive left sympathies on the margins of the Hollywood Ten) refused the kind of Dickensian simplification that too many modern viewers mistake as the essence of noir. The lumpen mob led by Slim is counterbalanced by hard-working proletarians like Steve's family. Bunker Hill is portrayed not as the heart of darkness but, more realistically, as a vibrant, hard-working neighborhood under siege from Slim's fascist bullies. The Rondo Club is alternately a depraved fleshpot and the friendly neighborhood bar in *Cheers*. And—at least on the days when Anna isn't floorboarding his testosterone level—Steve is a sentimental sort of fellow, almost a working-class hero.

Kiss Me Deadly

Ralph Meeker's Mike Hammer in *Kiss Me Deadly* (Robert Aldrich, 1955), by contrast, is a serial killer with a detective's license: the clear anticipation of later Bronson and Eastwood films where the "hero" is as violently psychotic as any of the villains. He also has a formidable co-star. After walk-on roles in innumerable B-movies, Bunker Hill once again has top billing in Robert Aldrich's adaptation of the 1952 Mickey Spillane bestseller. An even more acclaimed noir than *Criss Cross*, *Kiss Me Deadly* almost defies genre description. Like the infamous "thing-inside-the-box," it is a small apocalypse with no obvious precedent in film history.

The novel is set in New York. Mike Hammer is driving back from Albany to the city when a mysterious "Viking blonde" lurches in front of his headlights like a frightened deer. "Berga," wearing only an overcoat (whose contents are soon revealed in "a beautifully obscene gesture"), is a desperate dame on the run, and Hammer is sufficiently intrigued to lie their way past a police roadblock. Soon after, their car is run off the road by a gang and Hammer is sapped down. He awakes to the horror of Berga being slowly tortured to death ("the hand with the pliers did something horrible to her") in quest of some secret. Thinking he is still unconscious, the gang put him in the driver's seat next to the dead Berga and push his car off a cliff. Hammer manages to jump out the door as the car begins to fall into "the incredible void."

He next awakes in a hospital in New Jersey where his luscious girl Friday Velda and his cop friend Pat are keeping him out of reach of the angry upstate New York police. Hammer cajoles Pat into revealing that the killers were actually the Mafia and that some incredibly vast conspiracy is afoot. After ritual warnings from Pat and later the FBI to "stay out of it," Hammer returns to Manhattan to conduct a solo *jihad* against the Mafia—"the stinking, slimy Mafia"—who have not only murdered Berga but, more importantly, wrecked his car. He proceeds to rub the Mafia's face in forensic gore. Disarmed by police order, he discovers he enjoys ripping out eyeballs and crushing thoraxes even more than shooting punks in the belly. A long trail of corpses and blondes finally leads him to the Secret: a key in Berga's belly that opens a foot locker containing $4 million of narcotics stolen from the Mafia. Berga's former roommate, the seemingly sweet and terrified Lily Carver, is likewise unmasked in the last scene as a disfigured female monster. She considerably annoys Hammer by shooting him in the side and then demanding a kiss. As her lips approach his, he literally flips his Bic.

> The smile never left her mouth and before it was on me I thumbed the lighter and in the moment of the time before the scream blossoms into the wild cry of terror she was a mass of flame tumbling on the floor with the blue flames of alcohol turning the white of her hair into black char and her body convulsing under the agony of it.

Aldrich, we are told, despised the book, regarding Spillane's "cynical and fascistic" hero with "utter contempt and loathing." A much more serious fellow-traveler than Siodmak, he clearly understood that the Hammer series had become the popular pornography of McCarthyism. Although he later told François Truffaut that he should have refused the adaptation, his subversion of it was a much more powerful protest.[9] In collaboration with A. I. Bezzerides, the brilliant scenarist of *Thieves Highway* (1949), Aldrich transformed *Kiss Me Deadly* into a Cold War allegory that deliriously combines *grand guignol* hyperbole with subtle commentaries on the psychology of fascism. The Herculean machismo of Hammer in the novel is now exposed as the sadism of a small-time bully who routinely batters Velda and, in one chilling scene, wantonly destroys a precious phonograph record belonging to a harmless old opera buff. The Mafia, in turn, is replaced by an occult conspiracy led by the Himmler-like Dr. Soborin (Albert Dekker) and the ravenous Lily Carver (Gaby Rogers). Most masterfully, the sordid stash of street drugs becomes a mythic Pandora's box of stolen plutonium. In the incomparable final scene, the incautious Lily literally unleashes the fires of Hell.

The wages of fascism and greed, it seems, are nuclear holocaust.

And what better location for ground zero than a Malibu beach-house? The film version is unmoored from the novel's hackneyed Manhattan settings and reanchored in witchhunt-era Los Angeles. Although the action moves through an archipelago of sinister locales, the film is visually dominated by Hammer's repeated forays to Lily Carver's Bunker Hill tenement. Where Siodmak treated the Hill as a complex urban microcosm, Aldrich strips it down to the darkest layer of metaphor, removing all traces of vibrant normalcy. Its residents now cower behind their doors as killers (including Hammer) silently stalk their prey in the hallways. Whereas Siodmak filmed almost exclusively in full daylight, Aldrich prefers allegorical darkness and uses the deep-focus "3-D" technique of *Citizen Kane* (1941) to accentuate the shadowy vertigo of interior stairwells and landings, creating the ambience of a single, vast haunted house.

The Hill's visual and architectural antipode is Hammer's fastidiously modern (Wilshire Boulevard?) apartment with its Eames chairs, telephone-answering machine, and middlebrow art. Here Aldrich is accurately mapping the polar-ized social geography (and real-estate values) of L.A. in the mid 1950s. Mike is emphatically a "westside" guy in a period when Downtown was generally presumed to be dying, and the cultural and business life of Los Angeles was migrating down Wilshire Boulevard toward Westwood and the future Century City. If Chandler's Marlowe (like his creator) incarnated an old-fashioned petty-bourgeois work ethic, Aldrich's version of Hammer is primordial material boy, a sleazy hustler interested only in Jaguars and expensive gimmicks, who has man-aged to shoot his way into a petite version of the Westside *nouveau-riche* lifestyle and obviously wants more. He is a human type nauseatingly familiar to Aldrich in a city and industry where greed and betrayal packaged as patriotism had recently made a hecatomb of the careers of his best friends.

They Live!

A few years after the release of *Kiss Me Deadly*, the wrecking balls and bulldozers began to systematically destroy the homes of 10,000 Bunker Hill residents. After a generation of corporate machination, including a successful 1953 campaign (directed by the *Los Angeles Times*) to prevent the construction of public housing on the Hill, there was finally a green light for "urban renewal." A few Victorian landmarks, like Angel's Flight, were carted away as architectural nostalgia, but otherwise an extraordinary history was promptly razed to the dirt and the shell-shocked inhabitants, mostly old and indigent, pushed across the moat of the Harbor Freeway to die in the tenements of Crown Hill, Bunker Hill's threadbare twin sister. Irrigated by almost a billion dollars of diverted public taxes, bank towers, law offices, museums, and hotels eventually sprouted from its naked scars, and Bunker Hill was reincarnated as a glitzy command center of the booming Pacific Rim economy. Where hard men and their molls once plotted to rob banks, banks now plotted to rob the world. Yet history is sometimes like the last scene in *Carrie* (1976).

In John Carpenter's *They Live!* (1988), the old Bunker Hill suddenly rises from the grave to deal summary justice to the yuppie scum who have infested its flanks. Carpenter is, of course, notorious (and among some, well loved) for his use of right-wing plot elements—like the Bronson revenge-massacre formula or the city-as-penal-colony—to advance his progressive sympathies. The only Hollywood liberal with a bigger gun collection (I am being metaphorical) than Charlton Heston, he had the chutzpah in *They Live!* to suggest guerrilla warfare as a well-deserved response to Reaganomics and The Age of Me.

The opening of the film contains the bluntest imagery of class polarization since *Battleship Potemkin*. On now glamorous Bunker Hill, the *nouveaux riches* cavort in Armani splendor, while across the freeway, in the little valley (Beaudry Street) that separates it from Crown Hill, the Other America is camped out in homeless squalor. The Sears and Roebuck middle class has disappeared, and there are only yuppie princes and blue-collar paupers left. Still stunned by the enormity of their downfall, the workers (now racially integrated in catastrophe) sulk in their postmodern Hooverville under the watchful eye of the now openly fascist LAPD.

These scenes had authentic local poignancy since Carpenter's neighborhood—Temple-Beaudry—was deliberately turned into a wasteland in the late 1970s and early 1980s by corporate speculators based in Singapore, Dallas, and

Toronto who preferred to destroy the housing of an estimated 7000 residents than face future resettlement costs. As the time the film was shot, the area was mainly occupied by fugitive gang members, older homeless Black men, and young Mexican and Central American immigrants. Their homes were razed, but from the flanks of Crown Hill they could arrange old car seats and discarded sofas to enjoy the spectacle every dusk of the illumination of the downtown office towers. In an ironic, Dickensian reversal, the poor were now the voyeurs of the rich.

In the film, the passivity of the homeless workers explodes into resistance when John Nada (whom Carpenter himself has characterized as "an everyman/ working class character") discovers that the Reagan voters, when viewed through special glasses, are actually alien invaders who have hijacked the city and immiserated its common people.[10] This is, of course, Spillane's McCarthyite paranoia turned back against itself, and once Nada (Nothing) convinces his fellow-workers that their class enemies are monstrous extraterrestrials, there is no longer any moral scruple preventing a war of total extermination/liberation. From then on, it is urban renewal in violent reverse, and the Rainbow Coalition puts a serious hurt on the grotesque insects masquerading as *L.A. Law*. The concept of a second Battle of Bunker Hill is, of course, idiotic and breathtaking as the same time. But the return of the repressed is always that way.

2000

Notes

1. See the section "Sunshine or Noir" in my own *City of Quartz: Excavating the Future in Los Angeles*, London 1990.

2. Reyner Banham, *Los Angeles: The Architecture of the Four Ecologies*, London 1971.

3. Quoted in Frederick Brown, *Zola: A Life*, New York 1995, p. 347. Brecht would later boast that he was the Kipling of Berlin.

4. Charles Dickens, *American Notes*, London 1842; with an introduction by Christopher Hitchens. New York 1996.

5. E. Z. C. Judson, *Mysteries and Miseries of New York*, New York 1848; Mathew Hale Smith, *Sunshine and Shadow in New York*, Hartford, Conn. 1868.

6. See Pat Adler, *The Bunker Hill Story*, Glendale, Calif. 1963.

7. See Deborah Lazaroff Alpi, *Robert Siodmak: A Biography*, Jefferson, N.C. 1998; Hervé Dumont, *Robert Siodmak: le maitre du film noir*, Lausanne 1981.

8. Raymond Borde and Etienne Chaumeton, *Panorama du film noir americain, 1941–1953*, Paris 1975.

9. Edwin T. Arnold and Eugene L. Miller, *The Films and Career of Robert Aldrich*, Knoxville, Tenn. 1986.

10. Carpenter, as quoted on the *They Live!* website at www.sopuown.com/dorms/creedssonegase/shey/jcarp.hsm, 1998.

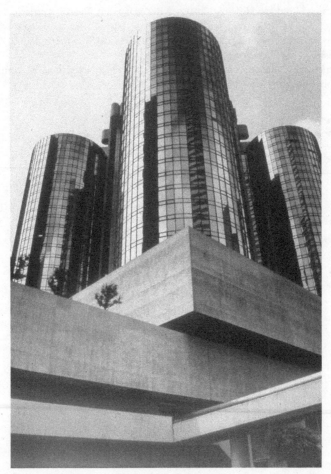

Bunker Hill (ca 1990)

8

The Infinite Game

On warm evenings, the homeless men who live furtively in the wastelands of Crown Hill like to set up old car seats and broken chairs under the scorched palms to watch the spectacle of dusk over Downtown Los Angeles. They have ringside seats to enjoy the nightly illumination of 26 million square feet of prime corporate real estate, half of it built in the last decade. This incomparable light show and the plight of the homeless themselves are the chief legacies of a generation of urban redevelopment. Thanks to over a billion dollars of public subsidies and diverted tax revenues, the "suburbs in search of a city" have finally found what they were looking for.

Despite Reyner Banham's disparaging 1971 "note" ("because that is all Downtown Los Angeles deserves") that it had become irrelevant, the center has held after all.[1] Indeed, since the arrival of Pacific Rim capital in the early 1980s, Downtown Los Angeles has grown at warp speed. The stylized crown on the top of Maguire Thomas's overweening new skyscraper, the 73-story First Interstate World Center, symbolizes the climax of redevelopment in the new financial core from Bunker Hill to South Park. Meanwhile, on every side of the existing corporate citadel, panzer divisions of bulldozers and wrecking cranes are clearing the way for a doubling or tripling of Downtown office space in the 1990s. The desolate flanks of Crown Hill itself (the Cinderella stepsister of Bunker Hill, across the Harbor Freeway) may become another glowing forest of office towers and high-rise apartments in a few years. And the homeless, their ranks swollen by the displaced from the redeveloped "West Bank," will probably be watching the nighttime special effects from Elysian Park or beyond.

The terrible beauty struggling to be born Downtown is usually called growth, but it is neither a purely natural metabolism (as neoliberals imagine the marketplace to be) nor an enlightened volition (as politicians and planners like to claim). Rather it is better conceptualized as a vast game—a relentless competition between privileged players (or alliances of players) in which the state intervenes much like a card-dealer or croupier to referee the play. Urban design, embodied in different master plans and project visions, provides malleable rules for the key players as well as a set of boundaries to exclude unauthorized play. But unlike most games, there is no winning gambit or final move. Downtown redevelopment is an essentially infinite game, played not toward any conclusion or closure, but toward its own endless protraction. The Central City Association's fairy-tale imagery of Downtown 2020 as a cluster of "urban villages" offering Manhattanized lifestyles and pleasures is bunkum for the hicks.[2] Downtown's only authentic deep vision is the same as any casino's: to keep the roulette wheels turning.

How the Game Started

Certain primordial facts organize the playing of the game. Above all, there is the ghost of sunk capital: a large part of the spoils of the suburban speculations of the early twentieth century—the subdivision of Hollywood and the Valley—were invested in Downtown high-rise real estate in the 1900–1925 period. But these investments (including the legendary patrimonies of the Chandlers, Lankershims, and Hellmans) were almost immediately imperiled by the revolutionary tendency of the automobile to disperse retail and office functions.[3] The old-guard elite resisted this decentralization (represented in the 1920s by the rise of Wilshire Boulevard as a "linear Downtown") by marshaling an ironic municipal socialism on behalf of the central business district.[4]

The first priority of this "recentering" crusade, led by the *Los Angeles Times* and

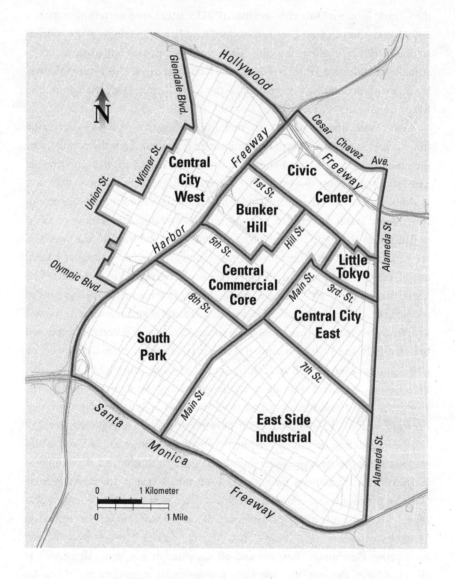

Hollywood

Glendale Blvd.

N

Witmer St.

Cesar Chavez Ave.

Freeway

**Central
City
West**

Union St.

1st St.

**Civic
Center**

Freeway

**Bunker
Hill**

Harbor

5th St.

Hill St.

Alameda St

Olympic Blvd.

**Central
Commercial
Core**

**Little
Tokyo**

3rd. St.

Main St.

8th St.

**Central City
East**

**South
Park**

7th St.

Main St.

Santa

**East Side
Industrial**

Monica

Alameda St

Freeway

| 0 | 1 Kilometer |
| 0 | 1 Mile |

the Central Business District Association (CBDA, later Downtown Businessmen's Association, then Central City Association [CCA]), was to reinforce the concentration of civic life within the core. Thus the public-private initiatives that constructed the Biltmore Hotel and Memorial Coliseum in the 1920s were followed in successive decades by the creation of the Civic Center, Dodger Stadium, the Music Center, and the Convention Center.[5]

At the same time, the CBDA also mobilized to keep the region's major traffic flows centered on Downtown. Redistributing tax revenue from the periphery to the center, the city subsidized a heroic program of transportation improvements. The Los Angeles River was bridged by a series of magnificent viaducts (1920–40), Downtown streets were widened and tunneled through Bunker Hill, the centralizing Major Traffic Street Plan was adopted (1924), rail commuters were taken underground through Crown Hill in a "Hollywood Subway" to the profit of the Chandlers and other investors in the Subway Terminal Building at Fourth and Hill (1925), and the main rail lines were finally persuaded to consolidate in a Union Station (1937–39).[6] Repeated campaigns by Downtown business groups to recapitalize and grade-separate the electric railroad and streetcar system (as well as extend it via monorail into the San Fernando Valley) were successfully opposed between 1920 and 1970 by suburban commercial interests.[7] With the support of city engineer Lloyd Aldrich and the Southern California Automobile Club, however, Downtown forces were successful in persuading the state highway department to accept a radial freeway grid that minimized "the destructive aspects of decentralization" and eventually made Downtown the hub of eight freeways.[8]

The recentering of L.A. is even better envisioned, however, as a succession of social struggles between different interest groups, classes, and communities. If Downtown landowners have always been pitted against the developers of Wilshire Boulevard and suburban retail and, later, office centers (now veritable outer cities), there is also a bitter legacy of resentment among San Fernando Valley homeowners, who believe that their tax dollars have been confiscated to improve Downtown. But most of all, Downtown has been "defended" at the expense of the working-class communities on its immediate periphery. An estimated 50,000 residents—Chinese, Mexican, and Black—were displaced to make way for such "improvements" as Union Station, Dodger Stadium, the Civic Center, industrial expansion east of Alameda, central business district (CBD) redevelopment on Bunker Hill, city and county jails, and especially, the eight freeways (always carefully routed to remove homes, not industry). Chronicling the story of Downtown's land grabs and landuse dumping east of the river, Rodolfo Acuña talks about a "community under a thirty-year siege."[9]

For a few years in the early postwar period, however, Downtown boosters had to face the challenge of an ambitious housing program that aimed to reconstruct, rather than displace, the working-class neighborhoods next to Downtown. Mayor Fletcher Bowron, supported by the CIO and civil-rights organizations, signed a contract with the federal government under the Housing Act of 1949, to "make Los Angeles the first slum-free city in the nation" by building ten thousand public housing units in areas like Chavez Ravine and, potentially, Bunker Hill. The Los Angeles Community Redevelopment Agency was established under state law to assist in the assemblage of land for this purpose. The vision of a stabilized, decently housed Downtown residential fringe roused vehement opposition, however, from CBD landowners. Bowron and public housing were defeated by hysterical red-baiting orchestrated by the Los Angeles Times and police chief William Parker in 1953.[10] Anything that even smacked of a socialistic rehousing strategy was henceforth excluded from discussions of Downtown renewal.

Early Game Plans

Infrastructural improvement alone, however—even in tandem with Cold War politics—could not prevent the relative decline of Downtown. Postwar Los Angeles continued to trade its old hub-and-spoke form for a decentralized urban geometry. Although Downtown remained the financial as well as governmental center of Southern California through the early 1960s, it inexorably saw its retail customers migrate outward along Wilshire Boulevard and eventually toward dozens of suburban shopping centers. Moreover, by 1964, as plans were completed to create Century City—a "Downtown" for Los Angeles's Westside—out of an old movie lot, the historic headquarters role of the central business district was suddenly put to question as well.

Embattled Downtown landowners were virtually unanimous that the CBD's great competitive disadvantage—even more than the age of its building stock (circa 1900–1930)—was the growing accumulation of so-called blight along Main Street (Skid Row) and in the old Victorian neighborhood of Bunker Hill.[11] The Hill, in fact, was a double obstacle, physically cutting off the Pershing Square focus of the business district from the Civic Center as well as preventing the CBD from expanding westward. Public discussion became riveted on images of dereliction, ignoring the simple fact that most of the Hill's eleven thousand inhabitants were, in fact, productive Downtown employees: dishwashers, waiters, elevator operators, janitors, garment workers, and so on. The role of city government in the redevelopment of the Hill had already been extensively debated before 1940. In 1925, Allied Architects, denouncing the Hill as "an unsightly landmark ... blocking business expansion to the west and north," envisioned rebuilding it as a "civic acropolis" of parks and public buildings.[12] In contrast, C. C. Bigelow simply wanted to obliterate the Hill by leveling it to the Hill Street grade, and engineer William Babcock in 1931 proposed a less drastic regrading to buckle the new Civic Center to Pershing Square.[13] Despite considerable political support, both the Allied Architects and Babcock schemes were defeated, and by 1938, the city council threw in the towel to let "the natural forces of economics do the job."[14]

The Bunker Hill debate resumed after World War II with the advent of Greater Los Angeles Plans, Inc. (GLAPI), sponsored by an elite group that included Norman Chandler and Asa Call (often described by his contemporaries as L.A.'s "Mr. Big"). GLAPI actually bought land on Bunker Hill for a music center, but found its plans thwarted by the reluctance of voters to approve the necessary bond issue (even with a sports arena appended). In the meantime, market forces were given a chance to transform Bunker Hill. An early 1950s insurance-company scheme to build upscale apartment towers on the Hill (along the lines of Park LaBrea on Wilshire) never managed to get beyond its directors' anxieties about investing in Downtown L.A. A few years later, GLAPI believed that it had convinced Union Oil to build its new headquarters on Bunker Hill, but at the last moment, the corporation instead chose Crown Hill.[15] In light of these failures, piecemeal private-sector redevelopment of Bunker Hill was abandoned.

Instead, the Community Redevelopment Agency (CRA)—in original intention a public housing agency—became simultaneously the largest developer
Downtown and the collective instrument of all the developers. Classically, like
other regulatory agencies, it was captured by the very interests it was supposed to
regulate. Its mayorally appointed board of seven was ideally shielded from direct
public scrutiny or electoral responsibility. Moreover, it possessed autonomous
financial authority, based on the use of diverted tax increments. After the failure
of various private initiatives, the CRA wrested the entirety of Bunker Hill from
its slumlords by invoking eminent domain. The city council approved the final
plan for Bunker Hill in the spring of 1959, and within eighteen months bulldozers
began demolishing the Hill's Gothic mansions and Queen Anne tenements.

The Hill's population, meanwhile, was simply dumped into other Downtown
areas. Although some ended up on Skid Row, most of the 10,000 ex–Bunker Hill
residents were displaced to the west bank of the Harbor Freeway, driving a salient
of "blight" and rack-renting across the Temple-Beaudry area well into the fash-
ionable Westlake district. Twenty years passed before the CRA bothered to establish a fund to rebuild the quarter of Downtown housing units it had abolished in
this single stroke.[16]

While the CRA was clearing, regrading, and assembling Bunker Hill into parcels suited for sale to developers, the major Downtown stakeholders (organized
as the Central City Committee [CCC]) were helping CRA chairman William
Sesnon and city planners create a master plan "to bring about the rebirth of [the
entire] Central City." The 1964 plan, titled *Centropolis*, was the first comprehensive design for redevelopment: the product of a series of studies that had begun
with an economic survey of Downtown in 1960.

Its core vision was the linkage of new development on Bunker Hill with the
revitalization of the fading financial district along Spring Street and the retail
core along Broadway and Seventh Street. Pershing Square, still envisioned as the
center of Downtown, was to be modernized with a large underground parking
lot, the beginning of Wilshire Boulevard was to be anchored with a dramatic
Wilshire Gateway, and El Pueblo de Los Angeles historical park around Olvera
Street was to be completed.

The outstanding innovation of the plan, however, was a proposal to link the major structures in the retail core by means of mid-block malls, with pedestrian circulation lifted above the street on "pedways." This superstructure would unify prime property, old and new, into a single vast Downtown mall. At the same time, it addressed department store concerns about an enhanced definition of social areas and the insulation of shoppers from "bums." Indeed, the "rollback" of Skid Row was one of the plan's major objectives. The idea was to deploy new or augmented land uses, including parking lots, a low-cost shopping precinct, and a light industrial strip along Main and Los Angeles Streets, to create an effective buffer zone between Skid Row and the born-again CBD.[17]

Just a year after the premiere of *Centropolis,* the Watts rebellion and the attendant white backlash almost completely vitiated the plan and the seven years of work that had gone into it. The flames of August 1965 had crept to within a few blocks of Downtown's southern perimeter, causing the establishment to lose its nerve. The McCone Commission predicted "that by 1990 the core of the Central City of Los Angeles will be inhabited almost exclusively by more than 1,200,000 Negroes," and the Los Angeles Police Department warned Downtown merchants against an "imminent gang invasion" by Black youth ("when encountered in groups of more than two they are very dangerous and armed").[18] Faced with such spectres, mortgage bankers and leasing agents started talking about a wholesale corporate defection to Century City and the Westside, even the "death of Downtown."[19] As a result, landowners and financiers jettisoned the central tenet of the *Centropolis* plan—the renovation of the historic core—and began to vote with their feet: leaving the Broadway–Spring Street corridor to decline and fall.

In the midst of crisis and flight, the Central City Association rallied to save Downtown by reinventing it. Rejecting as inadequate the 1969 CBD plan prepared by city planning director Calvin Hamilton, the CCA established its own planning committee, the Committee for Central City Planning, Inc. (CCCP, "a who's who of business power"), in substantial continuity with the tradition and membership of both Greater Los Angeles Plans, Inc. and the Central City Committee.[20] With the CCCP and the city contributing $250,000 each, an eminent planning firm, Wallace, McHarg, Roberts and Todd of Philadelphia, was hired to create a new urban design for the post-Watts reality.

The firm's *Central City L.A., 1972–1990* became universally known as the *Silverbook* because of its striking metallic cover. Replacing the dead letter of *Centropolis*, it adumbrated the political and design principles that have guided Downtown to the edge of the 1990s. For the purposes of analysis, these guidelines can be divided into two orders of importance: "dogmas" and "gadgets."

The dogmas, outlined below, gave new directions to the redevelopment process and established far-reaching goals for public–private cooperation.

1. First, the *Silverbook* categorically reasserted *contra*-Banham that Downtown was the center of the Los Angeles metropolitan area. As Robert Meyers pointed out at the time, this directly contradicted planning director Hamilton's laboriously constructed "Centers Concept": the keystone of a city master plan emphasizing polycentric development and the equality of major growth poles.[21]

2. The *Silverbook* also proposed a dramatic enlargement of the Community Redevelopment Agency's scale of planning and tax-increment authority to include virtually all of Downtown between Alameda Street (on the east) and the Harbor (on the west), Hollywood (on the north), and Santa Monica (on the south) freeways.

3. The defense of the old office core was abandoned in favor of resiting Downtown a few blocks further west in the frontier being cleared by the CRA on Bunker Hill and along Figueroa between Fifth and Eighth streets.[22] This was in essence a disguised corporate bailout using diverted tax monies. The chief role of the CRA was envisioned as *recycling* land value from old to new, as discounts on greenfield parcels (together with rapid appreciation after building) compensated stakeholders for the depreciation of their obsolete properties in the old core.

4. The new *growth axis* (supplanting the Wilshire–Seventh Street–West direction of the last wave of prewar Downtown building) was established along Figueroa and Grand, integrated at one end with the Civic Center and pointing toward the University of Southern California at the

other. The luxury apartment community on Bunker Hill was to be coun-
teranchored at redevelopment's prospective southern frontier by a South
Park Urban Village. This envisioned southward flow of Downtown fortuitously coincided with the personal strategy of CCA president and
Occidental Insurance executive Earl Clark, who had erected a solitary
skyscraper (today the Transamerica Center) at Olive and Twelfth Streets,
almost a mile south of the center of new highrise construction. The *Silverbook* plan, if implemented, would bring Downtown and soaring land
values to Clark's speculative outpost.

5. Even while rotating the axis of redevelopment ninety degrees from the
   west to the south, *Silverbook* premised its Downtown renaissance on the
   coordinated construction of a new rapid-transit infrastructure (Metro
   Rail) along the Wilshire corridor (with an ancillary line running toward
   South Central L.A.). At the same time, the neighborhoods immediately
   west of Downtown, across the Harbor Freeway, were reserved as a
   periphery for parking and CBD services.

6. *Silverbook* amended the corporate-center vision of *Centropolis* to a post-
   Watts rebellion corporate-fortress strategy. Rather than creating a pedestrian superstructure to unify the old and new in a single mall-like configuration as in the *Centropolis* plan, new investment was now massively
   segregated from old. In the CRA's actual practice—more drastic than
   the model—pedestrian access to Bunker Hill was deliberately removed,
   Angel's Flight (the Hill's picturesque funicular railroad) was dismantled,
   and Hill Street, once a vital boulevard, became a glacis separating the
   decaying traditional business district from the new construction zone.

7. Skid Row, circumscribed and buffered in *Centropolis*, was now scheduled
   for elimination, thus freeing up "Central City East" for redevelopment
   as a "joint university communications center and extension school."

In addition to these strategic dogmas, the *Silverbook* unveiled a number of gadgets to make the new Downtown cohere in an efficient working order. Most
important was the proposed "people-mover" to distribute office workers and

shoppers from mass transit terminals across the broad spaces of Bunker Hill megastructures, to individual buildings, and then, southward, to "South Park Village."[23] Similarly, the elevated, grade-separated pedways of the *Centropolis* plan were reintroduced in Bunker Hill as a preferred option to street-level pedestrian circulation. A second-level plaza and pedway complex ("Bunker Hill East"), again copied from the previous plan, was envisioned as a "five-way, 'pivotal interface'" connecting Bunker Hill, above street level, with the Civic Center, Little Tokyo, "Central City East" (the reclaimed Skid Row), and a corner of North Broadway. Downzoning was proposed throughout the central commercial core (excluding Bunker Hill) to create a "development rights bank" to be allotted or auctioned off according to priorities defined by a prospective Specific Plan. Finally, *Silverbook* had a whole toolchest of miscellaneous gadgets—ranging from a Downtown industrial freeway to an in-town industrial park—to stimulate new investment in the industrial salient between Los Angeles and Alameda Streets.

The political translation of the *Silverbook* concepts into a legally valid blueprint for the CRA–Central Business District Redevelomment Plan[24] encountered unexpected opposition. Although only the redoubtable Ernani Bernardi opposed passage of the plan through the fifteen-member city council in July 1975, the dissident councilman was soon reinforced by powerful allies, including the county board of supervisors, the county assessor, and State Senator Alan Robbins, a mayoral aspirant from the Valley.[25] They joined Bernardi in suing the council to prevent the CRA from diverting billions of dollars of future tax increments (the increase in assessments due to redevelopment) from general fund uses. As the debate grew increasingly nasty, the CRA and its council supporters (backed by new mayor Tom Bradley) argued that the increments were essential to renewing growth and jobs Downtown, whereas opponents insisted that a handful of large property owners—led by Security Pacific Bank, Prudential Life Insurance, and the Times Mirror Company—stood to reap a windfall at public expense. In the end, before the CBD plan was allowed to take effect, Bernardi and the county forced the CRA (in 1977) to accept a consent decree capping the tax-increment bond-issuing capacity of the project at $750 million.

Meanwhile, the CRA bureaucracy itself, under commission chairman Kurt

Meyer (a well-known L.A. architect) and administrator Edward Helfeld, balked
at the CCA's demand that the agency implement the *Silverbook* to the letter. Wallace, McHarg proposals for a large lake in South Park and the university complex
on Skid Row were rejected as "unfeasible" (privately, the CRA thought them
"preposterous"), and Meyer and Helfeld took a principled stand against a Charles
Luckman scheme to move the central public library to Broadway to serve as a
buffer between Latino small businesses and the remnant upscale shopping precinct on Seventh Street.[26] Most of all, they railed against the CCCP's attempt to
perpetuate itself as the CRA's "shadow government." Although the CCA, under
the urging of Franklin Murphy of Times Mirror, ultimately wound down its
parallel planning arm, Downtown leaders did not forget, or forgive, the disobedience of Meyer and Helfeld. After Meyer resigned (officially to return to his busy
architectural practice), he was replaced by a consummate wheeler-dealer and
CCA ally, construction trades' spokesman Jim Wood. A few years later, the CCA
combined forces with Helfeld foes on the planning commission and city council
to purge the controversial CRA administrator.

Japan Ups the Ante

Having cleared the initial hurdle of political opposition, however, the central business district plan still had to prove that it could command the requisite levels of
investment from private developers. The *Silverbook* had coincided with the epochal
transition in city hall from Sam Yorty to Bradley, and the Downtown old guard
was initially skeptical of what to expect from Los Angeles's first Black mayor with
his coterie of South-Central ministers and wealthy Westside liberals. But Bradley,
as his biographers emphasize, took great pains from the very beginning to con-
ciliate the powerful Downtown interests. Moreover, in the latter part of his first
term, a vice arrest—which most insiders believed was set up—led to the dismissal
of Maury Weiner, his liberal chief deputy and *bête noire* of conservative critics.
Weiner's replacement, to the chagrin of liberals, was a Pasadena Republican, Ray
Remy (later head of the Los Angeles Chamber of Commerce). The new deputy
mayor was instrumental in consolidating the rapprochement between the mayor
and the Central City Association. Bradley, supported by the powerful building-

trades wing of the local labor movement, became an aggressive proponent of the CBD plan and the strategy outlined in the CCA's *Silverbook*.[27]

With city hall (and a city council majority) routinely approving every request of the developers' lobby (or abdicating power to the CRA), new capital was encouraged to flow into Downtown's greenfields. If there were just five new high rises above the old earthquake limit of thirteen floors in 1975, there are now fifty. Moreover, as the game picked up pace, purely speculative trading also increased, with perhaps a third of Downtown changing hands between 1976 and 1982. Ironically, as the ante has risen, many of the original champions of Downtown renewal, including large regional banks and oil companies with troubled cash flows, have had to cash in their equity and withdraw to the sidelines. As Volcker-ism first created a superdollar and then destroyed it, the volatile US commercial real estate markets favored highly liquid investors and foreign capital.

Downtown simply became too big for local interests to dominate. Thus in 1979 the *Times* reported that a quarter of Downtown's major properties were for-eign-owned; six years later the figure was revised to 75 percent (one authority has even claimed 90 percent).[28] The first wave of foreign investment in the late 1970s, as in Manhattan, was led by Canadian real estate capital, epitomized by Toronto-based Olympia and York. The Reichmann clan, which owns Olympia and York, collects skyscrapers like the mere rich collect rare stamps or Louis XIV furniture. Yet since 1984, they, along with the New York insurance companies and British banks, have been swamped by a tsunami of Asian finance and flight capital.

What the Japanese call *zaitech*, the strategy of using financial technologies to shift cash flow from production to speculation, has radically restructured Down-town's investment portfolios and given a new impetus to the realization of the CBD redesign (indeed, they have become its major motive force). As the superyen and foreign protectionism slowed domestic industrial reinvestment in Japan, giant corporations and trading companies shifted black ink abroad in search of lucrative foreign assets. The liquid resources of other investors have simply been dwarfed by the sheer mass of the Japanese trade surplus, which has rapidly found its way from US treasury bonds to prime real estate. In the particular case of Downtown Los Angeles, the superyen of the late 1980s put the skyscrapers along Figueroa's "gold coast" at rummage-sale discounts compared to Tokyo real estate. A virtually unknown condominium developer, Shuwa Company Ltd., stunned the Downtown establishment in 1986 by purchasing nearly $1 billion of L.A.'s new skyline, including the twin-towered ARCO Plaza, in a single two-and-a-half-month buying spree. As local real estate analysts complained at the time, "The major Japanese companies are borrowing at very cheap rates, usually 5% or less. They borrow in Japan [in Shuwa's case, through ten L.A. branches of Tokyo banks], deduct it from their taxes in Japan, convert it to dollars, and invest in dollars in the United States."[29]

In singing praise to the miracle of the Pacific Rim economy, Los Angeles boosters in the 1980s generally avoided reference to the specific mechanism of the Downtown boom. But, to the extent that Japanese capital was now the major player, the Downtown economy had become illicitly dependent on the continuation of the structural imbalance that recycled US deficits as foreign speculation in American assets. In a word, it had become addicted to US losses in the world trade war, and bank towers on Bunker Hill were rising almost in direct proportion to plant closings in East Los Angeles and elsewhere in the nation. The Downtown renaissance had become a perverse monument to deindustrialization.

But the ironies of international geopolitics were scarcely noted by the Community Redevelopment Agency. Its concern was, rather, that the very success of Downtown redevelopment was imperiling the agency's *raison d'être*. By 1989–90, the CRA, working hand-in-glove with offshore capital, had reached the limits of the 1977 Bernardi cap, endangering its hegemony in the central business district and setting off a complex process of plan redesign and political negotiations. Before analyzing this new conjuncture, however, it is first necessary to draw a notional balance-sheet of redevelopment in the fifteen years since the creation of the CBD project. To what extent has the grand design, à la *Silverbook*, actually been realized, and how has it been further modified?

First, there have been some strategic setbacks. Skid Row, slated for demolition (or deinstitutionalization, in the Orwellian language of the *Silverbook*), has survived, however infernally, largely as the result of council members' fear of the spillover of the homeless into their districts. This has led Little Tokyo to expand eastward, along First Street toward the Los Angeles River, rather than southward as expected. And despite the deliberate siting of the Jewelry Mart on its eastern margin, the redevelopment of Pershing Square (a subsidiary goal of the *Silverbook*) languishes two decades behind schedule, with street people in occupation of the park and the developers squabbling among themselves. As a result, the Biltmore Hotel, in designing its recent tower annex, rotated its main entrance 180 degrees to face the library—the new focal point of Downtown. (The library, in turn, was left in place, *contra* earlier plans, because its air rights were used to add density to the huge Maguire Thomas projects across the street.)

More serious still are the transportation anomalies in the realized Downtown design. In the *Silverbook*, the viability of the new Downtown depended upon the articulation of Wilshire-axis mass transit with a pedestrian distribution system along the new Figueroa corridor. Although those in the CRA talk wistfully of reviving the scheme, federal funding for the people-mover—a proposed $250 million system of airportlike moving sidewalks—was vetoed by the Reagan administration after heavy lobbying by opponents from the San Fernando Valley. This has marooned pedestrians in the various megastructures Downtown and left a useless $30 million people-mover tunnel underneath Bunker Hill.[30]

The fate of Metro Rail has been stranger still. After loud protests from Westside homeowners, Metro Rail was diverted from Wilshire, at Western, to run north through Downtown Hollywood and then under the mountains to North Hollywood. This suits some CRA leaders and their developer friends, since it links three major redevelopment projects and creates a continuous corridor of real estate speculation.[31] The environmental impact report of the Southern California Rapid Transit District (SCRTD) forecasts a staggering 50 million square feet of new commercial development (virtually two new Downtowns) centered on eleven Metro Rail stations.[32]

But the current alignment also negates the original economic rationale for subway construction, since only the Wilshire corridor currently has the population density to generate an amortizing ridership for Metro Rail. As a result, Metro Rail faces a very likely danger of insolvency, while most Downtown commuters (coming from the Westside or, especially, the San Gabriel Valley to the east) will continue to rely on their cars. Metro Rail, at least in its current configuration, will act as an Archimedean lever to increase development densities in the CBD–Hollywood–North Hollywood corridor without mitigating current levels of congestion Downtown (but more on traffic in a moment).[33]

In vindication, the CRA and its supporters can claim that, whatever setbacks or anomalies may have occurred, the agency has triumphally achieved the central vision of the *Silverbook*. A new financial district has taken shape on the east bank of the Harbor Freeway, with its skyscraper pinnacle along Grand, focused on the library, and pointing southward toward the expanded Convention Center and USC. Because this successful recentering has been largely fueled by a land rush of Asian and Canadian capital, it has simultaneously transferred ownership to absentee foreign investors.[34] Yet there is little anxiety Downtown that the ultimate economic control panels are thousands of miles away. Although CBD Downtown office space remains a surprisingly small fraction of the total regional inventory, more power—in the form of financial headquarters and $400-per-hour firms—is now concentrated Downtown than at any time since the 1940s.[35]

Who Wins, Who Loses?

Creating this physical infrastructure for international finance has been unquestionably the chief policy objective—and accomplishment—of the Bradley administration since 1973. More than mere "urban renewal," Downtown redevelopment has also been the city's major economic strategy for creating jobs and growth. In the face of deindustrialization of its older, nondefense, branch-plant economy, the city has gambled on creating office jobs.[36] Has it worked? And who has benefited?

Certainly the major private-sector players have exploited a real estate bonanza. Speculators and developers have consistently realized large windfalls from Community Redevelopment Agency write-downs and the equity-raising effects of public investment. For example, the CRA bought sixteen rundown parcels at Fourth and Flower streets in the early 1960s for $3 million; in the early 1970s, despite the explosion in property values, it discounted the combined parcel to Security Pacific Bank for a mere $5.4 million. By 1975, the land alone was worth more than $100 million. In another instance, Richard Riordan, a prominent local speculator and mortgage banker, bought property in 1969 at Ninth and Figueroa for $8 per square foot; within a decade, it had soared to $225 per square foot. (Riordan's successes have attracted unusual attention because he is a major contributor to Mayor Bradley and a member of two city commissions.)[37] A veteran Downtown real estate and corporate-leasing expert has "guesstimated" that the $1 billion that the CRA has invested in Bunker Hill and the central business district has helped generate "at least one billion, perhaps two billion dollars' worth of sheer profit for Downtown players, above and beyond their own outlays."[38]

City hall—while in effect promoting Downtown redevelopment as "industrial policy"—has never bothered to collect accurate figures on the new employment generated by the high-rise boom. As a result, conventional cost-benefit analysis is impossible. Don Spivak, the CRA's manager for the entire CBD project during the 1980s, confessed in an interview that the agency had no idea how many jobs for women or minorities have been created, or what has been their per capita cost.[39]

Likewise, while city hall has been throwing $90,000 topping-off parties for new skyscrapers, it has paid no attention to the success of outlying areas in capturing the "back-office" jobs ("number crunching'" and data processing) that are such vital employment multipliers for entry-level clericals. Thus Glendale (a city that in the last census had 450 Black residents out of a population of nearly 130,000) has managed to snare 3 million square feet of secondary bank, insurance, and real estate investment—becoming as a result the third largest financial center in the state.[40] Other major back-office complexes have grown up in Chatsworth, Pasadena, City of Commerce, and Brea (the main base for Security First National). The CRA's indifference to the new geography of service jobs is disturbing since these are precisely the kind of compensatory jobs that East and South Central Los Angeles—hard hit by plant closings—desperately need, and which presumably might have located there if the city had linked front-office development rights Downtown with back-office investment in the surrounding inner city.[41]

The CRA's record in Downtown housing has also been considerably obfuscated in agency propaganda. Planners maintain that the creation of a "jobs–housing balance" Downtown—both to mitigate traffic congestion and to generate a residential base for a "24-hour Downtown"—is one of their major priorities. Yet the CRA, which defines Downtown objectives almost exclusively in terms of middle-class populations and needs, ignores the jobs–housing equilibrium that exists between the garment industry (Downtown's other major industry) and surrounding Latino neighborhoods. It is precisely this existent balance that is now threatened on every side by agency projects (for example, the removal of nearly four thousand people for the recent Convention Center extension) and other public-private initiatives (the potential 10,000 West Bank residents who may be forced out by the proposed specific plan in that area, for example).

The CRA was badly embarrassed in March 1989 when Legal Aid analysts proved that the agency had been deliberately misleading the public by counting cots in Skid Row shelters as "units of affordable family housing." Because neither the agency nor city hall has accurately monitored the destruction of housing by private action Downtown, it is virtually impossible to construct an overall balance sheet of the housing record of redevelopment.

A quarter-century after the clearance of 7310 units on Bunker Hill, the CRA claims to have finally constructed their replacements, although most are outside the Downtown area and only a quarter are "section 8," or "very low income," like those originally destroyed. Setting aside the rehabilitation of Skid Row hotel and shelter rooms, it would appear from the agency's tangled statistics that it has so far increased the city's net stock of "affordable" housing (after deducting units demolished by agency action) by slightly more than 1000 units. Much of this, however, is actually gentrification—that is, replacing lost "very low income" units with more expensive "moderate income" units (an income differential as great as $21,000). In conversations with CRA staff, it was apparent that they conceptualize "affordable housing" as integrating legal secretaries and school teachers, not garment workers or janitors, into the "new Downtown community."[42]

At the end of the day, and in lieu of any official cost-benefit assessment, the redevelopment game yields the following approximate scores:

1. A tripling of land values Downtown since 1975, thanks to public action.

2. Zero increment in property taxes available for general-fund purposes (schools, transportation, welfare).

3. Thirty-five to forty thousand commuter office jobs added to Downtown (presumably these jobs would have ended up somewhere in the region anyway—the CRA did not create them, but merely influenced their location).

4. A small net increment of "affordable" housing scattered around the city, which would probably be canceled out if statistics on private demolition were available.

5. A series of ineffable and questionable "public benefits" (for example, "Downtown culture," "being a World City," "having a center," and so on).

6. The yet uncalculated "negative externalities" generated by redevelopment (that is, the additional traffic load, pollution, neglected investments in other parts of the city, negative tax impact on other services, and so on).

In addition, a full balance sheet on redevelopment would have to estimate the corrupting impact of "centermania" on city politics. City hall and the Downtown development community interpenetrate to such a profound extent that it has become literally impossible to tell where private capital ends and the Bradley administration begins. The resulting trade in influence is a miniature mirror of the military-industrial complex. Just like retired Air Force generals rushing off to fat sinecures on the boards of the aerospace industry, the illuminati of city hall—Art Snyder (ex-councilman), Dan Garcia (former planning commissioner), Tom Houston (former deputy mayor), Fran Savitch (ex–mayoral lieutenant), Maureen Kindel (ditto), and now Mike Gage (another ex–deputy mayor), to give an incomplete list—inevitably seem to end up as lobbyists for bulldozers. With such an extreme concentration of Los Angeles's best minds on moving dirt (and thus creating lucrative second careers for themselves) it is not surprising that lesser priorities—like jobs, safety, health, and welfare in South Central Los Angeles—have been so neglected.

Disneyfying Downtown South

The social costs of Downtown growth will rise steeply in the next decade. But before analyzing the destabilizing impact of emerging "countergames" (the West Bank) and "side-moves" (Central City North, South, and East), let us first consider how the Community Redevelopment Agency proposes to play out the rest of its central business district hand. With construction in the new core in the mopping-up stages (including a controversial plan to demolish historic structures on Seventh Street), the focus of the CRA has shifted to the poles of CBD development: South Park and the Third Street corridor between Bunker Hill and Little Tokyo.

South Park, as we have seen, was a coinage of the 1972 *Silverbook*. The idea was to create a mixed-income "urban village" of clericals and professionals to "brighten" the face of Downtown around the Convention Center and to extend redevelopment in the direction of the University of Southern California campus.[43] Although the CRA reaffirmed a South Park plan in a 1982 rewrite of development guidelines (eventually extending area boundaries south of Seventh and west of Main to the two freeways), speculators had plenty of time to bid up land values to as much as $300 per square foot before the agency finally acted to assemble parcels. In the face of such land inflation, even luxury units in South Park now require large subsidies.

South Park's massive need for public financing is probably the major item on CRA's hidden agenda in the struggle to remove the cap on tax increments in the central business district.[45] The CRA sticks obdurately to the dogma that South Park's critical mass (a projected build-out population of 25,000) is absolutely necessary to transform Downtown into a "true community" (poor people evidently do not count) and to shore up street-level leases and overall CBD property values into the twenty-first century. Not surprisingly, housing activists have attacked the premise that the "yuppification" of South Park should be the city's top residential priority. Thus Michael Bodaken of Legal Aid (now Mayor Bradley's housing advisor), in a 1989 *Times* interview, denounced the $10 million subsidy that the CRA had furnished to Forest City Properties to build $1200-per-month apartments in South Park. "It is just unbelievable that the city is subsidizing developers with millions of dollars to lure yuppies Downtown. This city is the homeless capital of the nation. The money ought to be earmarked for homeless shelters and low-income housing."[46]

Housing advocates have also criticized the relocation of an entire residential community in order to expand the Convention Center, the other major component of the South Park plan. The $390 million expansion—the single largest bond issue in Downtown history—is headed for troubled waters as Calmark Holding Co., the developer of an adjacent super-hotel, collapses under the weight of its junk bonds. Without Calmark's $400 million Pacific Basin Hotel—the largest ever planned in Southern California—the expanded Convention Center would be left without a single hotel room within walking distance.[47]

On the rim of South Park (Eighth Street and the Harbor Freeway), a jocularly named monstrosity called Metropolis is being designed by Michael Graves, current house architect for the Disney Corporation and tongue-in-cheek author of the Disney World Hotel, decorated with giant swans, and the Burbank Disney headquarters with its columnar figures of the Seven Dwarves. Impervious, like most architects, to the social impact of his multimillion-square-foot project on surrounding streets and neighborhoods, his concern is instead focused on making Metropolis a "total experience" for its corporate users. As design critics have appreciated, the arrival of Graves marks a new era Downtown, a shift from stern skyscraper monoliths and fortresses to more "livable" and playful environments. He plans colorful glazed bands, "party hat" roof lines, and flashy octagonal pavilions atop some of his towers. A key decorative element will be a six-story base of turquoise terra cotta—intended, according to Graves, "to show where Daddy works."[48] Indeed.

"Reaganizing" the Historic Core

Gentrification is also the municipal objective in the area between Bunker Hill and Little Tokyo. Back in *Silverbook* days, as we have seen, Third and Broadway figured as "Bunker Hill East," a kind of urban universal-gear meshing Bunker Hill, the Civic Center, and Little Tokyo. The fortification of Bunker Hill, however, precluded such interaction, and Broadway became instead the premier Spanish-language shopping street in North America. Now, with the Hill fully secured and almost completely redeveloped, the Community Redevelopment Agency is reviving the idea of a "pivotal interface" (read: yuppie corridor) to allow the free circulation of white-collar workers and tourists in the northern part of the central business district.

The anchor of this gentrification strategy is the new Ronald Reagan State Office Building. The CRA spent more than $20 million in direct subsidies to induce the state to bring three thousand office workers to Third and Spring as the shock troops of the area's "uplift." Both the Broadway Spring Center, across from the state building, and the new *L.A. Times* parking garage, on Broadway north of Third, have been ingeniously designed with CRA and LAPD expertise to provide high-security pedestrian passageways, with surveillance cameras, private guards, and steel-stake fencing, to allay the anxieties of white-collar workers.

The direct beneficiaries of the "Reaganization" of the area are two chief Bradley backers: Ira Yellin, owner of the Bradbury Building, the Grand Central Public Market, and the Million Dollar Theater Building (all at the corner of Third and Broadway), and his friend and associate, Bruce Corwin, proprietor of various Broadway theaters and largest contributor to the recent defense fund for the embattled mayor. Yellin and Corwin have for a long time been the principal players in the CRA-financed "Miracle on Broadway" association. Now they plan to exploit the captive clientele from the Reagan Building (as well as from the *Times* and Bunker Hill) to create a "Grand Central Square" with upscale restaurants, condos, and offices. Restoration architect Brenda Levin has been hired to "weave together the historical fabric" of the Million Dollar Building with the market and a new ten-story parking garage. The CRA has buttered the way by allowing Yellin to cash in the air rights of the Bradbury and Million Dollar Theater buildings for $12 million (a complex subsidy that after sale to another developer will eventually be costed to the public as further traffic congestion). As Spivack of the CRA put it, the deal-making on Broadway was a "win-win situation," the real "miracle" being the CRA's extraordinary willingness to bankroll Yellin and Corwin.[49]

Another component necessary to complete the corridor between Bunker Hill and Little Tokyo is the removal of the Union Rescue Mission—and its crowds of homeless men—from Second and Main, next door to Saint Vibiana's Cathedral. It is rumored that relocation of the mission is part of the deal the CRA made with the state to get the Reagan Building. Moreover, Archbishop Mahony was reported to have lobbied the CRA (whose chief, John Tuite, is a former priest) to shift the eyesore away from his doorstep. Even so, there was some consternation when the CRA suddenly announced in September 1989 that it was offering the Mission $6.5 million to move—nearly four times the appraised value of the property. Councilman Bernardi (still the hammer of the CRA) decried a new conspiracy of the "moneyed interests," and his Westside colleague, Zev Yaroslavsky, complained about public subsidies to a fundamentalist body (the Mission) that refuses to hire non-Christians. Nonetheless, the council majority (without any debate about the implied subsidy to the other sectarian institution, St. Vibiana's) approved the CRA maneuver.[50]

As a result of the Reagan Building and the other CRA initiatives, land prices have skyrocketed in the Third Street corridor, but the revival of the rest of the Historic Core (as the area bounded by First, Los Angeles, Ninth, and Hill Streets is now officially called) remains in doubt. The flight of banks and department stores after the Watts rebellion left millions of square feet of upper-story office space in the core unoccupied. Much of it has sat vacant for twenty years (the city, of course, has never imagined conscripting it for housing for the homeless or other "radical" uses). The CRA has planned to gradually bring this office desert back to life with infusions of restoration money, improved security, the addition of "nightlife" (for example, the old Pacific Stock Exchange transformed into a disco), and so on.[51] Now, however, the fate of the area appears inversely hinged upon the success of a plan to bootleg a second Downtown, west of the Harbor Freeway. The emergence of the so-called Central City West has suddenly put the CRA's best-made plans in jeopardy.

The Countergame

Certainly, the possibility has always existed of a "countergame." The growing differential between land values in the growth core and its immediate periphery encouraged outlaw developers to gamble on attracting investment across the

Harbor Freeway. Indeed, already by the mid 1960s, a diverse group of speculators, large and small, were staking positions west of the freeway (an area that the *Silverbook* had primarily designated for peripheral parking and services). While awaiting redevelopment to come their way, they were permitted, criminally, to demolish entire neighborhoods in the Crown Hill and Temple-Beaudry areas. It was to their advantage to "bank" land in desolation rather than take the risk of tenant organization or future relocation costs.

But the frustrated speculators had to wait nearly a generation before they could compete against the central business district. With the exception of Unocal (a major Downtown corporation stranded on the wrong side of the Harbor Freeway), they were either foreigners (overseas Chinese and Israelis) or minor-leaguers outside the mainstream power-structures, opposed by an awesome combination of the old-elite Central City Association and the Community Redevelopment Agency. Moreover, the notional "West Bank" was balkanized by several city council districts and had no clear "patron."

This calculus of forces began to shift in the mid 1980s. As the Figueroa corridor started to top-out with new development and turn its face away from Pershing Square, the western shore of the freeway suddenly became inviting. Despite the notorious fiasco of the Chinese World Trust building (still half-empty today), structures like the new Pacific Stock Exchange (relocated from its magnificent home on Spring Street) seemingly proved the viability of the other bank. This led several major-league players—including Hillman Industries and Ray Watt—to migrate west with their awesome financial resources and political clout. Moreover, most of the West Bank was politically consolidated into a new district under Gloria Molina, who was eager to find a resource base for jobs and housing in her crushingly poor constituency.

With Molina's forceful backing, the area's largest stakeholders (organized since 1985 as the Central City West Association [CCWA]), germinated a plan to literally create a second Downtown. Despite the dire warnings of former CRA chief Ed Helfand that West Bank competition would undermine the entire logic of Downtown renewal, Molina accepted the offer of the CCWA in 1987 to privately fund a "specific plan" for the area. This partnership deliberately excluded

CRA (seen as the custodian of CBD interests) and greatly reduced the role of the city planning commission. In July 1989, after two years of study, the urban design firm headed by ex-CRA commission president Kurt Meyer submitted a first draft of the plan, detailing transportation and land use for a maximum build-out of 25 to 30 million square feet of commercial space (that is, roughly equal to all new construction Downtown since 1975, or to two-and-a-half Century Citys).

The transportation requirements of such a scale of development are stupefying, especially in the wake of Downtown's past policy of "starving" the West Bank of transport links in order to make it undevelopable. In the CCWA's conception, the Harbor Freeway, rather than Figueroa, would become the new "Main Street" of a bipolar Downtown. Although Caltrans officials staunchly maintain that the freeway—"double-decked" or not—will simply not be able to absorb the new traffic volume from the proposed Central City West, the draft plan provides for four new off-ramps, as well as an additional Metro Rail station at Bixel and Wilshire, a $300 million "transit tunnel" under Crown Hill, and a funneling of traffic down Glendale Boulevard that could have nightmarish consequences for the already congested Echo Park area. (Some of the transport planners involved also argue for the conversion of Alvarado into a high-speed freeway connector.)

Another breathtaking dimension of the plan is the proposal for 12,000 units of new housing gathered in a predominantly affluent "urban village" similar to the South Park plan, but with a marginally greater inclusion of low- and very-low-income units (25 percent). Housing advocates, however, like Father Philip Lance of the United Neighbors of City West, point out that there is already a housing emergency in the area as the arrival of the big guns accelerated scorched-earth land-banking: 2100 units have been demolished in the last decade.

Moreover, the draft specific plan provides for the replacement of only three-quarters of the low-income units it proposes to remove for development.[52] Other critics, pointing to the twenty-year timeline of development, have demanded immediate rehousing of the existing tenants and restitution for the housing destroyed in recent years.

While the larval Central City West plan gestates through a labyrinthine process of political negotiation, a land rush of Klondike proportions has broken out on the West Bank. In some cases, land values have increased nearly 4000 percent in a single decade.[53] Speculators, reinforced by new arrivals from offshore, are now concentrating on an "underdeveloped" mile-long strip of Wilshire Boulevard between the freeway and the new Metro Rail station at MacArthur Park. As CRA planners recognize with some trepidation, this flow of investment threatens to revive Wilshire Boulevard–westward as the major axis of Downtown growth—in competition to the Figueroa–southward target of the *Silverbook* strategy.

Meanwhile, with stakes rapidly increasing, developer Ray Watt has bum-rushed his way ahead of the CCWA pack to break ground. Although the city planning department's chief hearing examiner opposed the plan for a 1-million-square-foot "Watt City Center" tower on the west side of the Harbor Freeway at Eighth Street, Watt—in one of the most impressive power-plays in recent city history—ramrodded it through the city council with the help of lobbyist Art Snyder (former East L.A. councilmember) and Molina, chair of the Planning and Landuse Committee. Molina, in liaison with the United Neighbors of Temple-Beaudry, cut a consciously Faustian deal: accepting the Watt Center's additional traffic load in exchange for eighty units of immediate low-income housing.[54]

Downtown Every-Which-Way?

To many Downtowners, the Watt City Center is a massive symbol that crime (in this case, skyscraper hijacking) does pay after all. And to make matters worse, the West Bank example seems to be spurring other landowners on the central business district's periphery to package megaprojects for sale to interested members of the city council. Venting the Community Redevelopment Agency's anxiety at the dissipation of a Downtown focus, the agency's chief, John Tuite, recently outlined the competing vectors of development: "There is the Convention Center (South Park), Union Station, [councilmember] Bob Farrell's ideas for a strategic plan to link USC and the surrounding area to Downtown, as well as other CRA areas, City West and City North."

Union Station, especially, is a variable of unknown, perhaps huge, dimension in Downtown's future. When Caltrans tried to purchase the station under eminent domain in the early 1980s, its owners (the three transcontinental railroads:

Southern Pacific, Union Pacific, and Santa Fe) brought to court a Charles Luckman model showing the site built out to the proportions of Century City. In Luckman's conception, the elegant old station was reduced to minor detail in an overscaled nightmare that included two skyscrapers, two mansard-roof Vegas-type hotels, a vast shopping concourse, acres of parking, and a fantastic thirty-story glass "Arc de Triomphe" smiling over 20,000 office workers and shoppers. Overawed by the model, the judge ruled in favor of the station owners, tripling the value of the site and forcing Caltrans to abandon its purchase attempt.

In following years, as Santa Fe (whose largest shareholder is Olympia and York—the world's largest commercial developers) laboriously negotiated to buy out its partners, the megadevelopment potential of the station became the focus of Councilman Richard Alatorre's attentions. Alatorre, chairing the redistricting of the city council in 1987, allocated to himself the cusp of Union Station and Olvera Street with the specific purpose of making station redevelopment a "financial motor" to drive economic development in his Eastside district's "enterprise zone." Although his idea of linking community development to a rich redevelopment project mimics Molina's strategy on the West Bank, he has collaborated with, rather than excluded, the CRA, in the evident hope of integrating Union Station into the CBD game plan.

Accordingly, in the spring of 1988, the CRA, on behalf of Alatorre and the station owners, completed an in-depth study of the site's development potential (including the vast, nearly empty shell of the neighboring Terminal Annex Post Office). In essence, the CRA analysts endorsed Luckman's 1983 vision of a "new urban center," proposing at least 3 million square feet of mixed-use office, hotel, retail, and residential development, as well as architectural unification with La Placita/Olvera Street across the street. But the study continues to raise as many questions as it answers.[55] First, it is not clear whether the office potential of the station site can be fully realized in the face of the growing competition of Central City West. Second, Olvera Street merchants and East Los Angeles political foes of Alatorre fear that station redevelopment will inundate and destroy the popular character of the old Plaza area—a crucial public space for Spanish-speaking Los Angeles. And, finally, Union Station is the fulcrum of competing claims between

Little Tokyo (core of an emergent Central City East) and Chinatown (center of a hypothetical City North).

It has become the passion of planning commission chairman William Luddy to unify the area north of the Civic Center—including Chinatown, El Pueblo, and Union Station—as a single planning unit designed to reinforce CBD redevelopment by adding a dynamic, nighttime tourist quarter. Moreover, as the planning department's December 1989 City North charette emphasized, "If Los Angeles is to compete favorably with Vancouver and San Francisco as a market for real estate development by Hong Kong and Singapore dollars, for investment from Chinese-Asian money, it must bolster that part of its city which represents its strong Chinese heritage.... "[56]

But this concept of packaging City North, including Union Station, for sale to the Chinese diaspora ignores the competing interest of Japanese capital in establishing a strong link between Little Tokyo and the station. Little Tokyo's Main Street–like function for Los Angeles's Japanese-American community has been eclipsed by its new role as a luxury hotel and shopping precinct for Japanese businessmen. Now, however, its developers (in the words of the *Downtown News*) "are making a bold play to capture the tourist windfall expected from the Convention Center expansion." Over a million square feet of hotel, retail, and residential construction is under way near First and Alameda (extending Little Tokyo to the edge of the Los Angeles River), and developers are pushing for a mixed-use, high-rise rezoning of the industrial corridor east of Alameda. Union Station, revived by Metro Rail as Downtown's transit hub, is hungrily envisioned as an integral part of Little Tokyo's expanded sphere of influence.

As different forces contend for the future of Union Station, another major eruption of development may be ready to occur on the CBD's southern flank. Since *Silverbook* days, most observers have believed that the CRA's ultimate goal is to extend the corridor of high-rise redevelopment along Figueroa to the Jefferson or Exposition Street edges of the University of Southern California campus. With utter conviction in its inevitability, one landowner (an auto dealer) has spent twenty years patiently assembling most of the long, low-rise stretch between Jefferson and Adams for conversion into office and hotel-block developments. USC, on the other hand, has been preoccupied since the Watts rebellion with fortifying its borders (an impressive Maginot Line of shopping centers and parking lots) and promoting the gentrification of its Hoover Street fringe. Few doubt, however, that the university is nurturing a far more ambitious vision, linking its housing strategy to the commercial development of Figueroa under joint auspices with the CRA. The 1988 appointment of Gerald Trimble, high-powered former redevelopment director for Pasadena and San Diego, as USC's development director has fueled endless speculation about the university's game plan as well as stimulated local councilmember Robert Farrell to agitate for a link—à la Molina's West Bank and Alatorre's Union Station cash cows—between commercial and community development in the south Downtown–USC nexus.

Perestroika or End Game?

In summary, the West Bank countergame, together with the emerging moves on the north, east, and south faces of Downtown, is beginning to disorganize the Community Redevelopment Agency's CBD game plan. The casino is in chaos, the developers are seen shooting craps with politicians in every alley. Existential questions are raised: Can Downtown grow in every compass direction at once? Who will supply the demand for one, two, three, or many Downtowns?

For the CRA, the problem is even more complicated, since it must confront these centrifugal tendencies while simultaneously surmounting the 1977 Bernardi cap and renewing its mandate to orchestrate Downtown's expansion. Moreover, Mayor Bradley's position as the agency's patron has been made more delicate by a highly publicized ethics scandal as well as by charges of benign neglect from his own Black political base.[57] An atmosphere of quiet crisis has served to concentrate minds in the CRA's Spring Street headquarters.

The "solution" hammered out from above necessarily proposed both a political realignment and a new design for Downtown. In 1989, the CRA survived a close call when its opponents on the city council almost achieved a majority for a takeover of the agency.[58] In the aftermath, Mayor Bradley urged a sweeping concordant between the CRA and its most trenchant community critics. In return for supporting a huge increase of the CBD's tax-increment capacity to lion, the mayor offered to split the addition evenly between CBD redevelopment and citywide housing needs. He also dramatically co-opted two of the CRA's leading housing critics into his administration (one as his housing advisor, the other as CRA commissioner). Simultaneously, city hall instituted new housing linkage fees, taxing high-rise development to support affordable home construction. Unsurprisingly, the former united front of CRA foes—community groups, public-interest lawyers, progressive planners, and so forth—has disintegrated.[59]

The mayor has appointed a Downtown Strategic Plan Advisory Committee to wrestle with the challenge of Central City West in the framework of a new, twenty-year master plan for Downtown. Chaired by two veterans of the *Silverbook* taskforce, CRA commissioners Frank Kuwahara and Edwin Steidle, the committee is dominated by a two-thirds majority of developers and corporate lawyers, including such familiar suspects as Ira Yellin, Chris Stewart (former secretary of the CCA), and the irrepressible Art Snyder. Although the CRA's authority ends at the Harbor Freeway, the committee has been specifically encouraged to visualize the CBD's future in the "broadest context—that is, to work out some reconciliation of developer interests on both sides of the freeway (taking into account USC, Little Tokyo, Central City North, and possibly Hollywood as well). In the meantime, the mayor and the CRA are readying a proposal to drastically expand the CRA's domain Downtown: taking in City North, the USC area, part of the West Bank, and perhaps the area in transition east of Alameda. If adopted, the expanded project areas would allow the agency to deal with two problems at once: reconciliation of growth poles and the linkage of community and commercial redevelopment.

The CRA has encouraged the view that these various initiatives are the beginning of an authentic Downtown *perestroika* that will eventually transform redevelopment to please everybody, from Japanese developers to the homeless on Skid Row. Despite the "encouragement," however, a hard core of doubt remains. Indeed, in the view of many insiders, the end game has already begun, as Downtown plays against the clock—perhaps time bomb—of two insurmountable contradictions: overbuilding and the coming traffic apocalypse.

The smart money on both sides of the Harbor Freeway has ceased to believe in the Downtown–Pacific Rim perpetual motion machine, and, like Ray Watt, is racing to bring their projects in and stuff them with high-class tenants before the market crashes. Even before the official arrival of recession in summer 1990, the torrent of incoming Manhattan law firms and Japanese banks had slowed to a trickle. In Japan itself, where convulsed stock markets registered the overaccumulation of fictional capital, high interest rates were beckoning capital to stay home. The 1980s fantasy of an infinite supply of offshore capital to sustain Southern California's real estate boom seems increasingly like a psychedelic aberration.

If the trophy-class Downtown office market still purred sweetly at the end of 1990, it was only because existing Downtown tenants (like First Interstate and Unocal) have been vigorously "trading up." As they have bailed out of their old offices (usually circa 1960s–1970s structures like the First Interstate Tower), vacancies have soared in the corporate schlock, or "class two" market. New development, in other words, is devaluing old. This is slowing job creation, and potentially undermining the CRA's putative tax base as well.[60]

But, as happens in all business cycles, production drastically overshoots demand in the final, fervid phase of the boom. In a situation where even redevelopment's *eminence grise*, CRA commission president Jim Wood, is admitting that Downtown is overdeveloped and the Japanese are acting nervous,[61] science-fiction-like quantities of office space are scheduled for delivery over the next decade. In the flush conditions of the 1980s, the Downtown market absorbed about 1.4 million square feet of new space per year. With more than 12 million square feet already approved and in the construction pipeline and with the financial-services expansion ending, supply should easily meet demand through to the millennium. Yet a further 20–30 million square feet of projects are on drawing boards, chasing investors and mortgage bankers around the city. (Altogether, councilmember Marvin Braude estimates that sixty-four new projects creating 37 million square feet of office space.)[62] With Southern California diving into deeper recession, who will occupy this embarrassment of space? (And why should tax dollars subsidize its construction?) Even in Los Angeles, speculators cannot go on endlessly building space for other speculators.

But a Downtown depression may be the lesser of potential evils. Worse still is the specter of hyper-gridlock paralyzing Downtown and a large part of Los Angeles County. The traffic nightmare of the 1990s—regardless of an economic slowdown—will be the simple addition of current planning exemptions and special cases. For example, two recently approved megaprojects—the Watt City Center and, directly across the Harbor Freeway, the Metropolis—will each add *fifteen thousand* trips per day to overloaded Downtown streets. Total new development will generate an additional 420,000 trips per day, making "the existing Harbor Freeway [according to councilmember Braude] a parking lot and paralyzing the movement of traffic in the Downtown area."[63] Lest Metro Rail and Downtown "village living" be immediately wheeled in as a *deus ex machina*, it is sobering to observe that a recent survey discovered that only a tiny fraction of Downtown office commuters (just 5.4 percent) have both the means and the desire to live in Bunker Hill or South Park. Certainly the nightmare of perpetual gridlock will persuade a larger percentage of commuters to reluctantly abandon Pasadena or Studio City,[64] but these same horrors may also persuade Mitsui and CitiCorp to look afresh at Wilshire Boulevard, Long Beach, or Orange County's Golden Triangle. They may even convince shaken Los Angeles voters to reexamine the premises of the city's pharaonic and socially irresponsible redevelopment strategy.[65]

1990

Postscript: Play Resumed

The 1990–94 recession, which sent CBD vacancies soaring into the double digits, accompanied by the 1992 riot that engulfed most of the MacArthur Park district, forced a humiliating retreat of speculative capital from the West Bank. Simulaneously, the meltdown of the superyen led to the panic-striken evacuation of Japanese investment from downtown trophy properties, while a wave of mergers and buyouts produced a rapid demotion of Los Angeles's rank as an international financial center.

Amid so much carnage, the old rules of the game dramatically asserted themselves. Although a new Downtown Strategic Plan alloted some "new urbanist" trinkets to the natives on the periphery, the Figueroa corridor was definitively reestablished as the axis of Downtown growth. Corporate nerves, badly rattled by recession and riot, were soothed by the 1994 election of Richard Riordan as mayor. He quickly began to rebuild the CBD growth coalition, incorporating for the first time the Catholic Church (anxious to build a huge new cathedral in the civic center) and the Latino leaderships of the Downtown service and hotel unions (who were rewarded with Riordan's tolerance of their organizing campaigns). Moreover, a dubious messiah appeared in the person of Denver billionaire Philip Anschutz, the eleventh or twelfth richest man in the United States, depending on the fluctuating fortunes of his myriad oil, railroad, telecommunications, and sports subsidiaries (including the Lakers).

Although even Anschutz isn't rich or stupid enough to attempt to fill the void left by the flight of the Japanese and the downsizing of LA-based banks, he is big enough to have his shoes shined by most of the city council. In partnership with media mogul Rupert Murdoch (who owns the Dodgers) and real estate wheeler-dealer Ed Roski (who grew his fortune in the corrupt topsoil of the City of Industry), he extracted $12 million from the CRA in 1997/98 to help assemble the site for the trio's new Staples Center sports complex that brought the Lakers and Clippers downtown (and across the street from red-ink-gushing Convention Center).

Despite fervent support from the Riordan administration, the $350 million Staples scheme suffered some minor wing damage as it flew into a cloud of suburban flak. An original proposal, for example, for a long-term city subsidy of $70 million was defeated by public outrage whose epicenter was the Downtown-hating San Fernando Valley. In 1999, moreover, revelations about a secret profit-sharing arrangement between Staples Center and the *Los Angeles Times Magazine* led to a shareholder revolt orchestrated by retired publisher Otis Chandler that resulted in sale of Times-Mirror—Downtown's oldest and most influential stakeholder—to the Chicago Tribune Company. The Downtown Old Guard (except for USC at the CBD's south pole) now is extinct, but the new elites continue to play the redevelopment game by the old Silverbook rules.

Indeed, the Anschutz-Roski camp (which also includes billionaire Ron Burkle and Casey Wasserman, who owns the LA Avengers) seems to have decided to go for the South Park/Convention Center hat trick. With Downtown booster James Hahn in Riordan's old office, they muscled a new $2.4 billion in tax-incre-ment redevelopment authority through a pliant city council in the winter of 2002. Nothing could have been more ingeniously designed to incite neighborhood anger or to bolster the cause of secession in the San Fernando Valley. But city hall, as usual, could see little further than the end of its nose. In the meantime, the new CRA masterplan is truly Viagra for a wilting Downtown boom, and much of the stimulus is targeted at hotel and high-income residential development to support the Staples complex and, possibly, a new stadium for an Anschutz-owned NFL franchise. County Supervisor Zev Yaroslavsky, who counts as one of his less palatable responsibilities the closing of bankrupt county healthcare facilities, characterized the entire boondoggle as "taking money out of the mouths of poor people." So what else is new?

2002

Notes

1. Reyner Banham, *Los Angeles: The Architecture of Four Ecologies*, London 1971, p. 201.
2. See Central City Association, *Downtown 2000*, Los Angeles, 1985.
3. Cf. Scott Bottles, *Los Angeles and the Automobile: The Making of the Modern City*, Los Angeles 1987; and Robert Fogelson, *The Fragmented Metropolis: Los Angeles 1850–1930*, Cambridge, Mass. 1967.
4. For Downtown interests' zealous but ultimately unsuccessful crusade to use zoning against centrifugal development, see Marc Weiss, "The Los Angeles Realty Board and Zoning," chapt. 4 in *The Rise of the Community Builders*, New York 1987.
5. Especially for the role of the *Times*, see Robert Gottlieb and Irene Wolt, *Thinking Big*, New York 1977, pp. 152–55, 306–17. Big public projects have been repeatedly used to revive or recycle real estate values in declining sectors of Downtown. Thus the construction of the Civic Center in the 1930s bolstered the value of *Times* properties in the older, circa-1900 core area, which had been in decline after the westward migration of the Downtown center in the early 1920s.
6. Bottles, chapt. 4 and 5 of *Los Angeles and the Automobile;* Central Business District Association, *A Quarter Century of Activities: 1924–1949*, Los Angeles 1950; Mike Davis, "Tunnel Busters: The Strange Story of the Hollywood Subway," unpublished, 1988; and Steven Mikesell, "The Los Angeles River Bridges," *Southern California Quarterly* (Winter 1988).

7. See Sy Adler, "Why BART But No LART? The Political Economy of Rail Rapid Transit Planning in the Los Angeles and San Francisco Metropolitan Areas, 1945–57," *Planning Perspectives* 2 (1987).

8. See David Brodsly, *L.A. Freeway,* Berkeley, Calif. 1981, p. 96.

9. Rodolfo Acuña, *A Community Under Siege: A Chronicle of Chicanos East of the Los Angeles River 1945–1975,* Los Angeles 1980.

10. Frank Wilkinson, "And Now the Bill Comes Due," *Frontier* (October 1965).

11. In 1956, Los Angeles had one of the largest skid rows in the nation, with 15,000 residents "in everything from abandoned buildings to packing crates in alleys and from 306 hotels to eleven flop-houses." See Aubrey Haines, "Skid Row Los Angeles," *Frontier* (September 1956).

12. See "Civic Center Plan" in Municipal League of Los Angeles, *Bulletin* 2 (March 1925), p. 13.

13. Cf. William Babcock, *Regrading the Bunker Hill Area,* Los Angeles 1931; and Pat Adler, *The Bunker Hill Story,* Glendale, Calif. 1965.

14. Cited in William Pugsley, *Bunker Hill: The Last of the Lofty Mansions,* Corona del Mar, Calif. 1977, p. 27.

15. See Gene Marine, "Bunker Hill: Pep Pill for Downtown Los Angeles," *Frontier* (August 1959).

16. Cf. John Brohman, "Urban Restructuring in Downtown Los Angeles" (M.A. thesis, School of Architecture and Urban Planning, UCLA, 1983); and Joel Friedman, "The Political Economy of Urban Renewal: Changes in Land Ownership in Bunker Hill" (M.A. thesis, School of Architecture and Urban Planning, UCLA, 1978).

17. *Centropolis 1—Economic Survey,* December 1960: *Centropolis 2—General Development Plan,* January 1962; *Centropolis 3—Transportation Study,* January 1963; and *Centropolis 4—Master Plan,* November 1964. The Central City Committee, appointed by Mayor Norris Poulson in 1958, was chaired by Walter Braunschweiger.

18. *Los Angeles Times,* 4 November 1965 and 24 December 1972.

19. By 1967, the Wilshire corridor had seventy financial headquarters versus forty-seven in the CBD. Only oil companies maintained their high headquarters concentration Downtown. (See Abraham Falick, "Transport Planning in Los Angeles: A Geo-Economic Analysis" Ph.D. thesis, Department of Geography, UCLA, 1970, pp. 172–75.) Eugene Grisby and William Andrews, moreover, claim that the CBD lost 40,000 jobs between 1961 and 1967. (See "Mass Rapid Transportation as a Means of Changing Access to Employment Opportunities for Low-Income People" [paper for the fifteenth Annual Meeting, Transportation Research Forum, San Francisco, October 1974].)

20. Gottlieb and Wolt, *Thinking Big,* p. 431. They argue that the shadowy "Committee of 25," organized by Asa Call and Neil Petree with the support of the Chandlers, was the ultimate invisible government behind the CCCP and other epiphenomenal forms of elite organization (pp. 457–58).

21. See Robert Meyers, "The Downtown Plan Faces Open Rebellion," *Los Angeles*

(December 1975), p. 85.

22. Certainly the study played lip service to upgrading Broadway and Seventh Street retail as well as preserving Spring Street, but the greatest area of opportunity defined by the *Silverbook* was expansion southward, along the Figueroa axis, into the South Park area.

23. The proposed circulation system corresponded entirely to the envisioned Figueroa corridor of the southward-moving office and apartment construction, completely ignoring the needs of tens of thousands of existing workers in the garment center—a bias reproduced in every subsequent phase of Downtown transportation planning.

24. Although all the different proposed "action areas" of the *Silverbook* were combined into one overall plan and project area, Bunker Hill remained legally and administratively separate under its original 1959 plan.

25. Councilman Donald Lorenzen, who had briefly left the chambers to chat with an aide, later claimed that his vote had been faked by another member to support the plan. He subsequently endorsed the Bernardi suit. See Meyers, "The Downtown Plan Faces Open Rebellion," p. 82.

26. The Downtown leadership brazenly proposed to destroy the old library (now recognized as Downtown's most distinguished architectural landmark) in order to create a development greenfield while simultaneously using the new facility to roll back Latino "intrusions" in the vicinity of the May Company and Broadway department stores. A study commissioned by Meyer and Helfeld in 1976 brilliantly rebuked the myth of Broadway blight. See Charles Kober Associates, *Broadway/Central Library: Impact Analysis*, Los Angeles 1976.

27. See J. Gregory Payne and Scott Ratzan, *Tom Bradley: The Impossible Dream*, Santa Monica, Calif. 1986, pp. 142–43, 149–50.

28. See Dick Turpin in the *Los Angeles Times*, 21 September 1986—confirmed by the *National Real Estate Investor* (December 1986), p. 102: the higher estimate is from Howard Sadlowski in the *Los Angeles Times*, 17 June 1984.

29. Stephen Weiner of Bear Stearns quoted in *National Real Estate Investor* (December 1986), p. 132.

30. See Davis, "Tunnel Busters," pp. 34–38. In the summer of 1990, rumors began to fly that the CRA was considering a monorail system to link nodes of development Downtown.

31. The Southern California Rapid Transit District is proposing to impose $75 million in special taxes on the ninety-eight hundred commercial property owners who stand to profit most from proximity to eleven Metro Rail Phase II stations. In an ominous precedent for the plan, however, MCA Inc., which owns Universal City (an unincorporated enclave) in Cahuenga Pass between Hollywood and Burbank, seems to have found a technical loophole to avoid assessment, although the entertainment conglomerate is planning massive new development next to its Metro Rail station (see Los *Angeles Business Journal*, 15 January 1990).

32. SCRTD, *Los Angeles Rail Rapid Transit Project SE/S/SE/H*, draft (November 1987), Table 3–21, p. 3-2-13.

33. In my view, Los Angeles's emerging transportation infrastructure will restructure land uses (and social groups) without necessarily alleviating gridlock. Thus Metro Rail will be a powerful link between development nodes (strengthening their sales value to offshore capital), whereas Light Rail (Downtown to Long Beach) will allow Downtown's low-wage workers to commute from a more dispersed labor-shed (opening up new development space in the CBD's periphery), It is unclear, however, whether any of this mass transit development will actually reduce freeway usage by Downtown office workers and professionals commuting from the valleys and the Westside.

34. On the internationalization of Downtown redevelopment and the new political alliances created in its wake, see Mike Davis, "Chinatown, Part Two? The 'Internationalization' of Downtown Los Angeles," New Left Review 164 (July/August 1987).

35. There has been a double restructuring of power on the West Coast. On the one hand, San Francisco has been supplanted as a financial capital by Los Angeles. On the other hand, Los Angeles capital has increasingly been overshadowed by the arrival of the big Tokyo and New York banks, whose local headquarters are Downtown. Booster hyperbole about the ascendancy of Los Angeles typically focuses on the first of these phenomena and ignores the second.

36. As Edward Soja has pointed out, the Los Angeles case is a unique combination of industrial decline (unionized auto, tire, and steel plants) and revival (military aerospace and new sweatshop manufacturers). In local impact, however,. the loss of high-wage branch-plant jobs has had a devastating and long-term impact on Chicano and Black communities uncompensated by the addition of new high-tech jobs in the Valley or minimum-wage garment-making jobs Downtown. See Soja, "L.A.'s the Place: Economic Restructuring and the Internationalization of the Los Angeles Region," in Postmodern Geographies, London 1989.

37. Compare Brohman, "Urban Restructuring in Downtown Los Angeles," p. 111: and Friedman, "The Political Economy of Urban Renewal," p. 261.

36. My anonymous informant (interviewed in the fall of 1989 for the L.A. Weekly) was referring to give-aways and discounts, on one hand, and to "positive externalities" (public investments raising private equities) on the other. The total present value of post-1975 private investment in Downtown is probably around $8 billion.

39. Interviewed for the L.A. Weekly in the fall of 1989. A survey of other public agencies revealed a similar ignorance of the economic impact of redevelopment.

40. See Glendale Redevelopment Agency's myriad brag-sheets and press releases. Glendale, just a few miles north of Downtown L.A., is also planning to dramatically "upscale" its Galleria mall to "Rodeo Drive standards"—another move that will steal thunder (and customers) from CRA-backed retail development Downtown.

41. Again, CBD chief Spivak (in the interview noted in footnote 39) confessed that the CRA "had never considered or studied the question of 'back-office' investment as an opportunity in its own right."

42. *Downtown News*, 16 November 1987.

43. See Dick Turpin, "Downtown Expansion to Take Southerly Direction," *Los Angeles Times*, 9 February 1986.

44. Developers fought like tigers to rezone South Park for offices. Ultimately an Urban Institute Panel had to be brought in to adjudicate whether, in light of land values, it was still possible to develop a residential community in the area. See *Urban Land* (September 1987), pp. 13–17; also *Los Angeles Business Journal*, 19 October 1987.

45. An internal CRA memo secured by the *L.A. Weekly* indicates that the agency wants to spend a further $372 million on developing South Park. See Ron Curran, "The Agency at a Crossroads," *L.A. Weekly*, 28 March 1990. Curran has been the only journalist in Los Angeles to doggedly follow the CRA's footsteps over the last five years, and his many articles in the *Weekly* are essential reading for anyone interested in Downtown L.A. or the politics of redevelopment.

46. *Los Angeles Times*, 25 June 1989. See also, ibid., 10 January 1988.

47. Just as Glendale has waylaid back-office jobs, so too is Long Beach preparing to hijack Downtown's convention trade. With nearly as much convention space as Downtown and a new oceanfront cityscape under construction on the site of the former "Pike" (once the West Coast's Coney Island), as well as a potential Disneyland II next to the *Queen Mary*, Long Beach (like Anaheim in Orange County) is geared up for competition.

48. Quoted in Leon Whiteson, *Los Angeles Times*, 22 March 1990.

49. Cf. *Miracle on Broadway—Annual Report 1989*; *Downtown News*, 10 February 1990; *Los Angeles Business Journal*, 6 November 1989: *Los Angeles Times*, 10 and 27 February, 9 April, and 22 September 1989.

50. Cf. *Los Angeles Times*, 16 October; and *Downtown News*, 16 October 1989.

51. See CRA, "Memorandum: Historic Core Three-Year Work Program," 21 December 1988.

52. I have been fascinated to learn that even the CRA study team assigned to evaluate the draft specific plan (for in-house purposes only) regarded the housing element as a "crock ... not proposing to do anything at all." *L. A. Times* architecture critic Sam Hall Kaplan has repeatedly questioned the adequacy of its housing provision as well as condemned its "segregation of uses ... and office tower ghetto in the southeastern portion of the community, and the isolation elsewhere of schools, housing and streets."

53. "In 1979 a parcel of land was sold to Unocal for $11 a square foot. Towards the end of 1988, an adjacent parcel was sold to Unocal for $270 a square foot. And just last spring, Hillman Properties reportedly purchased the entire contiguous site for $370 a square foot," *Los Angeles Business Journal*, 29 January 1990. As land prices rise on the West Bank, it nonetheless retains the important comparative advantage, vis-à-vis the CBD, of being "parking rich"—that is, of having more generous onsite parking allowances—an increasingly important variable for developers and their tenants in Los Angeles.

54. Councilmember Gloria Molina illustrates the classic dilemma of an inner-city politician negotiating with international capital without the clout of an activist community

movement backing her up. Although a tireless advocate of housing for her low-income community, she has accepted developer ground rules (and campaign contributions) as a strategy for generating at least a modicum of decent replacement units. Friendly critics have suggested that she would have saved more housing—or at least wrestled a better deal—by mobilizing the largely Central American community of the West Bank in opposition to the CCWA's redevelopment strategy. For an interesting profile of Molina, see Ron Curran, "Gloria Molina—A Perennial Outsider Comes to Power and Now Plans to Run for Mayor," *L.A. Weekly*, 13–19 October 1989.

55. For Union Station redevelopment in the context of the restructuring of railroad land holdings, see Mike Davis, "The Los Angeles River: Lost and Found," *L.A. Weekly*, 1–7 September 1989.

56. Los Angeles Design Action Planning Team, *A Plan for City North*, 5 December 1989.

57. On "Bradleygate," see Mike Davis, "Heavyweight Contenders," *Interview*, August 1989.

58. The vote took place along a pork barrel divide. Historically, council attitudes toward the CRA have been shaped less by ideology than by whether or not the councilmember has a significant CRA project in his or her district. Even Ruth Galanter, the recent "environmentalist" addition to the council from Venice Beach, has had a change of heart about the agency after working with it to renovate the aged Crenshaw Shopping Center.

59. The mayor meanwhile has tried to mollify the county Board of Supervisors—Bernardi's major ally—by deregulating development rights for county properties Downtown and increasing their share of the tax flow from Hollywood redevelopment. This leaves only Bernardi and Valley homeowners' groups as intransigent opponents of lifting the cap.

60. See Morris Newman, "Old Buildings Square Off Against New in Los Angeles Office Battle," *Los Angeles Business Journal*, 23 October 1989: also ibid., 29 January 1990.

61. See interview in *Los Angeles Business Journal*, 16 October 1989. The *Downtown News* (22 January 1990) reported the growing dissatisfaction of Japanese owners with the advice they had been receiving from asset managers and brokers about the quality of the Downtown market.

62. See Chip Jacobs, "Braude, Saying Downtown Growth 'Out of Control,'" *Los Angeles Business Journal*, 9 October 1989.

63. Ibid.

64. Is it conceivable that some Downtown visionaries are actually counting on gridlock to make their Manhattanized urban villages work? As CRA chairman Jim Wood explained in an interview last year, "We *planned* for there to be lots of traffic Downtown; we *wanted* traffic because that would make Metro Rail work." For a preface to a "Green" counterplan for Downtown, see Mike Davis, "Deconstructing Downtown," *L.A. Weekly*, 1–7 December 1989, as well as Davis, "The Los Angeles River: Lost and Found," previously cited.

The Hole (Hollywood Boulevard)

9

The Subway That Ate L.A.

Until this summer, the corner of Hollywood Boulevard and Vermont Avenue was best known for Hollyhock House, the epic neo-Mayan mansion that Frank Lloyd Wright designed in 1919 for the wealthy socialist Aline Barnsdall. But on 22 June 1995 a more sinister landmark appeared. Passengers waiting for the overdue 6 a.m. bus at the foot of Barnsdall Park were startled by an eerie groan emanating from under their feet. Suddenly Hollywood Boulevard began to collapse in front of them. Several stories underground, twenty subway construction workers were sprinting for their lives, a few steps ahead of falling steel and debris.

Miraculously, everyone escaped serious injury, but the monstrous cavity, seventy feet deep and half a block long, refused to stop growing. All morning long it continued to suck down slabs of pavement, as well as what remained of the tarnished reputation of Los Angeles County's Metropolitan Transportation Authority (MTA). Officially christened the Hollywood Sinkhole, it has become the biggest transportation fiasco in modern American history.

The Last Pork Barrel?

L.A. County currently consumes one-quarter of the federal government's rail-transit construction budget. As originally planned in 1980, the MTA proposed to spend $183 billion over thirty years on a 400-mile network of subway and light rail lines. Although an implacable regional recession forced the MTA to scale back at the beginning of this year, rail construction remains budgeted at $70 billion over twenty years—the largest public-works program in fin-de-siècle America. The half-finished Red Line subway, which collapsed in Hollywood, has averaged $290 million per mile.

Although the Sinkhole is now plugged with several hundred tons of concrete, the political damage continues. Influential politicians, left and right, have demanded a halt to further construction. The Justice Department has intensified an investigation into fraud and corruption at all levels of the MTA. On 22 July, the FBI made its first arrest: a top MTA administrator was charged with operating a kickback scheme involving the agency's insurers. Meanwhile, a militant Bus Riders Union has emerged on the city's buses, whose profoundly poor and mostly nonwhite riders are at the losing end of L.A.'s transit schemes. The sky is rapidly darkening for the MTA: Southern California's last great pork barrel now that aerospace employment has crashed.

The comparison fits. With its huge budget and a bigger swarm of lobbyists (nearly 1100) than Sacramento, the MTA walks and talks like the military-industrial complex. And as with that other behemoth, the design of its "product" owes less to real need than to special-interest agendas. The Red Line, for example, intended to connect Downtown to the San Fernando Valley and, at $5.8 billion, the most expensive subway in history, is the culmination of a seventy-year crusade by Downtown business groups to use mass transit to recentralize investment and commerce. Although L.A.'s middle-class voters twice approved hikes in the sales tax to support the subway project, the system fails to tap any significant residential concentration of Downtown automobile commuters. For poor and minority communities, dependent on the bus system, it is a vast diversion of resources.

Between 1925 and 1974, the old Central Business District Association and its descendants placed half a dozen subway or monorail initiatives on the county ballot. All were shot down by the combined forces of outlying city neighborhoods and the suburbs. Then, in 1980, Downtown interests used the promise of an enlarged system and ample federal pork to bring the suburbs aboard. In 1987 the California congressional delegation flexed all of its muscle to override President Reagan's veto of the transit bill that contained the crucial funding.

If any consensus had emerged from decades of otherwise quixotic transit planning in Los Angeles, it was that only the Wilshire corridor, from Downtown to Westwood, had the minimum population density and concentration of activities to justify the cost of underground construction. Yet the Wilshire main line was stopped in its tracks by Congressman Henry Waxman after a methane explosion in his district in 1987 (unrelated to the subway) injured twenty-one of his constituents.

This should have been the death knell for the Red Line, but the route was quickly reconfigured to run north, through Hollywood to the Valley. Abandoned was any pretense of serving existing transit demand. The project became simply an aphrodisiac to attract real estate investment to the city's three largest redevelopment projects—which happened to be in the Downtown–Hollywood–North Hollywood corridor. In the MTA's environmental impact report, the current Red Line is envisioned as generating its own ridership, virtually *ex nihilo*, through land-use policies that would concentrate most of L.A.'s foreseeable commercial and residential construction above its eleven stations. The report projects a fantastic *50 million* square feet of such development over the next generation—roughly equivalent to two new Downtowns. Although construction on this scale is still science fiction, "transit-oriented development" has become the city's official growth strategy and the Red Line is already jacking up corporate land values in choice plots.

The Catellus Corporation, heir to the Santa Fe Railroad land empire, enjoys the sweetest of sweetheart deals at its Union Station property, where the MTA is building new headquarters—derided as a "Taj Mahal" by State Senator Tom Hayden. Meanwhile, Universal City, the unincorporated tax island owned by MCA in the Hollywood Hills, is getting its own deluxe subway stop as well as a new offramp and major street improvements—all gratis from its MTA friends. Nick Patsaouras, the chief gadfly on the MTA board, claims that the deal "sells out the public. ... MCA pays nothing; MTA pays for everything."

Theaters of Confrontation

The #204 Vermont line, which serves a fifteen-mile corridor from Gardena to Hollywood, is the most congested bus route in America. On an average day it carries 61,000 riders, four times the current volume of L.A.'s subway.

Martín Hernandez, a full-time organizer for the Los Angeles Bus Riders Union, practically lives on the overcrowded buses that some Black passengers refer to as "slave ships." Since breakfast, he has made five complete circuits, handing out leaflets ("Sinkholes or Buses") that urge riders to protest MTA plans to further reduce bus services. As Martín explains his postmodern theory of organizing, the buses are "factories on wheels," the new theaters of confrontation: "Since deindustrialization, buses are among the last public spaces where blue-collar people of all races still mingle."

Indeed, from front to rear the Vermont bus is an extraordinary kaleidoscope of the city's transit-dependent classes: an elderly Black woman wearing a "Pray for O.J." cap, a young Latina with two restless toddlers and a huge bag of groceries, a blind man with a white-tipped cane, buffed young men with gang tattoos and gold neckchains, a weary maid on her long trip home from Beverly Hills, and so on. No one pays much attention to Martín until he identifies himself, in English and Spanish, as a member of a group that last year won an injunction against fare increases and the elimination of the popular monthly bus pass. Interest immediately perks up, and even the blind man takes a leaflet.

Then the driver, a sassy young woman, pipes in: "You know, the Red Line is total bullshit. The MTA is just flushing your tax money down the subway." One of the homeboys mutters, "Yeah, fuck the choo-choo." Several passengers nod their heads and trade phone numbers with Martín. He gets off at Imperial Highway to work a northbound #204.

The Bus Riders Union, which draws about fifty members to its weekly meetings, is a project of the Labor/Community Strategy Center. The union concept first emerged during grass-roots protests in the summer of 1993 when the MTA board shifted discretionary revenue from buses to a Pasadena light-rail line (the Blue Line extension) favored by Mayor Richard Riordan. As the center's director, Eric Mann, pointed out at board hearings (before being hauled off by transit cops), rail commuters—6 percent of the total ridership—already monopolize over 70 percent of the budget. Moreover, for each rail passenger that the MTA has managed to add, it has lost six bus riders through fare increases and reduced services. "Incredibly," Mann said, "despite billions of dollars in new transit investment, overall ridership has plummeted and gridlock has increased."

In addition to opposing the Blue Line's diversion of funds, the Strategy Center proposed a halt to Red Line construction and a reorientation of MTA resources to buses. When the board ignored this proposal, the center put organizers on buses. As Mann explained, "Buses have been symbols of the civil rights movement since the days of Rosa Parks and the Montgomery boycott. But the issues have fundamentally changed. Then it was the right to sit at the front of the bus; now it is the right to have a seat on a bus, period."

The MTA arrogantly raised bus fares in 1994 while allocating another $123 million to the Pasadena project. The fledgling Bus Riders Union, represented by the NAACP Legal Defense Fund, promptly filed suit charging the MTA with violating the Civil Rights Act of 1964 and "intentionally and unnecessarily imposing these extreme hardships on minority poor bus riders." After detailing L.A.'s history of transit apartheid, the Fund's brief argued that the MTA used regressive sales taxes and lopsided subsidies to establish "separate and unequal" transit systems. Or, as the union's lawyer put it, "Third World buses for Third World people." Last September Federal Judge Terry Hatter stunned the MTA by issuing a restraining order followed by the injunction. "It is unacceptable," he wrote, "for the MTA to balance its budget on the backs of the poor." Judge Hatter's ruling has opened the road for a possible constitutional showdown—perhaps this fall—which the union hopes will affirm that "affordable transit is a fundamental human right."

'This Cursed Project'

If the competing interests of capital and the poor are the foundation of L.A.'s transit politics, the MTA is its superstructure, doling out favors—but none so extravagant as those it reserves for itself and its circle. The MTA's equivalent of the military industry's notorious cost overrun is the "change order," a reimbursement to contractors for "unforeseen" expenses. Change orders on the first two phases of construction exceeded $230 million, while internal audits show that prime MTA contracts typically end up costing taxpayers more than three times the original bid. MTA staff have shredded incriminating documents, rigged bidding, leaked secret data to contractors, and disguised conflicts of interest.

The thirteen politicians on the MTA board, meanwhile, reap huge campaign contributions from major contractors. A study by Hayden's office showed that ten contractors had expended $579,000 in campaign funding and lobbying over eighteen months from 1993 to 1994. The dividends for such investment are handsome: board appointees of County Supervisor Deane Dana voted twenty-nine times to increase payments to the construction firm headed by a key campaign contributor despite dangerous defects in its work.

The most glaring conflict of interest, however, involves Mayor Riordan, who directly controls four seats on the board and has been the recipient of the largest single share of campaign money from MTA contractors. Like a fat spider, Riordan is at the center of a complex web of lucrative connections. His old law firm, Riordan and McKinzie, has long acted as counsel for the MTA. It also represents Tetra Tech, the Pasadena-based engineering corporation in which Riordan is a major investor. As consultant on the Green Line, Tetra Tech used change orders to increase its income an incredible 118 times over the original bid. Moreover, it won a major contract on the controversial Pasadena light-rail extension—a deal that is now under investigation by the state.

Riordan's close friend John Shea is head of the consortium Shea-Kiewit-Kenny, which managed to engineer the Hollywood Sinkhole. On 11 July, FBI agents raided SKK's offices, collecting evidence of safety violations and inferior construction. SKK's recklessness is not new. In 1993 it bored straight into an underground river, flooding the tunnels and stopping construction for nearly six months. The next summer, after ignoring contract instructions, SKK undermined Hollywood Boulevard so badly that buildings began to sink, pipes burst, and the Walk of Fame buckled and cracked. Apartments and theaters had to be evacuated, and a nine-block segment of the boulevard was closed for fear of collapse.

Eleven hundred angry Hollywood property owners filed a $3 billion lawsuit against the MTA, charging the agency and its contractors with corruption, influence peddling and the persecution of whistleblowers. The Federal Transit Administration temporarily froze $1.6 billion in Red Line funds. Work resumed only after a supposedly contrite SKK accepted new safeguards. Yet the kamikaze contractors again ignored elementary precautions when, early on the morning of 22 June, they ordered their workers to remove the concrete liner that supported a huge column of wet, ungrouted sand. Suddenly Hollywood collapsed on their heads. Catastrophic as the cave-in was, SKK is not an exception when it comes to flouting safety standards. All of the major Red Line contractors, screened from effective oversight, have been allowed to run roughshod. Tunneling through hazardous subsoil conditions that include pockets of dangerous gas as well as treacherous water-saturated sands, contractors have routinely failed to build subway walls to the required thickness or to grout unstable soil. Plastic liners designed to protect tunnels from gas have been discovered full of punctures. In the few stations now operating, gas sensors have been installed above the passengers' heads, although deadly hydrogen sulfide is heavier than air.

Still, digging under the stars' footprints is the easy part. Technical difficulties will increase in the next phase of Red Line construction, as the MTA attempts to bore twin 2.5-mile tunnels through the unstable rock of the Hollywood Hills. Geologists have criticized the agency's "simplistic" studies for radically underestimating the ground water and seismic hazards of this Runyon Canyon segment.

After surveying the debacle in Hollywood, Tom Hayden denounced the Red Line as "criminal folly" and demanded an immediate halt to "this cursed project." Simultaneously, County Supervisor Mike Antonovich, a conservative from Glendale and former chairman of the MTA board, urged his friend Newt Gingrich to "pull the plug" on further federal aid. "The taxpayers are subsidizing their own suicide," he said. Across the political dial, Eric Mann and his organization have turned up the volume on their demand for radical reform. In the meantime, Martín Hernandez, like the hero of the old Kingston Trio song, is prepared to "ride forever" on the streets of L.A., organizing riders on the #204 Vermont bus.

1995

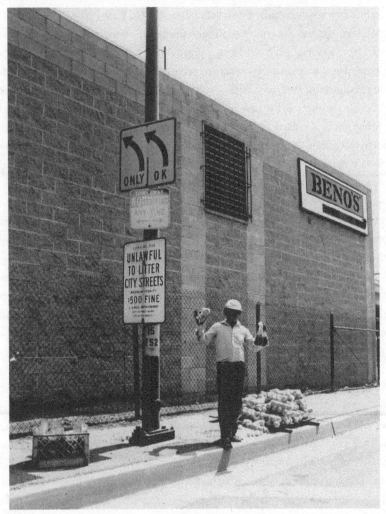

Bitter oranges

10

The New Industrial Peonage

The descending sun is temporarily eclipsed by a huge water tower emblazoned THE CITY OF VERNON. Shadows play off the concrete embankments of the Los Angeles River and dance across the shallow trickle of sewage in its channel. A locomotive shunts a dozen hazardous-chemical cars into a siding. A trucker somewhere pulls hard at his air horn. A forklift scurries across a busy road. We are only about five miles from Downtown Los Angeles, but have entered a world invisible to its culture pundits, the "empty quarter" of its tourist guides. This is L.A.'s old industrial heartland—the Southeast.

It's 4:30 p.m. Two workers are standing behind an immense metal table, partially shaded by a ragged beach umbrella. A portable radio is blasting rock-'n'-roll *en español*, hot from Mexico City. Each man is armed with a Phillips screwdriver, a pliers, and a ballpeen hammer. Eduardo, the taller, is from Guanajuato in north-central Mexico and he is wearing the camouflage-green "Border Patrol" baseball cap favored by so many of Los Angeles's illegal immigrants. Miguel, more slightly built and pensive, is from Honduras. They are unconsciously syncopating the beat as they alternate between hammering, prying, and unscrewing.

Towering in front of them is a twenty-foot-high mound of dead and discarded computer technology: obsolete word processors, damaged printers, virus-infected micros—last decade's state of the art. The Sisyphean task of Eduardo and Miguel is to smash up everything in order to salvage a few components that will be sent to England for the recovery of their gold content. Being a computer breaker is a monotonous $5.25-an-hour job in the Black economy. There are no benefits, or taxes, just cash in a plain envelope every Friday.

Miguel is about to deliver a massive blow to the VDT of a Macintosh when I ask him why he came to Los Angeles. His hammer hesitates for a second, then he smiles and answers, "Because I wanted to work in your high-technology economy." I wince as the hammer falls. The Macintosh implodes.

L.A.'s Rustbelt

The computer breakers have been in Los Angeles three years. Eduardo had a local network of contacts from his village; Miguel, who had none, lived feral on the streets for nearly two years. Now they share a tiny two-bedroom house in the nearby city of Maywood with four other immigrant workers (two of whom are also "illegal") for $1100 per month. All are married, but no one has yet managed to save enough money to bring his wife or kids *al otro lado*. Like so many of the 400,000 new immigrants who work and live in southeast Los Angeles County, they feel trapped between low wages and high rents. "Just like peons," Miguel broods, "like slaves."

It wasn't always this way in the Southeast. Twenty-five years ago the archipelago of bungalow communities that surrounds the factories and warehouses was predominately Anglo. The Southeast area contained most of Southern California's nondefense factories, including three auto plants, four tire plants, and a huge complex of iron and steel fabrication. The plants were unionized and they paid the mortgages on bungalows and financed college educations. On weekends well-dressed couples swung to Jimmy Wright's band at the South Gate Women's Club or rocked with Eddie Cochran at Huntington Park's Lyric Theater. For sons of the Dustbowl like Cochran (from the nearby "Okie suburb" of Bell Gardens), the smokestacks of Bethlehem Steel and GM South Gate represented the happy ending to *The Grapes of Wrath*.

But this blue-collar version of the Southern California dream was reserved strictly for whites. Alameda Street, running from Downtown to the Harbor, and forming the western edge of the Southeast industrial district, was L.A.'s "Cotton Curtain" segregating Black neighborhoods from the white-controlled job base. Blacks who crossed Alameda to shop in Huntington Park or South Gate risked beatings from redneck gangs with names like the "Spookhunters." In the early 1960s, as ghetto joblessness soared in an otherwise "full employment" economy, the situation became explosive. During the Watts rebellion, Black teenagers stoned the cars of white commuters, while Lynwood and South Gate police reciprocated by beating and arresting innocent Black motorists. In the wake of rioting on its borders, the white working class began to abandon the Southeast. Some entrenched themselves in the "white redoubt" of nearby Downey, but most moved to the rapidly industrializing northern tier of Orange County. Employers followed their labor force in a first wave of plant closings in the early 1970s.

This outward seepage of the Anglo population in the 1960s (–36,510) became an exodus in the 1970s (–123,812) and the 1980s (–43,734). Racial hysteria, abetted by "block-busting" in the city of Lynwood, was followed by a second wave of plant closings in the late 1970s. Much of the trucking industry, escaping gridlock and land inflation, migrated to new industrial zones in the Inland Empire, fifty miles east of L.A. And disastrously, within the short space of the "Volcker recession," local heavy industry—including the entirety of the auto-tire-steel complex—collapsed in the face of relentless Japanese and Korean competition. In most cases, plant closure followed within a few years of watershed Black and Chicano breakthroughs in shop-floor seniority and local union leadership. While white workers for the most part were able to retire or follow their jobs to the suburban periphery, nonwhites were stranded in an economy that was suddenly minus 50,000 high-wage manufacturing and trucking jobs.

Unlike Detroit or Youngstown, however, L.A.'s derelict industrial core was not simply abandoned. Almost as fast as Fortune 500 corporations shut down their L.A. branch plants, local capitalists rushed in to take advantage of the Southeast's cheap leases, tax incentives, and burgeoning supply of immigrant Mexican labor. Minimum-wage apparel and furniture makers, fleeing land inflation in Downtown L.A., were in the vanguard of the movement. Within the dead shell of heavy manufacturing, a new sweatshop economy emerged.

The old Firestone Rubber and American Can plants, for instance, have been converted into nonunion furniture factories, while the great Bethlehem Steel Works on Slauson Avenue has been replaced by a hot-dog distributor, a Chinese food-products company, and a maker of rattan patio furniture. Chrysler Maywood is now a bank "back office," while US Steel has metamorphosed into a warehouse complex, and the "Assyrian" wall of Uniroyal Tire has become a façade for a designer-label outlet center. (On the other hand, the area's former largest employer, GM South Gate, remains a ninety-acre vacant lot.)

A Family Dictatorship

Although on a site-by-site average, two high-wage heavy manufacturing jobs have been replaced by only one low-wage sweatshop or warehouse job, the aggregate employment level in the Southeast has been sustained at 80 to 85 percent of its 1970s peak by the infiltration of hundreds of small employers. The secret formula of this new, low-wage "reindustrialization" has been the combination of a seemingly infinite supply of immigrant labor from Mexico and Central America with the entrepreneurial energy of East Asia. Chinese diaspora capital has been particularly vigorous in such sectors as food processing, apparel, novelties, and furniture, which employ large minimum-wage workforces of Latino immigrants. As a result, much of southeast Los Angeles County has come to resemble a free-trade zone or manufacturing platform of ambiguous nationality. Along Telegraph Road in the aptly named City of Commerce, for instance, the yellow sun of the Republic of China (Taiwan) flies side by side with the Stars and Stripes, while billboards advertise beer and cigarettes in Spanish.

The nation-state, moreover, yields real sovereignty to the city-states, which contain most of L.A. County's industrial assets. Exploiting the prerogatives of California's promiscuous constitution, a half-dozen industrial districts have incorporated themselves, almost without residential populations, as independent cities, selfishly enabled to monopolize land use and tax resources. The oldest, and strangest, of these "phantom cities" is the city of Vernon—the industrial *hacienda* to which Eduardo, Miguel, and thousands of their *compañeros* commute each morning for work. The two most important facts about Vernon are: first, that it has a permanent residential population of only 90 adult citizens (70 of them municipal employees and their families living in city-owned housing) but a work force of more than 48,000—that is to say, a commuter-to-resident ratio of 600:1. Second, the city has been controlled by a single family, the Basque-origin Leonis dynasty, since its formation in 1905.

Originally established as a safe haven for "sporting" activities (boxing, gambling, and drinking) under attack by L.A.'s early municipal reformers, Vernon evolved during the 1920s into an "exclusively industrial" (official city motto) base for eastern corporate branch plants. Under the iron heel of John Leonis, city founder, existing housing was condemned or bought out in order to reduce the residential population to a handful of loyal retainers living in the literal shadow of the Hitler-bunker-like city hall. Elections in Vernon thereby became a biennial farce where the Leonis slate (now headed by grandson Leonis Malburg) is *unanimously* reelected by a micro-citizenry of Leonis employees. (Although civic officials are required to live in the city of their election, the mayor has for decades brazenly flouted state law by residing in a family mansion in Los Angeles.)

This family dictatorship has been routinely accepted by Vernon industrialists in exchange for exceedingly low tax rates and high standards of municipal police and fire services. Conversely, Vernon cityhood has been a disaster for the city of Los Angeles, which has lost perhaps 20 percent of its potential industrial property-tax base. Now Vernon is angling to establish a "community redevelopment" project that will authorize it to withhold $873 million in tax revenues from the county general fund (which would go to schools, welfare, and hospitals) over the next generation. The primary beneficiaries of this raid on the county treasury—to be used to "modernize" the city's older plant and warehouse sites—will be Vernon's major landowners: a list headed by the Santa Fe Railroad Land Company (now Catellus Development—forty-one parcels) and hizzoner Leonis Malburg (nineteen parcels).

The thousand-odd pages of documents used to argue Vernon's case for redevelopment inadvertently unmask an economy capitalized on poverty and pollution. A detailed survey of local wages, for example, reveals that 96 percent of

Vernon's 48,000 workers earn incomes so low that they qualify for public housing assistance. At least 58 percent of this largely unorganized work force fall into the official "very low income" category, making less than half the county median—a dramatic downturn from the area's union-wage norms twenty years before.

Moreover, this low-wage army is working under conditions of increasing toxicity. Vernon has long been the worst air polluter in the county, but public health officials in nearby communities are most worried by the 365 hazardous material use or storage sites within the city. A recent investigation by the California Public Interest Research Group revealed that Vernon annually emits, processes, or stores 27 million pounds of toxins—more than three times as much as the entire city of Los Angeles. As part of its early 1980s "replacement" industry—especially apparel and furniture—begins to relocate to Mexico or East Asia, Vernon is becoming increasingly dependent on an odd couple of growth industries: ethnic-food processing and toxic waste disposal. In Vernon there is nothing unusual about a Chinese frozen-shrimp processor being located on the same block with a company recycling battery acids or treating industrial solvents. Vernon's neighbors have been especially alarmed by the city's plans to build an incinerator for infectious hospital debris, as well as by a proposal to locate a plant processing up to 60,000 gallons per day of deadly cyanide on the city border, less than a thousand feet from overcrowded Huntington Park High School. Local activists nervously wonder if hypertoxic Vernon might not become the Bhopal of Los Angeles County.

Rent Plantations

Like Gaul, the Southeast is divided into three parts: 1) the industrial incorporations, Vernon and City of Commerce, containing nearly 100,000 jobs; 2) three "normal" suburbs, Huntington Park, South Gate, and Lynwood, that retain some semblance of downtowns as well as significant industrial land use; and 3) the more or less exclusively residential and very poor cities of Bell Gardens, Cudahy, Maywood, and Bell, which have almost no industry and lack recognizable business centers. Jim McIntyre is city planner for Bell Gardens, the third poorest suburb in the United States (nearby Cudahy is the second). He is denouncing the slumlords "who turn up at meetings in torn T-shirts pleading poverty but actually own scores of units." He is particularly incensed that the worst offenders include such sanctimonious types as the head of a local realty board, the pastor of a prominent church, and the president of the Korean-American Chamber of Commerce.

Once the "workingman's paradise," where a Dust Bowl immigrant could buy a home and garden for $20 down and $10 per month, Bell Gardens has become a "rent plantation," controlled by absentee landlords, where Mexican immigrants (88 percent of the population) are forced to squeeze as many as fifteen people into a unit to afford housing. McIntyre explains that local landlords can charge much higher rents than their counterparts in nearby white middle-class Downey because Bell Gardens is a "totally cash economy," catering to Mexican blue-collar workers, many of them "illegal," without bank balances or credit lines. In exchange for not requiring credit checks or deposits, landlords routinely demand extortionate rents for units that they fail to maintain or repair. Tenants adapt to the high cost of housing by overcrowding.

As a result, population densities in Bell Gardens as well as Huntington Park, Cudahy, and Maywood are beginning to approach the threshold of New York (26,000 people per square mile). Since officially, by the 1990 census, the popula-tion in eight residential Southeast cities has grown by 85,145, while the hous-ing stock has declined by 1120 units, the person-per-dwelling ratio has had to increase by nearly a third. In effect, this has been accomplished by a variety of strategies. In tiny Maywood, where the computer breakers live, "hotbedding" (rotating occupancy of the same bed) is common. In South Gate, once the richest of Southeast communities, every other garage has been illegally converted into a rental unit. In Cudahy, the poorest and densest of all, "victory garden" lots, 60 feet wide and as much as 390 feet deep, which were designed for a bungalow, a chicken house, and an orchard, now accommodate "six-pack" stucco tenements three or four deep—in effect, continuous barracks housing as many as 125 people on a former single-family site.

Southeast local governments have reacted to this crisis by trying to restrict the supply of illegal or tenement housing. Rather than seeking assistance to meet the demand for low-income housing, they have torn down cheap housing in order to build strip malls or upscale townhouses. The explicit aim has been to reduce (or "stabilize") the number of poor renters while importing new tax resources in the form of retail businesses and middle-class residents. Thus Bell Gardens destroyed a poor neighborhood on its west side in order to build a shopping center and poker casino, while Huntington Park, the area's most ardent "gentrifier," has actually reduced its housing stock by 3 percent. Even the apparently laudable local effort to enforce building codes against slumlords and the owners of illegal garage conversions is fundamentally a stratagem to limit the growth of the renter population, not a serious crusade to improve housing conditions.

Meanwhile, a sea of children has simply overwhelmed the aging physical plant of the Southeast schools (Region B of Los Angeles Unified). Over the last quarter-century, as 204,000 aging Anglos have been replaced by 328,000 young Latinos, the median age in the Southeast has fallen from 30–32 to 18–20. There are 125,000 children in the area, almost double the number in 1960. Although local secondary schools are among the largest in the country and have long operated on staggered year-round schedules, they have failed to absorb the burgeoning teenage population. New students in the Southeast are now immediately bused across Los Angeles, as far as the San Fernando Valley. The shortfall in school capacity was aggravated in 1988–89 by the closures of elementary schools in Cudahy and South Gate after the discovery of dangerous chemical contamination in their playgrounds. (In the case of Park Avenue Elementary, kindergarteners found toxic tars and methane gas bubbling under their swings.)

Nor has the school district been able to adequately address the needs of students who, in nine cases out of ten, now come from Spanish-speaking households. There have been numerous battles between majority-Anglo teaching staffs and Latino parents incensed over low test scores and 50 percent high school dropout rates. Despite the Southeast's population, there is no nearby community college, and per capita recreational space is a tiny fraction of the county average. It is not surprising, therefore, that many local kids—betrayed by bad schools and predestined for sweatshop jobs—choose instead to join the flourishing gang subculture. Although the Southeast is not yet a full-fledged war zone like South Central L.A., it offers a broad array of *las vidas locas,* from traditional local gangs

like Florentia and Clara (Street), to newer affiliations like the all-around-town
18th Street supergang.

"Law and order" in the Southeast long meant keeping Blacks out of the
area. Now, more than anything, it has become *carte blanche* to terrorize Latino
youngsters, especially suspected gang members. A few sensational cases hint at
widespread practices. Several years ago two Huntington Park policemen were
prosecuted for "torturing" a gang member with an electric cattle prod. In 1990,
twenty-two sheriff's deputies based at the Lynwood station were accused of
illegal beatings and shootings in a federal class-action lawsuit filed by eighty-one
local victims, most of them Latino or Black. In spring 1991, while the world was
watching the LAPD club Rodney King on video, a superior court jury convicted
three officers for a savage beating of a Latino suspect in the Maywood jail the
previous year.

Rotten Boroughs

Unfortunately, police brutality, like sweatshop wages and overcrowded housing, has become routine fare in the communities of the Southeast. The Latino
majority (mainly immigrants from west-central Mexico, especially Jalisco) is as
effectively disfranchised as were Blacks in Mississippi a generation ago. Although
the population of the Southeast has increased by 174,000 since the 1960s, the
active electorate has *shrunk* by 40,000 votes (from one-half of the population to
less than one-sixth). A geriatric Anglo residue, only 6 percent of the total population, remains an electoral majority, dominating politics in a region that is now
90 percent Latino. In Bell Gardens (population 42,000), for example, 600 aged
Anglos outpolled 300 Latinos in April 1990 to reelect an all-white city council.
Indeed, until the April elections, only three out of thirty councilpeople in a bloc
of six Southeast cities had Spanish surnames (the number has now increased, by
election and appointment, to seven).

Latino powerlessness in the Southeast is a collusion of demography, citizen
status, and benign neglect by the Democratic Party. The restrictive electoral
mathematics work as follows: first, deducting the large number of children on
the Latino side of the ledger reduces the population ratio from 9:1 to 5:1 (adults).

Second, a large percentage of adult Latino residents are noncitizens. Accepting the high-range figure ventured by Jim McIntyre in Bell Gardens of 60 percent, an overwhelming Latino majority is further whittled down to a ratio of 2.5:1 or even 2:1 (potential electors). At this point, the actual balance of power is a function of electoral mobilization, which in turn depends on issue salience and political resources. Given the partisan propensities of blue-collar Latinos, it seems logical to assume that Democrats would make major efforts to activate this electorate.

To date, however, they have failed to make a significant investment in voter registration. To understand why not, it is necessary to evoke the peculiar geopolitical evolution of the region. Until reapportionment in the early 1970s, the Southeast had a strongly reinforced political identity from overlapping state and federal legislative boundaries, making it a key prize in long struggles between liberal and conservative Democrats, and infusing local councilmanic politics with the larger passions of assembly or congressional rivalries. In the 1940s and 1950s, for example, the Southeast's Twenty-third Congressional District was the crucible of repeated attempts by large CIO locals to create a liberal-labor political majority. By the 1960s, however, racial hysteria eclipsed class consciousness, and the region became the bedrock of the statewide anti–school integration and anti–fair housing campaigns led by reactionary assemblyman Floyd Wakefield of South Gate. Partially to punish Wakefield and to break up his political base, the Democratic state legislature redrew electoral boundaries to distribute fragments of the old Twenty-third CD among districts anchored by Black majority votes in South Central L.A. The legislature's nullification of the Southeast as a distinctive political entity, together with the reliability of the Black vote, vastly diminished Democratic attention to the fate of nonpartisan local politics. At the same time, Latino civil rights and community-power movements in Los Angeles County have almost totally concentrated on mobilizing neighborhoods with high percentages of Chicano (second-through-fourth-generation Mexican-American) homeowners and middle-income earners. Only desultory resources have been made available to such port-of-entry areas for Mexican and Central American immigrants as the Southeast, Boyle Heights (east of Downtown), or the Westlake District (west of Downtown). Although the recent election of Gloria Molina as the first Spanish-surnamed L.A. County supervisor in the twentieth century has rekindled rhetoric about the "Latino decade," grassroots activists in the Southeast have yet to see even Chicano Democrats focus adequately on the problems of their rotten borough.

To the extent that power sharing—which is, of course, different from empowerment—is occurring in the Southeast, it is being instigated from the right. The two Latinos elected last year to the Huntington Park City Council were both Republicans, and there is evidence that the county Republican campaign fund operated by ex-supervisor Peter Schabarum is playing a significant role in encouraging conservative Latino businessmen to run for office. In Bell Gardens, community activists criticized the appointment of businesswoman and police advocate Rosa Maria Hernandez to fill a council vacancy as a crude attempt by the Anglo leadership to placate the community symbolically without surrendering any real power. Others worry that as elderly white voters die off, the old Anglo business elite will simply be replaced by a look-alike clique of conservative Latino realtors and shopkeepers elected by an attenuated and unrepresentative electorate.

Poker Faces

Meanwhile, the still-dominant Anglo establishment has enjoyed a generation of uncontested control over municipal affairs. The shrinkage of the electorate in the early 1970s and the disappearance of the big union locals as political actors, combined with Anglo solidarity and paranoia, consolidated the power of local chamber of commerce types, ruling from smoke-filled rooms in relaxed unaccountability. In Huntington Park, for example, the same five Anglo businessmen (with one substitution) composed the city council from 1970 to 1990 (three still remain). In Bell Gardens, Claude Booker, the city manager who has also been a councilman, represents a continuity of power from the 1960s. In other cities, power has alternated between conservative cliques with only the rare reformist candidate (like UAW leader Henry Gonzales in South Gate) to brighten the picture.

The chief interest of these "good ole boys" has been to manipulate the lucrative pork barrel of redevelopment. California law allows municipalities to fight "blight" with urban renewal financed by tax increments withheld from the general fund. In smaller cities, the city council acts as the executive of the redevelopment agency, which typically means that the same local realtors and businessmen who will benefit directly from renewal are also its administrators. "Blight," moreover, is a conveniently elastic category, encompassing everything from too many railroad spurs in Vernon to too many poor renters in Bell Gardens and Huntington Park.

In the Balkanized landscape of the Southeast, redevelopment has devolved into a crazy zero-sum competition between impoverished municipalities. While the Lilliputian towns of Maywood, Bell, and Cudahy wager scarce tax resources in a "war of supermarkets" along Atlantic Boulevard, the larger cities—Huntington Park, South Gate, and Lynwood—struggle to resurrect central shopping districts wiped out by the competition of regional malls in Downey and Carson. The result is excessive strip development, a redundancy of franchises, and profligate sales-tax abatement. Never having recovered from the twin blows of plant closure and Proposition 13 in the late 1970s, most Southeast communities have only exacerbated their fiscal distress with overly ambitious and poorly targeted redevelopment projects. Huntington Park, for example, has almost ruined itself in the course of revitalizing its Pacific Avenue shopping corridor, while South Gate lost a quarter of its sales-tax income when one heavily subsidized car dealership suddenly went bankrupt.

In the face of the unforeseen costs of retail modernization, gambling on redevelopment has literally become redevelopment by gambling. For decades, Southeast citizens' groups, led by the Methodist Church, had successfully resisted attempts by outside gambling interests to establish poker parlors (a constitutional local option in California). But with the attrition of the electorate in the late 1970s, pro-gambling forces managed to legalize card casinos in Bell, Commerce, Bell Gardens, Huntington Park, and Cudahy. Using their redevelopment powers to discount land to casino developers, these five cities have attempted to utilize gambling as a tax generator to keep housing and retail development alive.

The record is mixed. Bell Gardens' Bicycle Club, the largest card casino in the country (and perhaps the world), turns over $100 million of gross profits each year, $10 to $12 million of which become local tax revenue—most of it immediately loaned to the city's aggressive redevelopment program. The giant 120-table Commerce Casino generates almost as much cashflow as the Bicycle Club and has long been the city's chief source of revenue. In Bell, Cudahy, and Huntington Park, on the other hand, smaller-scale casinos have collapsed in bankruptcy (although a new Bell casino has emerged). Whether solvent or not, however, each card club has offered seductive opportunities to councilmen and their cronies.

In the Bell case, Mayor Pete Werrlein, a council member for sixteen years and archtypical "good ole boy," convinced anxious Anglo voters that without the additional revenue from poker, the police would be unable to protect them from an influx of "dangerous Mexican gangs." Shortly after the opening of the California Bell Club in 1980, however, the county grand jury revealed that Werrlein, together with his city administrator and the former police chief of Huntington Park, had been engaged in "sex orgies" in a Cudahy warehouse with eighteen-year-old prostitutes supplied by Kenneth Bianchi—*aka* the Hillside Strangler, L.A.'s most notorious serial murderer. Two years later, in 1984, Werrlein and the city administrator were indicted for bribery, fraud, and racketeering when their secret ownership of the Bell Club was exposed.

Hard on the heels of the Bell scandal, the ex-mayor of Commerce and two councilmen were arrested in a similar hidden shareholding conspiracy, this time in partnership with Vegas mobsters. Perennial rumors that the casinos are laundries for drug money were partially confirmed in 1990 when the Feds seized the almighty Bicycle Club after evidence emerged that the casino had been financed with coke-tainted dollars. Finally, throughout the spring of 1991, the old-guard three-Anglo majority on the Huntington Park council has waited nervously for the district attorney to issue possible indictments in an ongoing investigation that links them to a profit-skimming operation at the city's now bankrupt card club.

Super-Pan Nine

As the sun plunges into the ocean off Venice Beach the yellow-brown smog over Southeast L.A. suddenly turns pastel in a cheap imitation of an Ed Ruscha painting. The two square-block parking lots of the Bicycle Club are already full, and several thousand Friday paychecks have been converted into poker chips. Around the seven-card stud and Texas hold-'em tables the professional card cheats are sizing up an excited mob of bikers, housewives, truck drivers, hippies, and senior citizens—all of them wearing their best poker face.

But the real action is happening at the other half of the casino devoted to Asian card games. On the super-pan nine tables there is no bluff or bluster, just silence and intense concentration as the dealers fire cards from 432-card stacks and players respond with volleys of $100 bills. Over half of the club's income now comes from the high-speed, high-stake play of the Asian games. For the intrigued but uninitiated player, there are convenient instruction sheets in English, Chinese, Korean, Spanish, Vietnamese, and Cambodian. In the exclusive inner sanctum of the Asian Room, off-limits to ordinary punters, wealthy Chinese and Korean businessmen, some of whom own nearby factories and warehouses, venture a year's sweatshop wages in a single hand. They accept their losses with icy composure.

Meanwhile, a few miles away, a man is standing in the twilight on the curb of Alameda Street selling oranges. Behind him is the vast empty lot, overgrown with jimsonweed and salt grass, that used to be General Motors' South Gate assembly plant. He has been standing on Alameda, gagging on smog and carbon monoxide, since seven in the morning. A bag of plump but second-rate navel oranges costs $1. He has sold twenty-five bags and will be allowed to keep half his earnings by the boss who drops him off in the morning and picks him up in the evening. He is anxious because he still has three bags left to sell.

There are at least a thousand other men—most of them destitute recent arrivals from Central America—selling oranges on freeway ramps and busy street corners in L.A. I ask him in broken Spanish if he is Salvadoran or Guatemalan. He replies in abrupt English that he was born in San Antonio and until 1982 was a machinist making $12 per hour in nearby Lynwood. When his shop went out of business, he was certain that he could easily find another job. He hasn't, and at age fifty-two he is selling oranges for $12 a day. So that's his story, and if my curiosity is satisfied, how about buying three bags of oranges? I pay him, he shuffles off, and when I get home I discover that the oranges have a strangely bitter taste.

1992

PART III

RIOT CITY

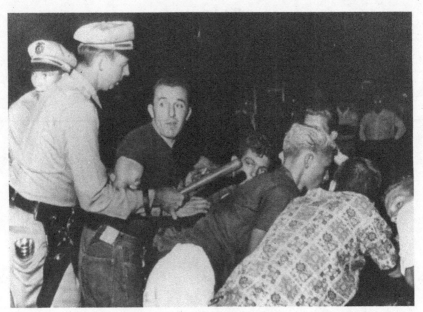

El Cajon Boulevard riot (1960)

11

'As Bad as the H-Bomb'

"It is time," Dr. Fred Schwartz urged residents of the nation's most patriotic city, "to lay aside secondary and unimportant issues.... Americans must renounce complacency, ignorance and greed, and dedicate themselves to hurling the gauntlet of freedom into the face of godless Communism." The expatriate Australian physician, executive director of the Christian Anti-Communist Crusade, was in San Diego to open the biggest session yet of his School for Anti-Communists. More than eleven hundred enthusiastic local anti-subversives were registered to hear Schwartz and his co-star, famed FBI counterspy Herbert ("I Led Three Lives") Philbrick, explain why "the free world is losing the battle with the Reds." In five days of workshops and expert testimony, Schwartz promised to unravel the arcane secrets of Communism's appalling success. "To think like a Communist you've got to develop a mind like a corkscrew. You'll never do it without an understanding of dialectical materialism."[1]

If, a thousand years from now, archaeologists from another planet wanted to understand the culture of ancient Cold War America, the perfect stratigraphic section in time and space might be San Diego, California in late August 1960, on the eve of Schwartz's anti-Communist revival. No other major city had so exclusively capitalized its future on what President Eisenhower (in a rare moment of apprehension) had recently described as the "military-industrial complex." Seen from the top of the El Cortez Hotel, the city's only skyscraper, the martial land-scape was overwhelming, even sublime: a gorgeous blue sky streaked with Navy and Marine jets; a perfect harbor crowded with massive gray flattops, a waterfront lined for miles with military storage depots, boot camps, and bomber assembly lines; and, crowding a mesa in the far distance, a vast complex where General Dynamics was building the first Atlas missile. San Diego reciprocated this defense bonanza with a civic cargo cult that extolled the Marines, the Pacific Fleet, and ICBMs. Decades earlier, the city fathers had chased Emma Goldman and the Wobblies out of town with the warning that nothing even remotely smacking of un-Americanism would ever be tolerated in San Diego. With the ultra-conserva-tive *San Diego Union* (an official sponsor of Schwartz's Anti-Communist School) as its watchdog, the city's intolerance of dissent was legendary. In short, this capital of the Cold War was the least likely place in America to be the scene of one of the first major youth riots of the 1960s.

Internal Combustion or Red Conspiracy?

While Black students in the summer of 1960 were boldly assailing Southern seg-regation with heroic lunch-counter sit-ins, white kids in San Diego were preoc-cupied by the seemingly trivial question of whether or not they could continue to burn rubber on a local drag-strip. On Friday night, 19 August, youthful patrons at San Diego's most popular drive-in restaurants received handbills along with their burgers and cherry cokes. The leaflet invited "all drag racing fans" to a "mass protest meeting" the next evening at the intersection of El Cajon Boulevard and Cherokee Street. Two weeks previously, following an accident that injured three bystanders, the Navy had shut down the last drag strip in San Diego County: an old auxiliary runway known as Hourglass Field where formally illegal weekend competitions under the adult auspices of the San Diego Timing Association (the parent body of twenty-two hot-rod clubs) had been tolerated since the closure of the association's original drag strip on Paradise Mesa in 1959. The Navy's action delighted the conservative triumvirate of Police Chief Jansen, Mayor Dail, and Supervisor Gibson, who had long denounced drag racing, sanctioned or not, as a stimulant to "recklessness and disorder." (Under Jansen's orders, the police had been conducting an intensive crackdown on street-racing and illegal "speed equipment" since the beginning of the year.) Conversely, local car clubs like the Vi-Counts and Roman Chariots, who had been cooperating with police efforts to suppress dangerous street-racing in exchange for being able to run dragsters at Hourglass Field, were infuriated by what was, in effect, the criminalization of the sport. "If we don't get [a new] strip," some of the Vi-Counts warned, "cars will start dragging in the street."[2]

Indeed, when the San Diego Police dispatcher began frantically calling in reinforcements at 1 a.m. on Sunday morning (21 August) an estimated 3,000 teenagers and young adults had blocked off a long section of El Cajon Boulevard (the city's major east–west thoroughfare) and were cheering on racers in a miscellany of vintage hot-rods and customized family sedans. "The cars, of all models and shapes," reported the *Union*, "raced two abreast for about three blocks down El Cajon Boulevard. Thousands of spectators lined the sidewalk and center island, leaving almost no room for cars to pass." It took police wielding batons, lobbing tear gas, and driving their patrol cars onto the sidewalks almost three hours to disperse the crowd.[3]

Veteran cops, accustomed to local teenagers' deference, were shocked by the crowd's angry defiance. One contingent of about a hundred stubbornly held their ground on a gas station parking lot, answering tear gas and police charges with volleys of "soft drink bottles, glasses and rocks," slightly injuring two officers. "Uniforms of several others were torn, and guns were stolen from some. Patrolman M. Addington was struck on the head with a rock, and Patrolman W. Pfahler suffered leg burns from a tear gas grenade. Sgt. J. Helmick was saved by fellow officers after a group of rioters attempted to overturn his patrol car."[4] The *Los Angeles Examiner*, melodramatizing the danger, quoted a "terrified homeowner": "They were like wild dogs, racing up and down the street at high speed and gunning their motors. I don't own a gun, but I armed myself with a knife and just hoped that no one would try to break into my house."[5]

Eventually 116 "demonstrators," including 36 juveniles, were hauled away in paddywagons. The adults were booked on suspicion of rioting, refusal to disperse, and conspiracy, then interrogated by homicide detectives. Police Chief Jansen reassured the City Council that as soon as the "ringleaders of the conspiracy" were identified, they would be charged with felonies. The names and addresses of the young adult arrestees, punctually published in Monday's *San Diego Union*, revealed that the hard core of the crowd, at least, came from the city's most typical blue-collar neighborhoods and suburbs. The largest contingent, not surprisingly, was from the tough, car-crazy town of El Cajon in the east county—also a major center of biker subculture. Another group may have been affiliated with the popular East San Diego car club, the Unholys. Others were from equally working-class Linda Vista, Lakeside, Spring Valley, Chula Vista, and Imperial Beach. A dozen young Marines and sailors gave only fleet addresses. Only one defendant had a Spanish surname, and there was a striking absence of outlaw street racers from upscale areas like Point Loma or La Jolla where Daddy controlled the keys to the T-Bird.[6]

Yet the El Cajon Boulevard Riot, as it became officially known, electrified teenagers of all classes, if not of all races. (I can personally testify that among my crowd of El Cajon fourteen- and fifteen-year-olds this was unanimously the "bitchinest" event of our lifetimes and the older rioters—with their ducktails and James Dean insouciance—were our Homeric heroes.) San Diego braced for the unknown. On Monday night, after one councilman had warned that the kids were "trying to run the town," police reserves were called up and issued riot sticks and tear gas. Instead of a single mob, they found themselves playing "motorized tag" with long convoys of protestors who alternately slowed down and speeded up, but never exceeded the speed limit.[7] Their unofficial anthem was the Ventures' thrilling punched-to-the-floorboard instrumental, "Walk, Don't Run."[8] Many of the cars displayed handlettered signs: "Wipe Out Tear Gas" and "We Want a Drag Strip."

In addition to El Cajon Boulevard, where several hundred hot-rodders taunted authorities in a tense confrontation at a popular drive-in, police and Highway Patrol officers struggled to keep up with the large contingents cruising Clairemont, Linda Vista, and Pacific Beach. In El Cajon, where Chief Joseph O'Connor had vowed that "we will resist mob rule down to the last man," the police blocked

Main Street and ticketed protestors for real or spurious "equipment violations." Meanwhile, San Diego police, aided by the Shore Patrol, impounded cars and arrested more than 100 juveniles and adults: many of them, it seems, for the sole purpose of interrogation about supposed ringleaders.[9]

On Wednesday, Herbert Sturdyvin, a twenty-year-old printer, was charged with conspiracy for having printed the "riot flyers" and warrants were issued for two other reputed instigators. But the clearly headless protests continued to annoy police with nightly convoys (centered on what police described as the "East San Diego trouble spot") and, later, with a furtive attempt, defeated by the Navy and Highway Patrol, to stage nocturnal drag races at Hourglass Field. As arrests (more than 200 by Friday) and impoundments sapped energy from the movement, Chief Jansen voiced apprehension that local teenage disobedience was only a prelude to a full-scale "invasion of hot-rod gangs" from Los Angeles. All through the weekend of 27–28 August, the San Diego police—again aided by the Navy—mobilized for the arrival of the L.A. hordes: a menace, it was clear by Monday, that had only existed in Jansen's fervid imagination.[10]

City leaders, however, were loathe to concede the obviously spontaneous and local context of teenage protest. Councilman Chester Schnieder was scoffed at by colleagues and county supervisors when he suggested that officials should work with car clubs and the Junior Chamber of Commerce to restore the safety valve of a legal drag strip.[11] The majority preferred to assimilate the week's disturbances to a dark constellation of conspiracies—including Southern sit-ins, San Francisco student demonstrations, and the previous month's riot at the Newport Jazz Festival—whose ultimate origin was the Politburo in Moscow. "This type of incident," Chief Jansen explained, "coupled with the jazz riots on the East Coast and the recent San Francisco riot provides *Pravda* with propaganda material to support their claims that this country is a lawless nation." Similarly, the *Union*, in an editorial tirade against "anarchy," found that "the sit-down strike, the lynch mob, the violence on picket lines, and the student riots all have a family relationship with what happened on El Cajon Boulevard."[12]

To underscore the deep, possibly satanic forces at work, the *Union* printed its "Drag Strip Riot—This Cannot Be Tolerated" editorial in tandem with an article by syndicated antisubversive George Sokolsky headlined "Communists Aiming at US Youth as Target." "Just as at another period," Sokolsky explained, "the Communist party of the United States devoted itself to infiltrating the Negro population, so, for 1960, the program is youth. The Communist party, which went underground during the early years of the cold war and peaceful competitive co-existence, is now coming out into the open again."[13]

Meanwhile, at the *Union*-sponsored Anti-Communist School, speaker after speaker expanded upon this equation between Communist resurgence and teenage disrespect for authority. Schwartz showed footage of the recent student demonstrations against House Un-American Activities Committee hearings in San Francisco, identifying what he claimed were Communist instigators in the crowd. W. Cleon Skousen, former assistant to J. Edgar Hoover and author of the best-selling *The Naked Communist*, excoriated the schools for teaching such "dangerous ideas" as "free world trade, disarmament and Russian peaceful coexistence." A Navy chaplain meanwhile warned that "if the nation goes to war, 50% of the 17-year-olds won't go with us," and, explicitly referring to the young sailors and marines arrested the previous weekend, "those that do come through into the service cause plenty of trouble." Philbrick, finally, explained that the current epidemic of delinquency was directly attributable to a Red Chinese plot to flood the country with drugs and pornography.[14] (The next day, in a packed meeting on measures to tighten the city's already draconian anti-obscentity ordinance, the city council gravely discussed a paperback entitled *High School Sex Club*, which reputedly "gave detailed instructions on how to start such a club.")[15]

Gidget Goes to the Riot

Kremlin-endorsed hot-rodders and Maoist high-school sex clubs? If the hallucinations of the high Cold War period seem utterly bizarre today, their official espousal nonetheless had the tonic effect of discrediting knee-jerk ideology among those who had been so nonsensically branded as dupes and subversives. By asserting a conspiratorial affinity between hot-rods, Negro rights, sexual freedom, and radical dissent, Chief Jansen and Fred Schwartz unwittingly planted a tiny seed of the later sixties. Incipient anti-authoritarianism, moreover, was deepened by the stepped-up police harassment of teenage motorists and party-goers in the weeks after the riot. Yet any larger meaning to the San Diego events was difficult to sustain after the nonevent of the L.A. hot-rod "invasion." As fall semester came and went, the police kept a tight lid on streets and drive-ins, and San Diego teenagers, seethe as they might against relentless curfews and traffic stops, seemingly fell back into line.

Then, in the spring of 1961, Southern California suddenly erupted from the valleys to the beaches in angry generational conflict. There were ten so-called "teen riots" in six months; three of them, including Griffith Park in May, Zuma Beach in June, and Alhambra in October, involved thousands of youth. If largely forgotten today, at the time these confrontations generated worldwide headlines and national controversy. Despite the disparate sociological and geographical characteristics of the individual events, civic and law enforcement leaders con-flated them as a single sustained outburst, an unprecedented insurrection against adult authority. And, following the causal chain back to El Cajon Boulevard the summer before, some of the country's leading anti-Communists once again dis-cerned a conspiratorial pattern in youthful defiance. As Los Angeles's acting chief of police warned the nation: "The eruption of violence and disorder directed at society's symbols of authority could be more devastating to America's hopes for the future than rockets and the 100-megaton bomb."[16]

The first explosion occurred—according to subversive schedule?—on 1 May. Although the national news was dominated by the huge "Splash Day" riot in Galveston, Texas, where 800 youth were arrested, Los Angeles County sheriff's deputies had to mount an amphibious landing to save the island resort of Avalon from its own teenage hordes. The city's third annual "Buccaneer Ball" celebration was disrupted by hundreds of rowdy high school and college students who "lit-tered streets with beer cans and wine bottles, climbed over cars, trampled flower planters, ripped fire hoses and extinguishers from hotel walls and sprayed cor-ridors." Panicky local authorities called in deputies from the mainland who even-tually arrested fifty-seven of the "mob."[17] The next weekend in Long Beach, in a fracas variously described in the local press as a "riot" or "near riot," 400 youths "all in bathing suits, swarmed out of the Bayshore Recreation Area ... they halted cars on Bayshore Avenue and Ocean Boulevard, tussled with drivers, yanked out ignition keys and flung water-filled balloons." When the police arrived, they too were jeered and bombarded with water balloons. Later in sentencing one of the participants, a flustered municipal judge complained: "I wish we had a whipping post. The youth of this country has absolutely no respect for authority. I just don't understand it."[18]

The more significant Memorial Day (30 May) riot in Griffith Park was a direct, if unplanned challenge by Black youth to de facto segregation in Los Angeles's public spaces. Although after the event, the local Hearst paper, the *Examiner,* would sermonize that "there is no segregation in the use of public facilities ... [and] there is no Negro group of comparable size anywhere in the world, including the continent of Africa, which has available and unopposed the opportunities of the half-million colored citizens of this region"—this was nonsense.[19] Faced with a radical shortage of parks and recreational facilities in South Central Los Angeles, Black residents, like Chicanos from the equally ghettoized Eastside, were systematically harassed by police when they attempted to freely enjoy Los Angeles's famed outdoor amenities. Only a small portion of the county beaches, for example, were integrated, and older folk recalled with bitterness how Black residents had been burnt out of their homes by the KKK in several beach communities during the 1920s. Likewise, Griffith Park, the city's largest public space, had an ugly history of racial exclusion which Black youth had recently begun to challenge.

A major focus of contestation was the park's famed merry-go-round: a natural magnet to teenagers of all races. Blaming "the publicity coming out of the South in connection with the Freedom Rides," Los Angeles Police Chief William Parker would later insinuate a Black conspiracy to take over the merry-go-round area. "We have been aware," he told the press, "of a potential problem ... for some time ... [because] that part of the park has been preempted by Negroes for the last year."[20] On Memorial Day there was palpable tension as Blacks arrived to find the LAPD deployed throughout the park. The riot erupted around 4 p.m. when the carousel operator accused a teenager of boarding without paying. When the youth denied the allegation and refused to leave, he was wrestled to the ground by white police officers with billy clubs. The sight of the youth being violently pulled off the merry-go-round enraged the several thousand Black picnickers in the vicinity. A teenage crowd followed the officers, surrounded the squad car, and demanded the release of the prisoner. When he bolted from the car, all hell broke loose. One officer opened fire; the crowd replied with bottles. Five police were injured and forced to seek refuge in a park office. As LAPD reinforcements arrived with their sirens screaming, Black teenagers shouted back: "This is not Alabama!"[21]

There were many cameras in the park that afternoon and images of the Griffith Park melee were soon reproduced around the world by the wire services. Under a particularly sensational photograph of hundreds of Black youth rushing a policeman as he manhandled the original arrestee, a *U.S. News & World Report* caption asked (following Chief Parker): "Aftermath of 'Freedom Rides'?"[22] There was a brief premonition in the media that as the freedom movement came northward into those ghettoes of "incomparable opportunity" (sic), nonviolence might be left behind. Griffith Park symbolized the emergence of an audacious "New Breed," as James Brown would call them, ready to fight the police, if need be, to claim their civil rights. It was the first skirmish on the road to the Watts rebellion four years later.

Yet while Chief Parker was still fuming over "Negro hoodlums," Gidget and 25,000 of her beach-blanket friends were pelting sheriff's deputies and highway patrolmen with sand-packed beer cans. The weekend after the Griffith Park battle, Los Angeles's most popular rock-and-roll station invited listeners to a "grunion derby" at Zuma Beach, near Malibu. KRLA expected about 2000 to arrive on Saturday night; "Instead, 25,000, a conservative estimate, showed up."[23] County Parks and Recreation officials prevented the sponsors from erecting a planned dance floor and bandstand, so the huge crowd was left to organize its own amusements. At midnight, the official beach closing time, sheriff's deputies ordered the revelers to leave. The response was a fierce fusillade of beer cans and bottles. Fifty additional patrol cars had to be called in before the crowd dispersed.[24] Although KRLA disputed the hair-raising accounts of mayhem and near-rape promulgated by county officials, the general perception was that the

deputies had narrowly prevented "an uncontrolled riot of frightening proportions." "Only by great good fortune," claimed the *Los Angeles Examiner,* "the fracas
did not result in fatalities."[25]

By any measure it was a busy night for the Los Angeles County Sheriff's
Department. Some of the deputies speeding toward Zuma Beach were diverted
instead to quell a second "riot" in the blue-collar San Gabriel Valley suburb of
Rosemead. Several hundred teenagers—perhaps inflamed by radio reports about
the Zuma melee—had gathered at the corner of Garvey and River Avenues and
were reportedly stoning passing cars. Sheriff's deputies arrested 47 juveniles on
charges of rioting, battery, and curfew violation. Meanwhile police in the southeast industrial suburb of Bell were breaking up a street fight involving some 300
teenagers outside a wedding reception.[26] Sheriff Peter Pitchess was at a loss to
identify a root cause for these white riots. He could only observe that defiance
of authority "had moved beyond the point where blame can be placed solely on
juveniles or adults, minority or majority groups."[27]

The next weekend (11 June) several deputies were slightly injured when
they came to the aid of San Gabriel police attempting to enforce an archaic law
against Sunday dancing at a local wedding celebration. Fifty officers battled more
than 300 teenagers and young adults outside a rented hall on Del Mar Avenue.
At one point, a policeman fired a warning shot in the air. Several of the rioters were charged with "lynching" after they rescued a seventeen-year-old from
police custody.[28] As temperatures of all kinds soared in July, the *Los Angeles Times,*
conflating traditional street gangs and car clubs, worried that mobile teenage
hoodlums now owned the streets.[29] In response, Sheriff Pitchess announced that
his elite Special Enforcement Detail would be deployed to help regular deputies
stringently enforce 10 p.m. juvenile curfew ordinances throughout Los Angeles
County. Local police departments followed suit in a massive regional crackdown
on drive-ins, cruising strips, beach parking lots, and other nocturnal nodes of
teenage culture.[30]

The law enforcement mobilization seemed to work. After the lurid headlines
of the early summer, Southern California survived without commotion a notorious Labor Day weekend that was celebrated across the East with fire hoses,

police dogs, and tear gas. As headlines screamed "Youth Mobs Riot in Five States; Hundreds Jailed," high school and college students ended the summer with riots in Clermont (Indiana), Lake George (New York), Wildwood (New Jersey), Ocean City (Maryland), Falmouth and Hyannis (Massachusetts), and Hampton Beach (New Hampshire). Overwhelmed local cops had to be reinforced, variously, by police dogs, state troopers, and even civil defense officials.[31] But the Los Angeles area remained quiet ... for a few days.

Autumn Anarchy

The second weekend in September, as usual for the summer's finale, was a scorcher in Los Angeles, and the largest crowds of the season packed the beaches to escape the 100-degree-plus temperatures in the valleys. Six thousand fans were lucky enough to have tickets to hear Ray Charles perform at the Hollywood Bowl Sunday evening. The blind rhythm-and-blues genius was at the height of an extraordinary, "cross-over" popularity that brought huge racially mixed audiences together everywhere outside the South. His latest tour, however, had been plagued by logistical snafus and disputes with local authorities. A week earlier, police had turned firehoses on a thousand angry fans in Portland after Charles's plane had been grounded in Seattle. The crowd, in turn, wrecked the Palais Royale Ballroom and smashed car and office windows in the downtown area, the first riot in the city's history.[32]

The Hollywood Bowl concert began without a hitch under the vigilant eye of LAPD music critics. As the tempo increased, hundreds of teenagers—Black, white, and Latino—found the beat irresistible. "Some of the screaming youngsters," the Examiner reported the next day, "organized a dance group and put on an impromptu performance of what the police said were objectionable dances, including the popular 'Jungle Bunny.'" Whether the dancing was too "dirty," too interracial, or both, the police decided to stop the concert. The lights were turned on, and when the "screaming gyrating fans" protested, reserves were summoned from the LAPD's Hollywood Division. The ensuing "teen riot," spilling out into the parking lot and adjacent Griffith Park, involved an estimated 500 to 600 members of the audience; ten were arrested.[33]

Three weeks later the suburban west end of the San Gabriel Valley (just east of downtown Los Angeles) exploded in teen violence that severely taxed the combined resources of the Sheriff's Department, the Highway Patrol and twelve local police departments. The proximate cause was a football game, but other anxieties may have been involved. A column by *Pasadena Star-News* writer Russ Leadabrand suggests the weird, fearful, even apocalyptic atmosphere in many Valley homes during October 1961:

> The telephones at the Pasadena Civil Defense headquarters have been busy during these last few weeks—since the latest Berlin crisis and the resultant increase in the chance of nuclear war. The calls come from members of the general public who are concerned now more than ever before about the possibility of awful, sudden, searing death. The people of Pasadena seek the answer to one main question. Should they build a home bomb shelter?

Leadabrand interviewed local civil defense (CD) director Ted Smith who warned readers that they should be in their backyards digging for family survival in face of the holocaust that "could happen now more easily than ever before." "One frightening thing," however, "stands in the way of effective recovery from an attack."

> Smith is chillingly frank about this. It is sabotage. Smith is convinced that there are Russian agents in the Pasadena area who are not only actively engaged in trying to wreck CD, but who would, in the event of attack, try to scuttle recovery programs.[34]

While some of their parents on the advice of Leadabrand and Smith were shopping for geiger counters, hundreds of carloads of teenagers were converging as they always did after Friday night football games (in this case, 7 October) on their favorite Valley Boulevard drive-ins. Around midnight, insults were exchanged between the gloating victors (Monrovia High) and badly humiliated losers (Alhambra High), and the ensuing scuffle quickly grew into a "whirlwind of fist fights that spread over a five-block area," blocking traffic for four miles, east and west, along Valley Boulevard. From its mobile transmitter, a local radio station broadcast a vivid blow-by-blow account of the melee, which police claimed "drew hundreds of others to the scene, all of them itching to join in the brawling." While attempting to arrest a powerfully built youth whom they accused of "mob-raising," Alhambra police were themselves overwhelmed and roughed up. "They were pushing and shoving," reported the watch commander, "attempting to grab guns from officers' holsters, jerking off their hats, jumping on their backs and trying to knock them to the ground." Alhambra's desperate "999" (the code for riot) appeal was answered by more than one hundred police and sheriff's deputies from other jurisdictions. They blockaded access to Valley Boulevard and ordered the estimated 1000 to 1200 rioters to disperse. The common response was "Go to Hell." After an hour of further melee, thirty-one young adults and sixty juveniles were in custody. It was officially characterized as "one of the worst examples of civil disobedience" in Los Angeles County since the 1943 Zootsuit Riots.[35]

It was followed by further weekends of teen–police clashes in Los Angeles's suburban valleys. On the evening of 14 October, Monrovia and Arcadia police dispersed a "mob of more than 100 teenagers ... come armed with clubs" in the parking lot of Santa Anita racetrack. The next weekend, a posse of sheriffs and Highway Patrol officers in South El Monte "broke up an incipient teen-age riot ... with the arrest of sixteen armed suspects. The youths carried spike-studded baseball bats, wire flails, brass knuckles and nail toothed chains." They were charged with "rout" ("behavior leading to riot"). Finally, on 17 November, the chain reaction of teen violence culminated on the football field at Monroe High School in Van Nuys when a crowd of 300 youths fought with school guards and the LAPD.[36]

As Southern California recaptured national leadership in juvenile anarchy, the region's pols and pundits were both dumbfounded and furious. "These are not childish incidents," Los Angeles Mayor Yorty told the press, "but serious revolts against the law." He conceded, however, that "I don't know where the failure lies, whether in the schools, at home, or where."[37] The Los Angeles Times saw a "frightening picture" in the rising arc of teenage defiance from Griffith Park to Alhambra. It warned that "mob violence and attacks on policemen threaten to grow into full-scale terror" and hinted that there might be an underlying coordination to the outbursts ("a favorite weapon of the 'cop fighters' in this city is a nail-stud-ded plank.")[38] The *Examiner* claimed that "demagogic and subversive elements welcome these disturbances as a means of promoting public support for their own ambitions," and published an interview with a leading police official under the bizarre headline, "[teen] Violence as Bad as H-Bomb." Finally, a local colum-nist confidently assured his readers that behind such "apparently unexplainable instances of mob action against the police" as the Alhambra riot, the "FBI sees a deliberate Communist pattern of attack."[39]

Whatever the causes of the teen riot epidemic, the *Examiner* was certain that the only cure was for the police to take off their kid gloves. The voice of citizen Hearst applauded Sheriff Pitchess for ordering his deputies in the wake of the Alhambra riot "to carry truncheons in addition to their side arms, and use them whenever necessary." It also commended a superior court judge for sentencing two of the Griffith Park defendants to a year in prison. "It is time to meet force with force, and pleas for tenderness on the pretext of youth or sex with judicial sternness." In the past, liberal and minority complaints about police brutality had only tied the hands of law enforcement. "The police were induced to adopt an attitude of mousy meekness that often proved an invitation to disrespect, con-tempt of the law and finally armed resistance as exemplified by the riots at Zuma Beach and Griffith Park."[40]

Revising the Sixties

In Southern California, the wild summers of 1960 and 1961 were a prelude to a series of famous youth insurrections: the Watts riot of 1965, the so-called "Hippie riots" on Sunset Strip between 1966 and 1970, and the Eastside high school "blow-outs" of 1968–69. Although the street-racing mania had subsided considerably by 1964, adolescent challenges to police control of the Night and Street became elaborately institutionalized in the renown "cruising" subcultures of Van Nuys Boulevard (white Valley kids), the Sunset Strip ("hippies"), Whittier Boulevard (Chicano eastsiders), and, much later, Crenshaw Boulevard (Black kids). Yet in what sense did early sixties teenage insubordination directly nurture or condition the politicized outbursts after 1964? And to what extent did these racially segmented youth rebellions share any common ethos or sensibility?

The most dramatic genealogy is the spiraling progression of protest and consciousness that links the Griffith Park riot of Memorial Day 1961 to the Los Angeles ("Watts") riot of August 1965. An extraordinary story remains to be told. Frustrated in their ability to integrate or access the larger city, Black youth in Los Angeles and elsewhere began to fight spontaneously for substantive control over community space—a thrust that would later become enshrined in the Black Panther Party's program for "self-determination." Although historians are at last producing fine accounts of the ordinary heroes and grassroots activism of the Southern civil rights movement, we still know little about the generational cultural revolution in Northern Black communities, or the patterns of defiance that link coming-of-age in the late 1950s and early 1960s to the near-revolutionary uprisings of the later 1960s.

The real engine room of the sixties, both politically and culturally, was not the college campus but the urban ghetto, and the transformation of young transplanted Southerners into a militant "New Breed" was the decisive event. 1961, moreover, seems to have been the watershed year in this process of generational self-definition. The counterpart to the highly organized protests in the South was the sudden wave of violent resistance to police racism in the North. Following the Griffith Park riot, there were major confrontations in Harlem (31 May and again on 29 July), Chicago (14 July after Black youths attempted to integrate a "white only" beach), and Newark (27 September). In October, the rightwing *U.S. News & World Report* claimed that 48 had been killed and 9261 injured in dozens of ghetto outbreaks. Indeed the magazine raved that an "epidemic" of "cop fighting" was engulfing the nation's big cities. Clearly, here is a powerful antecedent of the Black militancy to follow.[41]

The social trajectory of the white teenage riots, and their possible contribution to the later appeal of the New Left, is of course far less clear. Indeed most historians of the 1960s ignore the wave of teen unrest at the beginning of the decade that created so much anxiety among police chiefs and professional anti-Communists. The few who do acknowledge a premonitory upheaval typically focus on the Newport Jazz riot of 1959 or evoke "affluent adolescents" who "flirted with the harmless part of the culture of delinquency." But the hot-rod and beach riots in Southern California for the most part involved a far different social stratum of youth than Ivy League college kids at Newport or the typical Spring Break crowd of yesterday and today. The published addresses of arrestees confirm the contemporary perception that the teenagers and young adults who fought the police on El Cajon Boulevard in 1960 or Valley Boulevard in 1961 were from working-class neighborhoods and suburbs. Likewise, the riotous crowd at Zuma Beach was most likely dominated by kids from monotonous San Fernando Valley subdivisions and flatland L.A. neighborhoods, not by Malibu movie spawn.

My own recollection of the time was of almost unbearable, claustrophobic tension between the perception of teenage lands of Cockaigne and the reality of growing up blue-collar. My friends and I were mesmerized by beatniks, surfers, easy riders, and other free spirits who seemed to live an Endless Summer of libidinal adventure without the constraints of after-school jobs, the draft, and pre-programmed futures in the same ruts as our fathers and mothers. The foretaste of utopia on high school Friday nights made prospective lifetimes of punching Monday morning time-clocks even less endurable. We seethed in jealousy against everyone who lived at a beach, spent their nights in a coffee house, or went to an elite university. Todd Gitlin is correct to assert that the "marketplace sold adolescent society its banners," but not all who were seduced by the vision could participate in it.[42] With the mirage of unattainable cornucopia in the distance, it became all the more urgent to wrest as much freedom, exhilaration, and sheer mileage from the Night as possible.

I am claiming, in other words, that the white teen riots of the early 1960s were largely driven by the hidden injuries of class colliding with an overweening ideology of affluence: an affluence, that is, that we reinterpreted with the help of beatniks and surfers as the possibility of free time and space beyond the program of Fordist society. This reinterpretation was a radical seed, made all the more compelling by nuclear showdowns and Cold War apocalypticism. This quest for freedom, however inarticulate and inchoate, gave a dignity and historical purpose to our small rebellions, and, in conflict with the suburban police state, generated a powerful revulsion against arbitrary authority. Indeed anti-authoritarianism, trending toward a new romanticism of revolt and disobedience, was the vital cultural substate of the sixties. And it was inevitable that the most courageous and intransigent anti-authoritarians—Black ghetto youth—would become potent models for everyone else.

In the end, the paranoid belief of Fred Schwartz and Chief Parker that white youth rebellion was somehow instigated by Southern sit-ins and Freedom Rides proved to be a self-fulfilling prophecy. For example, in the long struggle against curfews and crowd control on the Sunset Strip in the late sixties (parodied in the teenxploitation film *Riot on the Sunset Strip*), white youth increasingly were persuaded that their resistance to a violent sheriff's department was a second front to the battle then being waged by the Black Panther Party in South Central L.A. The culminating showdown between thousands of white kids and the sheriff's deputies in 1969 was mobilized by a psychedelic leaflet demanding "Free the Strip! Free Huey!" The battle over the urban Night had joined forces with the Revolution.

2001

Notes

1. *San Diego Union*, 24 August 1960.
2. *San Diego Union*, 20 August 1960.
3. *San Diego Union*, 22 August 1960.
4. *Los Angeles Times*, 22 August 1960.
5. *Los Angeles Examiner*, 22 August 1960.
6. *San Diego Union*, 22 August 1960.
7. *San Diego Union*, 23 August 1960.
8. The top-five hit parade in San Diego during the riot weekend: (1) "It's Now or Never" [Elvis]; (2) "Walk, Don't Run" [Ventures]; (3) "Twist" [Chubby Checkers]; (4) "Itsy, Bitsy, Teeny-Weeny Yellow Polka Dot Bikini" [Bryan Hyland]; and (5) "Only the Lonely" [Roy Orbison]: *San Diego Union*, 20 August 1960.
9. *San Diego Union*, 22–23 August 1960.

10. *San Diego Union*, 24, 28, 29 August; and *Los Angeles Examiner*, 24 August 1960.

11. *San Diego Union*, 24 August 1960.

12. *San Diego Union*, 23 and 24 August 1960.

13. *San Diego Union*, 23 August 1960.

14. *San Diego Union*, 25–26 August 1960.

15. *San Diego Union*, 26 August 1960.

16. *Los Angeles Examiner*, 15 October 1961.

17. *Los Angeles Times*, 2 May 1961.

18. *Los Angeles Examiner*, 8 and 9 May 1961

19. *Los Angeles Examiner*, editorial, 1 June 1961.

20. *Los Angeles Examiner*, news story, 1 June 1961.

21. *Los Angeles Times*, 31 May 1961. The embattled police were able to make only three arrests, but took vengeance by charging two of the defendants with "attempted murder" and "lynching." Charges were later reduced to felony assault and the two were sentenced to a year in jail (*Los Angeles Examiner*, 31 May, 2 June, and 25 October 1961).

22. 12 June 1961. The same photo had appeared in the *New York Times* on 1 June.

23. *Los Angeles Examiner*, 5 June 1961.

24. Ibid, and *Los Angeles Times*, 5 June 1961.

25. *Los Angeles Examiner*, 10 and 11 June 1961.

26. *Los Angeles Times*, 5 June 1961.

27. *Los Angeles Examiner*, 11 June 1961.

28. *Los Angeles Examiner*; and *Los Angeles Times*, 12 June 1961.

29. *Los Angeles Times*, 5 July 1961.

30. *Los Angeles Times*, 16 July 1961.

31. *Los Angeles Examiner*, 5 September 1961; and *U.S. News & World Report*, 18 September 1961.

32. *Los Angeles Times*, 5 September 1961.

33. *Los Angeles Times*, 11 September 1961.

34. *Pasadena Star-News*, 12 October 1961.

35. *Pasadena Star-News*, *Los Angeles Times*, and *Los Angeles Examiner*, 7–8 October 1961; *Los Angeles Examiner*, 15 October 1961.

36. *Los Angeles Examiner*, 15 and 27 October; 17 November 1961.

37. *Los Angeles Examiner*, 12 October 1961.

38. *Los Angeles Times*, 16 October 1961.

39. *Los Angeles Examiner*, 12 October 1961; *Pasadena Star-News*, 11 October 1961.

40. *Los Angeles Examiner*, 29 October 1961.

41. "Where Even Police Are Not Safe: 48 Killed, 9261 Hurt in U.S. Cities," *U.S. News & World Report*, 9 October 1961.

42. Todd Gitlin, *The Sixties: Years of Hope, Days of Rage*, New York 1987, pp. 26–29. This is a brilliant memoir (impersonating a synoptic history) of Gitlin's political cohort: the SDS

"old guard" who came from affluent, progressive families and attended elite universities. A former valedictorian of the famous Bronx High School of Science who boasts that he played hookey only once, Gitlin is hardly predisposed to understand the relationships between delinquency, anti-authoritarianism, and revolt in the larger youth culture.

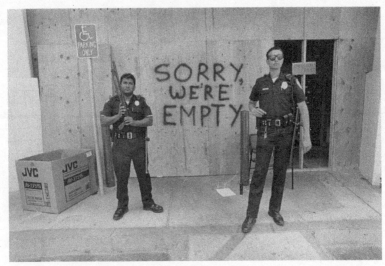

Sunset Boulevard and La Brea, Hollywood, 2 May 1992

12

Burning All Illusions

The armored personnel carrier squats on the corner like *un gran sapo feo*—"a big ugly toad"—according to nine-year-old Emerio. His parents talk anxiously, almost in a whisper, about the *desaparecidos*; Raul from Tepic, big Mario, the younger Flores girl, and the cousin from Ahuachapan. Like all Salvadorans, they know about those who "disappear"; they remember the headless corpses and the man whose tongue had been pulled through the hole in his throat like a necktie. That is why they came here—to ZIP code 90057, Los Angeles, California.

Now they are counting their friends and neighbors, Salvadoran and Mexican, who are suddenly gone. Some are still in the County Jail on Bauchet Street, little more than brown grains of sand lost among the 17,000 other alleged *saqueadores* (looters) and *incendarios* (arsonists) detained after the most violent American civil disturbance since the Irish poor burned Manhattan in 1863. Those without papers are probably already back in Tijuana, broke and disconsolate, cut off from their families and new lives. Violating city policy, the police fed hundreds of hapless undocumented *saqueadores* to the INS for deportation before the ACLU or immigrant rights groups even realized they had been arrested.

For many days the television talked only of the "South Central riot," "Black rage" and the "Crips and Bloods." But Emerio's parents know that thousands of their neighbors from the MacArthur Park district—home to nearly one-tenth of all the Salvadorans in the world—also looted, burned, stayed out past curfew, and went to jail. (An analysis of the first 5,000 arrests from all over the city revealed that 52 percent were poor Latinos, 10 percent whites, and only 38 percent Blacks.) They also know that the nation's first multiracial riot was as much about empty bellies and broken hearts as it was about police batons and Rodney King.

The week before the riot was unseasonably hot. At night the people lingered outside on the stoops and sidewalks of their tenements (MacArthur Park is L.A.'s Spanish Harlem), talking about their new burden of trouble. In a neighborhood far more crowded than mid-Manhattan and more dangerous than downtown Detroit, with more crack addicts and gangbangers than registered voters, *la gente* know how to laugh away every disaster except the final one. Yet there was a new melancholy in the air.

Too many people have been losing their jobs: their *pinche* $5.25-an-hour jobs as seamstresses, laborers, busboys, and factory workers. In two years of recession, unemployment has tripled in L.A.'s immigrant neighborhoods. At Christmas more than 20,000 predominantly Latina women and children from throughout the central city waited all night in the cold to collect a free turkey and a blanket from charities. Other visible barometers of distress are the rapidly growing colonies of homeless *compañeros* on the desolate flanks of Crown Hill and in the concrete bed of the L.A. River, where people are forced to use sewage water for bathing and cooking.

As mothers and fathers lose their jobs, or as unemployed relatives move under the shelter of the extended family, there is increasing pressure on teenagers to supplement the family income. Belmont High School is the pride of "Little Central America," but with nearly 4500 students it is severely overcrowded and an additional 2000 students must be bused to distant schools in the San Fernando Valley and elsewhere. Moreover, fully 7000 school-age teenagers in the Belmont area have dropped out of school. Some have entered the *vida loca* of gang culture (there are a hundred different gangs in the school district that includes Belmont High), but most are struggling to find minimum-wage footholds in a declining economy.

The neighbors in MacArthur Park whom I interviewed, such as Emerio's parents, all speak of this gathering sense of unease, a perception of a future already looted, the riot arrived like a magic dispensation. People were initially shocked by the violence, then mesmerized by the televised images of biracial crowds in South Central L.A. helping themselves to mountains of desirable goods without interference from the police. The next day, Thursday, 30 April, the authorities blundered twice: first by suspending school and releasing the kids into the streets; second by announcing that the National Guard was on the way to help enforce a dusk-to-dawn curfew.

Thousands immediately interpreted this as a last call to participate in the general redistribution of wealth in progress. Looting spread with explosive force throughout Hollywood and MacArthur Park, as well as parts of Echo Park, Van Nuys, and Huntington Park. Although arsonists spread terrifying destruction, the looting crowds were governed by a visible moral economy. As one middle-aged lady explained to me, "Stealing is a sin, but this is like a television game show where everyone in the audience gets to win." Unlike the looters in Hollywood (some on skateboards) who stole Madonna's bustier and all the crotchless panties from Frederick's, the masses of MacArthur Park concentrated on the prosaic necessities of life like cockroach spray and Pampers.

Now, one week later, MacArthur Park is in a state of siege. A special "We Tip" hotline invites people to inform on neighbors or acquaintances suspected of looting. Elite LAPD Metro Squad units, supported by the National Guard, sweep through the tenements in search of stolen goods, while Border Patrolmen from as far away as Texas prowl the streets. Frantic parents search for missing kids, like mentally retarded fourteen-year-old Zuly Estrada, who is believed to have been deported to Mexico.

Meanwhile, thousands of *saqueadores*, many of them pathetic scavengers captured in the charred ruins the day after the looting, languish in County Jail, unable to meet absurdly high bails. One man, caught with a packet of sunflower seeds and two cartons of milk, is being held on $15,000; hundreds of others face felony indictments and possible two-year prison terms. Prosecutors demand thirty-day jail sentences for curfew violators, despite the fact that many of those are either homeless street people or Spanish-speakers who were unaware of the curfew. These are the "weeds" that George Bush says we must pull from the soil of our cities before they can be sown with the regenerating "seeds" of enterprise zones and tax breaks for private capital.

There is rising apprehension that the entire community will become a scapegoat. An ugly, seal-the-border nativism has been growing like crabgrass in Southern California since the start of the recession. A lynch mob of Orange County Republicans, led by Representative Dana Rohrabacher of Huntington Beach, demands the immediate deportation of all the undocumented immigrants arrested in the disturbance, while liberal Democrat Anthony Beilenson, sounding like the San Fernando Valley's Son-of-Le-Pen, proposes to strip citizenship from the US-born children of illegals. According to Roberto Lovato of MacArthur Park's Central American Refugee Center, "We are becoming the guinea pigs, the Jews, in the militarized laboratory where George Bush is inventing his new urban order."

A Black Intifada?

"Little Gangster" Tak can't get over his amazement that he is actually standing in the same room of Brother Aziz's mosque with a bunch of Inglewood Crips. The handsome, 22-year-old Tak, a "straight up" Inglewood Blood who looks more like a Black angel by Michelangelo than one of the Boyz N the Hood, still has two Crip bullets in his body, and "they still carry a few of mine." Some of the Crips and Bloods, whose blue or red gang colors have been virtual tribal flags, remember one another from school playground days, but mainly they have met over the barrels of automatics in a war that has divided Inglewood—the pleasant, Black-majority city southwest of L.A. where the Lakers play—by a river of teenage blood. Now, as Tak explains, "Everybody knows what time it is. If we don't end the killing now and unite as Black men, we never will."

Although Imam Aziz and the Nation of Islam have provided the formal auspices for peacemaking, the real hands that have "tied the red and blue rags together into a 'Black thang'" are in Simi Valley. Within a few hours of the first attack on white motorists, which started in 8-Trey (83rd Street) Gangster Crip

territory near Florence and Normandie, the insatiable war between the Crips and
Bloods, fueled by a thousand neighborhood vendettas and dead homeboys, was
"put on hold" throughout Los Angeles and the adjacent Black suburbs of Compton and Inglewood.

Unlike the 1965 rebellion, which broke out south of Watts and remained
primarily focused on the poorer east side of the ghetto, the 1992 riot reached its
maximum temperature along Crenshaw Boulevard—the very heart of Black Los
Angeles's more affluent Westside. Despite the illusion of full-immersion "actual-
ity" provided by the minicam and the helicopter, television's coverage of the riot's
angry edge was even more twisted than the melted steel of Crenshaw's devastated shopping centers. Most reporters—"image looters" as they are now being
called in South Central—merely lip-synched suburban clichés as they tramped
through the ruins of lives they had no desire to understand. A violent kaleidoscope of bewildering complexity was flattened into a single categorical scenario:
legitimate Black anger over the King decision hijacked by hard-core street crimi-
nals and transformed into a maddened assault on their own community.

Local television thus unwittingly mimed the McCone Commission's summary judgment that the August 1965 Watts riot was primarily the act of a hood-
lum fringe. In that case, a subsequent UCLA study revealed that the "riot of the
riffraff" was in fact a popular uprising involving at least 50,000 working-class
adults and their teenage children. When the arrest records of this latest uprising are finally analyzed, they will probably also vindicate the judgment of many
residents that all segments of Black youth, gang and nongang, "buppie" as well as
underclass, took part in the disorder.

Although in Los Angeles, as elsewhere, the new Black middle class has socially
and spatially pulled farther apart from the deindustrialized Black working class,
the LAPD's Operation Hammer and other antigang dragnets that arrested kids at
random (entering their names and addresses into an electronic gang roster that is
now proving useful in house-to-house searches for riot "ringleaders") have tended
to criminalize Black youth without class distinction. Between 1987 and 1990, the
combined sweeps of the LAPD and the County Sheriff's Department ensnared
50,000 "suspects." Even the children of doctors and lawyers from View Park and

Windsor Hills have had to "kiss the pavement" and occasionally endure some of the humiliations that the homeboys in the flats face every day—experiences that reinforce the reputation of the gangs (and their poets laureate, the gangster rappers like Ice Cube and NWA) as the heroes of an outlaw generation.

Yet if the riot had a broad social base, it was the participation of the gangs—or, rather, their cooperation—that gave it constant momentum and direction. If the 1965 rebellion was a hurricane, leveling one hundred blocks of Central Avenue from Vernon to Imperial Highway, the 1992 riot was a tornado, no less destructive but snaking a zigzag course through the commercial areas of the ghetto and beyond. Most of the media saw no pattern in its path, just blind, nihilistic destruction. In fact, the arson was ruthlessly systematic. By Friday morning 90 percent of the myriad Korean-owned liquor stores, markets, and swapmeets in South Central L.A. had been wiped out. Deserted by the LAPD, which made no attempt to defend small businesses, the Koreans suffered damage or destruction to almost 2000 stores from Compton to the heart of Koreatown itself. One of the first to be attacked (although, ironically, it survived) was the grocery store where fifteen-year-old Latasha Harlins was shot in the back of the head last year by Korean grocer Soon Ja Du in a dispute over a $1.79 bottle of orange juice. The girl died with the money for her purchase in her hand.

Latasha Harlins. A name that was scarcely mentioned on television was the key to the catastrophic collapse of relations between L.A.'s Black and Korean communities. Ever since white judge Joyce Karlin let Soon Ja Du off with a $500 fine and some community service—a sentence which declared that the taking of a Black child's life was scarcely more serious than drunk driving—some interethnic explosion has been virtually inevitable. The several near-riots at the Compton courthouse this winter were early warning signals of the Black community's unassuaged grief over the murder of Latasha Harlins. On the streets of South Central Wednesday and Thursday. I was repeatedly told, "This is for our baby sister. This is for Latasha."

The balance of grievances in the community is complex. Rodney King is the symbol that links unleashed police racism in Los Angeles to the crisis of Black life everywhere, from Las Vegas to Toronto. Indeed, it is becoming clear that the King case may be almost as much of a watershed in American history as Dred Scott, a test of the very meaning of the citizenship for which African Americans have struggled for 400 years.

But on the grassroots level, especially among gang youth, Rodney King may not have quite the same profound resonance. As one of the Inglewood Bloods told me: "Rodney King? Shit, my homies be beat like dogs by the police every day. This riot is about all the homeboys murdered by the police, about the little sister killed by the Koreans. Rodney King just the trigger."

At the same time, those who predicted that the next L.A. riot would be a literal Armageddon have been proved wrong. Despite a thousand Day-Glo exhortations on the walls of South Central to "Kill the Police," the gangs have refrained from the deadly guerrilla warfare that they are so formidably equipped to conduct. As in 1965, there has not been a single LAPD fatality, and indeed few serious police injuries of any kind.

In this round, at least, the brunt of gang power was directed toward the looting and destruction of the Korean stores. If Latasha Harlins is the impassioned pretext, there may be other agendas as well. I saw graffiti in South Central that advocated "Day one: burn them out. Day two: we rebuild." The only national leader whom most Crips and Bloods seem to take seriously is Louis Farrakhan, and his goal of Black economic self-determination is broadly embraced. (Farrakhan, it should be emphasized, has never advocated violence as a means to this end.) At the Inglewood gang summit, which took place on 5 May, there were repeated references to a renaissance of Black capitalism out of the ashes of Korean businesses. "After all," an ex-Crip told me later, "we didn't burn our community, just *their* stores."

In the meantime, the police and military occupiers of Los Angeles give no credence to any peaceful, let alone entrepreneurial, transformation of L.A.'s Black gang cultures. The ecumenical movement of the Crips and Bloods is their worst imagining: gang violence no longer random but politicized into a Black *intifada*. The LAPD remembers only too well that a generation ago the Watts rebellion produced a gang peace out of which grew the Los Angeles branch of the Black Panther Party. As if to prove their suspicions, the police have circulated a copy of an anonymous and possibly spurious leaflet calling for gang unity and "an eye for an eye. ... If LAPD hurt a Black we'll kill two."

For its part, the Bush Administration has federalized the repression in L.A. with an eye to the spectacle of the President marching in triumph, like a Roman emperor, with captured Crips and Bloods in chains. Thus, the Justice Department has dispatched to L.A. the same elite task force of federal marshals who captured Manuel Noriega in Panama as reinforcements for LAPD and FBI efforts to track down the supposed gang instigators of the riots. But as a veteran of the 1965 riot said while watching SWAT teams arrest some of the hundreds of rival gang members trying to meet peacefully at Jordan Downs in Watts, "That ole fool Bush think we as dumb as Saddam. Send the Marines into Compton and get hisself reelected. But this ain't Iraq. This is Vietnam, Jack."

The Great Fear

A core grievance fueling the Watts rebellion and the subsequent urban insurrections of 1967–68 was rising Black unemployment in the midst of a boom economy. What contemporary journalists fearfully described as the beginning of the "Second Civil War" was as much a protest against Black America's exclusion from the military-Keynesian expansion of the 1960s as it was an uprising against police racism and de facto segregation in schools and housing. The 1992 riot and its possible progenies must likewise be understood as insurrections against an intolerable political-economic order. As even the *Los Angeles Times*, main cheerleader for "World City L.A.," now editorially acknowledges, the globalization of Los Angeles has produced "devastating poverty for those weak in skills and resources."

Although the $1 billion worth of liquor stores and minimalls destroyed in L. A. may seem like chump change next to the $2.6 trillion recently annihilated on the Tokyo Stock Exchange, the burning of Oz probably fits into the same Hegelian niche with the bursting of the Bubble Economy: not the "end of history" at the seacoast of Malibu but the beginning of an ominous dialectic on the rim of the Pacific. It was a hallucination in the first place to imagine that the wheel of the world economy could be turned indefinitely by a Himalaya of US trade deficits and a fictitious yen.

This structural crisis of the Japan-California "co-prosperity sphere" threatens, however, to translate class contradictions into interethnic conflict on the national and local level. Culturally distinct "middleman" groups—ethnic entrepreneurs and the like—risk being seen as the personal representatives of the invisible hand that has looted local communities of economic autonomy. In the case of Los Angeles, it was tragically the neighborhood Korean liquor store, not the skyscraper corporate fortress downtown, that became the symbol of a despised New World Order.

On their side, the half-million Korean-Americans in L.A. have been psychologically lacerated by the failure of the state to protect them against Black rage. Indeed, several young Koreans told me that they were especially bitter that the South Central shopping malls controlled by Alexander Haagen, a wealthy contributor to local politics, were quickly defended by police and National Guard, while their stores were leisurely ransacked and burned to the ground. "Maybe this is what we get," a UCLA student said, "for uncritically buying into the white middle class's attitude toward Blacks and its faith in the police."

The prospects for a multicultural reconciliation in Los Angeles depend much less on white knight Peter Ueberroth's committee of corporate rebuilders than on a general economic recovery in Southern California. As the Los Angeles Business Journal complained (after noting that L.A. had lost 100,000 manufacturing jobs over the past three years), "The riots are like poison administered to a sick patient."

Forecasts still under wraps at the Southern California Association of Governments paint a dark future for the Land of Sunshine, as job growth, slowed by the decline of aerospace as well as manufacturing shifts to Mexico, lags far behind population increase. Unemployment rates—not counting the estimated 40,000 jobs lost from the riot and the uprising's impact on the business climate—are predicted to remain at 8 to 10 percent (and 40 to 50 percent for minority youth) for the next generation, while the housing crisis, already the most acute in the nation, will spill over into new waves of homelessness. Thus, the "widening divide" of income inequality in Los Angeles County, described in a landmark 1988 study by UCLA professor Paul Ong, will become an unbridgeable chasm. Southern

California's endless summer is finally over.

Affluent Angelenos instinctively sensed this as they patrolled their Hancock Park estates with shotguns or bolted in their BMWs for white sanctuaries in Orange and Ventura Counties. From Palm Springs pool-sides they anxiously awaited news of the burning of Beverly Hills by the Crips and Bloods, and fretted over the extra set of house keys they had foolishly entrusted to the Latina maid. Was she now an incendiarist? Although their fears were hysterically magnified, tentacles of disorder did penetrate such sanctums of white life as the Beverly Center and Westwood Village, as well as the Melrose and Fairfax neighborhoods. Most alarmingly, the LAPD's "thin blue line," which had protected them in 1965, was now little more than a defunct metaphor, the last of Chief Gates's bad jokes.

1992

Postscript

The embers of April/May 1992 still smoulder. Despite delirious declarations that "LA is back!" there was significantly greater poverty in Los Angeles County in 1999—at the very height of "New Economy" prosperity—than in 1992. Since then, unemployment in flatland immigrant neighborhoods has soared while local government, especially the overwhelmed county health system (40 percent of the population has no medical coverage), faces the worst fiscal crunch in a generation. Even more ominously, the gang truce movement that preceded the uprising and brought about the social miracle of unity between the Crips and Bloods is dying, just as its organizers long ago predicted, from official hostility and the lack of employment resources. Teenage funerals are once again an almost daily ritual in the neighborhoods that the dot-com boom forgot.

City Hall, meanwhile, has heroically ignored the lessons of 1992. During the eight years of Mayor Richard Riordan's reign—a veritable "L.A. renaissance" according to his supporters—municipal politics was downsized to the two issues of paramount concern to the Anglo middle class in the beaches and hills: enlarging the police force and restoring business confidence. More recently, the political agenda has been hijacked by the secessionists in the San Fernando Valley: a motley alliance of exclusionary homeowner associations, Republican businessmen, and some opportunist Latino Democrats. The Valley confederates are led by the rightwing Sherman Oaks Homeowners Association—the same group that over the years has spearheaded the Jarvis-Gann tax rebellion (the original "white riot"), organized opposition to school busing, and most recently championed anti-immigrant state Proposition 187.

Valley secession is a direct aftershock of the 1992 uprising; a racist attempt to redraw the colorline against the new majority. Although Anglos even in the Valley are now only 40 percent of the population, they remain more than three-quarters of its electorate. Splitting up Los Angeles would allow white homeowners and business interests in the Valley to sustain their political domination for another decade, postponing the inevitable arrival of a majority Latino vote. Secession is white power on artificial life support. It is also a symptom of how little has been conceded to social justice in the last decade. And inequality, as we all know, is Southern California's most famous fire ecology.

2002

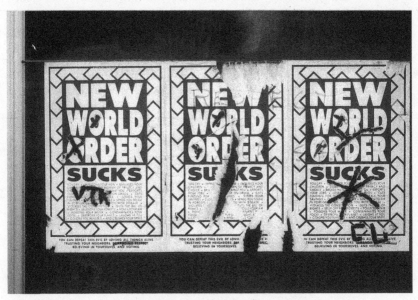

Popular sentiment (1992)

13

Who Killed L.A.?
A Political Autopsy

It was the most extraordinary conjuring feat in modern American political history. The spring presidential primary season had barely opened when a volcano of Black rage and Latino alienation erupted in the streets of Los Angeles. Elite Marine and Army units fresh from the Gulf War had to be landed to restore order to the bungalows of Compton and Watts. While the world press editorialized apocalyptically about the "decline of America," a grim-faced procession of inner-city leaders from Oakland to Bedford-Stuyvesant warned that their neglected neighborhoods too were tinderboxes awaiting a spark. They recalled the 164 major riots—the "Second Civil War" some warned at the time—that spread through urban ghettoes like wildfire for three summers after the original "Watts rebellion" in 1965.

The presidential candidates, meanwhile, jostled each other for the photo opportunity of squaring their jaws amid the smoking ruins of New Jack City. President Bush found meetings with residents "very emotional, very moving," and vowed that government had "an absolute responsibility to solve inner-city problems."[1] As the campaign promises flowed like honey, political columnist William Schneider reassured local leaders that "hundreds of millions of dollars will be funneled into L.A." The *Los Angeles Times*, meanwhile, applauded President

Bush and the House Democrats for joining together to take "swift action to bring relief to the nation's cities."[2]

The City Vanishes

Yet within weeks, and before a single scorched minimall had actually been rebuilt, the Second Los Angeles Riot, as well as the national urban-racial crisis that it symbolized, had been virtually erased from political memory banks. The Bush Administration's "new fervor" for urban reform quickly recooled into glacial indifference. When the US Conference of Mayors, for example, brought 200,000 marchers to the Capitol on 16 May under the banner "Save Our Cities, Save Our Children!" White House Press Secretary Martin Fitzwater simply shrugged his shoulders and complained, "I don't know anything about it. We have marches every weekend." The major palliative that Bush offered distressed cities in his stump speeches was an authoritarian "Weed and Seed" plan to place job training and community development funds under the jurisdiction of the Justice Department's war on gangs. Vice President Quayle, meanwhile, haughtily advised Mayor Bradley that if he really wanted to rebuild Los Angeles he should raise money by selling off the city's international airport.[3]

Among the Democrats only Jerry Brown remained an outspoken, if late-in-the-day, advocate of the big city mayors and their constituencies. His defeat in the June California primary ended, for all intents and purposes, further debate on urban poverty or the future of the cities. In the sharpest break yet with New Deal ideology, the 1992 Democratic Platform, drafted by Clinton supporters under new rules that eliminated formal amendment and vote-taking, scrapped traditional rhetoric about urban needs in favor of Republican-sounding emphases on capital formation and tax breaks for entrepreneurs. Clinton himself carefully "tiptoe[d] around the issues of urban problems and race." Every direct question about the Los Angeles uprising or the cities' fiscal crisis was met with neutered technobabble about "micro-enterprise zones" and "infrastructure."[4]

Listening to the fall presidential debates, it was almost impossible to avoid the suspicion that all three camps, including Perot redux, were acting in cynical concert to exclude a subject that had become mutually embarrassing. The word "city"—now color-coded and worrisome to the candidates' common suburban heartland—was expunged from the exchanges. Thus the thousand-pound gorilla of the urban was simply and consensually conjured out of sight. Indeed, if the verdict of the 1992 election is taken seriously, the big cities, once the very fulcrum of the Rooseveltian political universe, have been demoted to the status of a scorned and impotent electoral periphery.

For the weary populations of Detroit or Buffalo this may be old news. But in Los Angeles, which until recently had been preoccupied with fantasies of becoming the Byzantium of the Pacific Rim, it was a brutal shock. To the incredulity of local observers, the spring riots proved to be almost non-negotiable political currency outside Southern California. If the "white backlash" turned out to be more subdued than predicted, there was, symmetrically, very little national sympathy with Los Angeles's problems or its quest for state and federal aid. Its own affluent suburbs helped sabotage a bipartisan "urban rescue" bill in Congress, while the governor and legislature in Sacramento figuratively burned down the city a second time with billions of dollars of school and public sector cutbacks.

Unexpectedly left to their own devices in the thick of the worst economic crisis since 1938, Los Angeles's elites have invested inflated hopes in Rebuild LA (RLA), the corporate coalition headed by Peter Ueberroth, the czar of the 1984 Olympics who was given a virtually messianic mandate by Mayor Tom Bradley to save Los Angeles. When it became clear that the city would not receive any significant aid from either Sacramento or Washington, Ueberroth dramatically announced that RLA's corporate sponsors had already pledged more than a billion dollars in new investment for Los Angeles's inner-city neighborhoods. RLA was immediately embraced by everyone from Jimmy Carter to The Economist as the paradigm for a new corporate voluntarism that would save declining American and British cities. Quick-witted reporters, however, immediately interviewed Ueberroth's corporate Good Samaritans, half of whom emphatically denied making any such commitment. In the eyes of many, RLA was thereby exposed as the philanthropic equivalent of the classic Ponzi scheme; mendaciously pyramiding false promises into purely fictitious community "rebuilding."

This relentless melting of illusions in the context of the national nondebate about the urban crisis has left a refractory residue consisting of only the most base elements. In City Hall, for example, substantive discussion of community-level reform has been supplanted by monomaniac anxiety over police prepared-ness to deal with the new riot that virtually everyone now concedes is probable. Similarly, the current (spring 1993) mayoral race—arguably the most important in the city's history—has been largely reduced to a tawdry auction between com-peting schemes to lay off public employees in order to afford more cops. Even more depressingly, an embittering competition over shrinking resources, which RLA has only enflamed, has brought the Black and Latino communities to the brink of open street warfare. Local pundits now talk ominously about the city's "Yugoslav disease" as it Balkanizes into intercommunal strife.

In sum, the national and local responses to the 1992 Los Angeles uprising have revealed a doom-ridden inertia and shortage of reform resources at every level of the American political system. It is certainly different from 1965 when the John-son Administration steamrolled its huge Model Cities bill through Congress soon after the first Los Angeles riot. From an offshore perspective, moreover, the cur-rent situation must seem inscrutable if not incredible: what other affluent nation, much less planetary superpower, would tolerate such high levels of disorder in its second largest city? Is it conceivable that a suburban political majority is actually prepared to write off the future of Los Angeles (and possibly New York)? Won't the new Clinton Administration have to ride to the rescue?

Analysis of the political responses to the uprising reveals the formidable obstacles in the path of any resumption of urban reformism in the 1990s, as revealed by the debacle of "riot relief" legislation in Washington and Sacramento, where fiscal crisis has been the forcing house of a new anti-urban federalism. On Capitol Hill, Gramm-Rudman, Perot, and the international bond markets have tied a Gordian knot around urban policy that Clinton probably dare not cut. At the same time, a new version of the old congressional conservative coalition has emerged that unites suburban and rural representatives in both parties against any federal reinvestment in the minority-dominated big cities. Meanwhile, a less visible, but equally consequential, counterrevolution has been taking place on the state level since 1989. Key industrial states, including California, Ohio, Michigan, and Illinois, have radically reduced traditional welfare and educational entitlements with devastating results for their major urban cores.

For example, Los Angeles's formerly vaunted school system now compares unfavorably to Mississippi's, while community health standards have fallen to Third World levels. Although nativists have tried to blame last year's disturbances on the impact of promiscuous immigration, it is the accelerated decay of the public sector that best explains the rising tensions between different ethnic communities—in Los Angeles and elsewhere.

The Republican Wilderness

Flying home from Houston on Air Force One the day after his defeat, George Bush had the perverse consolation of vetoing the urban aid bill that he had helped launch six months before. Originally designed as a streamlined rescue package for riot-damaged Los Angeles and flood-stricken Chicago that combined federal emergency finance with enterprise-zone tax exemptions, the bill had become so grotesquely ornamented with expensive amendments that, according to Bush, it was now just a "Christmas tree." He bitterly blamed the Democrats for abandoning Los Angeles to a "blizzard of special interest pleadings."[5]

In fact, Bush himself had decorated the tree. The White House was directly responsible for attaching most of the ornaments, including amendments to help the urban poor by repealing luxury taxes on boats, planes, furs, and jewelry, as well as new tax breaks for real estate investors. What actually upset Bush most about the final form of the bill was a Democratic rider that proposed ending tax deductions for club dues: an unfair burden on the rich that might be interpreted as a stealth tax increase.

Historians may someday debate why the Republicans' callous refusal to help Los Angeles did not provoke a national scandal or at least give the Democrats valuable campaign ammunition. (The Clintonians deliberately declined the gift.) In its major features, the Bush Administration's response to the Second Los Angeles Riot was an inverted mirror image of the Johnson Administration's response to the First. In 1965 the LAPD's Chief Parker (assisted by the National Guard) retained total control over law enforcement while the federal government provided massive financial aid through its new urban programs. This time around, however, repression was immediately and dramatically federalized, while the rebuilding was left to shoestring local efforts and corporate charity.

There is, of course, an eerie indistinguishability in all the military interventions, "humanitarian" or exterminist, of the Reagan-Bush era. The fuzzy video images of the Marines or 82nd Airborne in the streets of Panama City, Miami, Los Angeles, Grenada, or Mogadishu all look alike and the prone figures on the ground are always Black. But the rapid deployment of federal combat troops to South Central L.A. was only one leg of the tripod of policies—an iron-fisted "Bush Doctrine" for troubled US cities—unveiled last May. Wielded into action with equally impressive speed, for example, was an unprecedented taskforce of federal law-enforcement agencies mandated to track down and prosecute riot felonies. The large FBI and INS components of the task force were later reorganized as permanent anti-gang units in line with Attorney General Barr's dictum that the Crips and Bloods, together with criminal illegal aliens, have replaced Communism as the major domestic subversive threat. This is also the official legitimation for the third leg of the tripod: the "Weed and Seed" program that ties neighborhood-level federal spending (the "seeds") to active collaboration with the war against the gangs (the "weeds").

If, in tendency, "Weed and Seed" prefigures the ultimate absorption of the welfare state by the police state, the parsimony of federal seed money has ensured that the actual results are less dramatic. Non-law-enforcement aid to Los Angeles has amounted to scarcely more than smoke and mirrors. A month after the riot, for example, HUD Secretary (and presumed 1996 Republican presidential front-runner) Jack Kemp showed up with a fat press entourage at Nickerson Gardens Housing Project in Watts to announce that his department was giving Los Angeles $137 million in housing assistance. The national press recorded the local elation at this unexpected windfall, but generally neglected to report the subsequent rage when Kemp's gift turned out to be nothing more than funds already in the pipeline.[6] Likewise the White House (which had previously blamed the riot on the legacy of the Great Society) established an awesomely named "Presidential Task Force on Los Angeles Recovery," headed by obscure deputy secretaries of Housing and Education, whose sole function was the repackaging of existing programs as dynamic Bush initiatives.

These sleights of hand allowed Republican campaign publicists to portray huge fictitious sums of assistance to Los Angeles when, in fact, the administration was blocking small business loans and food stamps to tens of thousands of needy residents in riot-impacted neighborhoods. According to city officials, fully 60 percent of eligible riot victims were denied disaster assistance and the Federal Emergency Management Agency even balked at reimbursing $1 million to the state for establishing disaster application centers.[7] In contrast, affluent Florida suburbs damaged by Hurricane Andrew and seen as crucial to a November Bush victory received massive, fast-tracked relief.[8]

In the end, the White House's only decisive response to Los Angeles's pleas for federal help, apart from Marines and FBI agents, was an audacious scheme to loot the city's major public assets. Behind Dan Quayle's seemingly off-handed remark about selling LAX was a concerted effort by advocates of radical privatization to force an urban fire sale. These latter-day privateers were led by Robert Pool Jr., founder of Los Angeles's rightwing Reason Institute; John Girudo, former counsel to President Reagan's privatization commission; and C. Boyden Gray, the chief counsel to President Bush. With Quayle riding roughshod over budget chief Richard Darman's scruples, Bush issued an executive order that paved the way for hard-pressed cities to sell off $220 billion worth of federally financed public works, ranging from sewage treatment plants to turnpikes and airports. An exultant Pool hailed the order as little short of a "Magna Carta for privatization."[9]

It was also another striking instance of how closely Washington's policy toward the cities has come to resemble the international politics of debt. In the Reagan-Bush era the big cities have become the domestic equivalent of an insolvent, criminalized Third World country whose only road to redemption is a combination of militarization and privatization. Otherwise, the Republicans have been absolutely adamant over the last twelve years in embargoing aid to the cities. Indeed, this de facto war against the cities has been one of the strategic pillars of modern conservative politics, embodying profound electoral and economic objectives.[10]

On the one hand, from the moment of victory in 1980, Republican ideologues were urging a "Thatcherite" offensive against core Democratic constituencies. The American Enterprise Institute, in particular, promoted the "win-win" logic of "blow[ing] up the political infrastructure of the urban Democratic Party" by killing programs like the Urban Development Action Grants (UDAGs) "that buy power for people who walk around with a capital D."[11] By savagely cutting back urban aid, they hoped to bury the remains of the Great Society and deepen the schism between Black inner-city and white suburban Democratic constituencies.[12]

On the other hand, federal disinvestment in the big cities was also supposed to liberate the animal spirits of urban capitalism. As Barnekov, Boyle, and Rich point out, this canonical Reagan-era precept—like so many others—was actually incubated in the second half of the Carter Administration. It was Carter's Commission for a National Agenda for the Eighties that rejected "centrally administered national urban policy" as "inconsistent with the revitalization of the larger national economy." According to the commission, Washington had to "reconcile itself" to the decline of older industrial cities and not interfere with the rise of a new "postindustrial economy" by directly aiding distressed communities.[13]

Within the Reagan Administration this resurgent Social Darwinism was given an even more implacable edge by Emmanuel Savas, the assistant secretary of HUD in charge of policy development. In various articles and official reports, Savas argued that federal urban policy had been a complete failure and that cities had to be weaned, however brutally, from their artificial dependence upon Washington. Casually admitting that "not all cities will benefit equally, and some may not benefit at all," Savas—supported by Budget Director David Stockman—advocated a competitive acceptance of the new discipline of the world economy and a thoroughgoing privatization of local government services. It was time for cities to stop being welfare cases and learn to become lean, mean entrepreneurs. Thus the 1982 National Urban Policy Report drafted by Savas envisioned an inter-urban war of all against all as cities were advised to "form partnerships with their private sectors and plan strategically to enhance their comparative advantages relative to other jurisdictions."[14]

Returning the cities to the Darwinian or Hobbesian wilderness, however, required massive Democratic complicity. The Republicans shrewdly calculated that Southern and suburban Democrats, given a suitable pretext, were ready to help put the knife in the backs of their big-city brethren. (Carter, after all, had already frozen urban spending in 1978.) This is exactly what happened in 1985–86 when the congressional Democratic leadership allowed general revenue sharing to be killed in committee and exposed urban grants to across-the-board cuts under the Gramm-Rudman deficit-reduction process.[15] The stiletto was given another vicious twist in 1988 when three-quarters of Southern Democratic representatives voted to eliminate UDAG in order to finance a major funding increase for NASA's Space Station.

Two years later, while cutbacks in federal aid were driving cities into their worst financial crisis since the Depression, the House Democratic leadership negotiated a budget compromise with the White House that precluded an urban bailout in the foreseeable future. Although Washington had twice invented "fiscal emergencies" to side-step Gramm-Rudman and finance the Gulf War and the S&L bailout, it simply "yawned" in face of the US Conference of Mayors' urgent plea for a domestic Marshall Plan.[16] Indeed, in passing the Budget Enforcement Act of 1990, with its moratorium on social spending, the Democratic majority abdicated any remaining pretence of committed opposition to the city-killing policies of the Bush Administration. It was the last nail in the coffin of the New Deal.

The Body Count

Abandoned by the national Democratic Party to the ill winds of "post-industrialism," the big cities have faced massive federal disinvestments at the very moment that deindustrialization and the epidemics of the 1980s (AIDS, crack, homelessness) were imposing immense new financial burdens. In an important study, Demetrios Caraley estimates that the 64 percent cutback in federal aid since 1980 had cost cities $26 billion per year (in constant 1990 dollars). For cities with more than 300,000 inhabitants, the average federal share of the municipal income stream has plummeted from 22 percent in 1980 to a mere 6 percent in 1989.[17]

Since state aid, nationally averaged, has remained constant at 16 percent, cities have had to make up the shortfall with local resources: usually very regressive sales taxes and user fees.

Table 1 shows the scale of the federal retreat from ten of the biggest cities. If Los Angeles has endured the steepest decline in budget share, the Republican war on the cities has probably inflicted the greatest absolute damage on Philadelphia and New York. Despite being placed under a virtual receivership by the state legislature, Philadelphia has lurched from one giant deficit to another since 1990. From Harlem to Flatbush, meanwhile, the missing federal aid has spelled the difference between the preservation of New York's La Guardian legacy and Mayor Dinkins's current "doomsday budget" with its 20,000 lay-offs. Forced to abandon redistributive programs and too broke to pave streets or modernize sewer systems, America's pariah big cities struggle simply to pay their financial creditors and keep a thin blue line of cops in uniform. As Ester Fuchs somberly notes, the coincidence of protracted recession and federal disinvestment ensures that "the prospects for urban America in the 1990s are in many ways worse than they were during the Depression era."[18]

The figures in Table 1, moreover, may substantially understate the real social impact of the Reagan revolution on urban finance. As James Fossett pointed out in a 1984 Brookings Institution study, federal grants and revenue sharing in the 1970s provided a much larger share of cities' operating expenditures than they did of their total budgets inclusive of capital outlays. By this alternative measure, federal aid to Los Angeles (42 percent of the operating budget in the peak year of 1978) may have been twice as significant as Table 1 suggests.

Even more importantly, federal funds constituted the predominant public resource for many, if not most, poor inner-city neighborhoods. Fossett estimated, for example, that 91 percent of federal grants to Los Angeles benefited poor and moderate-income groups.[20] Needless to say, these grants also greased the wheels of community politics. As we shall see later, the Republican dynamiting of the federal aqueduct to the inner cities has forced an important political realignment. Deprived of the funds and patronage that formerly flowed from Washington, many local politicos and organizers drifted back during the 1980s—just as the

Reagan ideologues intended—to Booker T. Washington-like dependencies on corporate paternalism. Similarly, most community organizations have had to "entrepreneurialize" themselves and their programs to survive the long drought of federal aid.

Table 1

Federal Contribution to Budgets, Selected Big Cities[19]

		1977	1985
1.	New York	19%	9%
2.	Los Angeles	18%	2%
3.	Chicago	27%	15%
4.	Philadelphia	20%	8%
5.	Detroit	23%	12%
6.	Baltimore	20%	6%
7.	Pittsburgh	24%	13%
8.	Boston	13%	7%
9.	Cleveland	33%	19%
10.	Minneapolis	21%	9%

In sectoral terms, meanwhile, the national urban programs that have suffered the most pitiless retrenchment since 1980 have been subsidized housing (–82 percent), economic development assistance (–78 percent), and job training (–63 percent).[21] Again, as ideologically designed, federal aid has been cut off from cities precisely as they have confronted the most wrenching restructuring since the industrial revolution. Like the Irish tenantry during the Potato Famine of the 1840s, the contemporary American urban poor have been doomed by the state's fanatical adherence to laissez-faire dogma. The decline in housing subsidies, for example, has helped put more urban Americans out in the cold than the Great Depression, while the evaporation of job training funds and the termination of the Comprehensive Employment Training Act (CETA) have consigned myriads more to the underground drug economy. The United States is the only major industrial nation to respond to the international competitive regime of the 1980s by ruthlessly eliminating structural adjustment assistance to workers and cities.

Federal policy has also pummeled city labor forces in other ways. Since the first wave of urban deindustrialization in the early 1970s, the local public sector and the US military have provided the most important compensatory employment opportunities for Black and Latino workers who, unlike their white counterparts, have been unable to move laterally into new suburban job slots or rise upward into the financial center professional-managerial strata. From the mid 1980s, however, the reduction of federal aid has accelerated the wage and workforce downsizing of local government that began tendentially during the Tax Revolt of 1978–79. The security of city and county employment has been undermined by the massive privatization of everything from sanitation to jails and schools. Takebacks, contracting out, and wage deflation are now as common in the local public sector as they were in the private sector during the 1980s.

More recently, the end-of-Cold-War shrinkage of the conventional military has closed the single most important employment option for ghetto and barrio youth. Since 1986 the percentage of young Blacks entering the armed forces has plummeted from 20 percent to 10 percent, while the overall nonwhite proportion of the military has fallen from one-third in 1979 to just one-quarter today.[22] Minorities have also suffered disproportionately from the closure of domestic military bases, like San Bernardino's Norton Air Force Base, the largest employer of Blacks in Southern California's "Inland Empire."

But Washington's culpability in the current urban crisis extends far beyond the mere cutoff of financial aid. The Republicans have also blown up city budgets by deliberately shifting the costs of many national problems onto Democrat-dominated localities. They have imposed mandates to provide new services without providing additional funding. New York and Los Angeles, for example, are the principal ports of entry for the greatest immigration wave since the early 1900s, but the Bush Administration has refused to pay them (or their state governments) the compensatory funds promised under the Immigration Reform and Control Act of 1988 (IRCA). Although several Southern California–based studies have shown that immigrants, undocumented as well as legal, contribute more in taxes than they consume, the federal government siphons off the net surplus via payroll taxes, leaving cities and counties with substantial deficits for services provided. Not surprisingly, this federal refusal to reimburse local government for its role in national immigration policy only exacerbates anti-immigrant prejudice at a local level, which is then politically harvested by nativists and conservatives.

The War on Drugs, of course, is the other Reagan-Bush initiative that has imposed crushing costs upon cities. Editorially endorsing the findings of a recent Rand Corporation report that examined the Los Angeles riots in the context of national policy, the *Los Angeles Times* conceded that the War on Drugs had "devastate[d] minority communities without significantly impairing narcotics distribution."[23] The Rand researchers had shown that the exponential increase in drug offenders arrested (over one million every year) and imprisoned was simply money and lives wasted. Despite federal subsidies to local law enforcement, the criminalization of drug use is accumulating huge long-term social costs that will fall primarily on insolvent city and county governments.[24]

Finally, as Table 2 demonstrates, the Reagan-Bush era's various anti-urban policies, combined with huge tax subsidies to suburban retail and office development, have opened a new chasm of inequality between core cities and their suburban rings. Over the last decade traditional urban areas have lost a staggering 30 percent of their job base while suburbs have seen employment soar by 25 percent.[25] In some cases, like Washington, D.C., the outer suburbs have accumulated fifteen times more per capita tax capacity than their dying cores.[26] A new, often startling, economic geometry has emerged as corporate headquarters and business services, following factories and shopping malls, have relocated to the highrise nodes strung like pearls on outer freeway rings, twenty to seventy-five miles from the old urban centers.

Table 2

Central Cities Compared to Suburban Ring (= 100%)[27]

	1980	1990
1. Households in poverty	360%	650%
2. Per capita income	90%	59%

Note: Per capita income figures are for 1980 and 1987.

The New Spatial Apartheid

Much of what Joel Garreau and other authors have celebrated as the rise of the "Edge City"—"the biggest change in a hundred years in how we build cities"[28]—is the artifact of the vastly different federal policies toward metropolitan centers and peripheries. While Reaganism was exiling core cities into the wilderness, it was smothering suburban commercial developers and renegade industrialists with tax breaks and subsidies. Most of the capital gains windfall of the 1980s that was supposed to technologically rearm corporate America for competition in the world market was actually siphoned into a vast overbuilding of office and retail space along the circumferential beltways and intercity corridors. Or, put another way, the "spatial trickle-down" from national economic growth that Savas and Stockman promised would eventually return to the chastened, entrepreneurial city has actually been centrifuged off to Edgeland.

In effect, these policies have also subsidized white flight and metropolitan resegregation. In the ideal world of neoclassical economics, the best option for workers in decaying, uncompetitive center cities is simply to follow the migration of jobs to the new edge cities. This is, of course, exactly what millions of white urbanites have done since the ghetto uprisings of the late 1960s. Table 3 (a–c) summarizes the ethnic recomposition of the fourteen cities (24 million people) that constitute the cores of the ten largest US metropolitan regions (76 million people).

At slightly greater magnification, it is possible to make important further distinction between the urban itineraries of whites, Blacks, Latinos, and other groups. To take Los Angeles as an example, almost the entire white working class of the older southeast industrial belt—some 250,000 people—moved to the job-rich suburban fringe during the 1970s and early 1980s.[30] They were replaced by 328,000 Mexican immigrants, primarily employed in nonunion manufacturing and service jobs. Indeed, in Los Angeles the counterpoint to the Latinization of manual labor has been the virtual disappearance of traditional Anglo blue-collar strata from the urban core. A cartoon of the city's resident workforce would depict a white professional-managerial elite, a Black public sector workforce, an

Asian petty-bourgeoisie, and an immigrant Latino proletariat. Table 4 shows the
distribution of political power that has accompanied that shift in urban ethnic
composition.

Table 3

a. Ethnic Shifts in Cores of 10 Largest Metropolises

- − 8,000,000 whites
- + 4,800,000 Latinos
- + 1,500,000 Asians
- + 800,000 Blacks
- − 900,000 total population

b. Ethnic Composition, 1970 versus 1990 (%) [29]

	1970	1990
white	70.0	39.9
Black	27.6	31.4
other	2.4	28.7
Asian		6.8
Latino		21.9

c. 10 Largest Metropolitan Cores: Percentage white [29]

		1970	1990
1.	New York	75.2	38.4
2.	Los Angeles	78.3	37.2
3.	Chicago	64.6	36.3
4.	D.C. area	41.4	33.0
5.	Bay Area	75.1	42.9
6.	Philadelphia	65.6	51.3
7.	Detroit	65.6	20.3
8.	Boston	81.7	58.0
9.	Dallas	75.8	49.8
10.	Houston	73.4	39.9

Although second- and third-generation Mexican-Americans do not move as freely within the Southern California metro area as working-class or middle-class Anglos, their mobility rate is surprisingly high. One of the major ethnic-political shifts of the last decade, for instance, has been the explosion of Chicano political power in the suburban San Gabriel Valley east of Los Angeles.

African Americans, by contrast, have been trapped in place in Los Angeles, as elsewhere in urban America. Dramatic figures that purport to show the suburbanization of Black Los Angeles primarily represent the territorial expansion of the traditional South Central ghetto into adjacent, but separately incorporated cities: for example, Lynwood on the east, Inglewood and Hawthorne on the west, and Carson on the south. When this quotient of "ghetto shift" is deducted from 1990 Census figures, what remains of Black suburbanization in Southern California is a mono-trend movement to blue-collar suburbs (principally Fontana, Rialto, and Moreno Valley) in the Inland Empire of western San Bernardino and Riverside Counties.

Certainly this is a significant phenomenon, and there are indications that the Black exodus to the Inland Empire may have accelerated since last spring's rebellion. But it must be emphasized that "Black flight" has been restricted to a handful of outer suburbs with dramatic deficits in their jobs-to-housing ratios. Compared not only to blue-collar whites, but especially to Chicanos, there has been, at best, only a desultory diffusion of Blacks within Southern California's wider housing and job markets.

The color bar, in other words, remains alive and well in Southern California's growth-pole exurbs like Simi Valley, Santa Clarita, Temecula, Irvine, Laguna Hills, and Rancho Bernardo. Between 1972 and 1989 Los Angeles's suburban rim gained more than two million new jobs while its Black population languished at less than 2 percent. (Blacks are 11 percent of Los Angeles County's population as a whole.)[31] Whatever the precise combination of class and racial discrimination involved, African Americans have been systematically excluded from the edge-city job boom. Conversely, they have become as a result more dependent than ever on center-city public employment, the cornerstone of the Black communal economy.

Table 4

Largest Ethnicities (1990) and Mayors (1992) [29]

		Plurality		Next Largest		Mayor
1.	New York	white	38.4	Black	29.8	Black
2.	Los Angeles	Latino	39.3	white	37.2	Black
3.	Chicago	Black	40.8	white	36.3	white
4.	D.C. area	Black	62.2	white	33.3	2 x Black
5.	Bay Area	white	42.9	Asian	22.8	2 x white, 1 Black
6.	Philadelphia	white	51.3	Black	39.9	Black
7.	Detroit	Black	75.7	white	20.3	Black
8.	Boston	white	58.0	Black	25.6	white
9.	Dallas	white	49.8	Black	27.2	2 x white
10.	Houston	white	39.9	Black	28.1	white

Note: D.C. includes Baltimore; Bay, San Francisco, Oakland, and San Jose; and Dallas, Fort Worth.

With minimal nuance or exception, this pattern of spatial apartheid (often, mistakenly, called "spatial mismatch") has been recapitulated in every metropolitan area in the United States during the 1980s. In the Bay Area, for instance, San Francisco's financial industry has ignored Black-governed Oakland's desperate efforts to attract white-collar employment, preferring instead to export tens of thousands of back-office jobs over the Berkeley Hills to the white edge cities of Contra Costa County. Greater Atlanta and Detroit, meanwhile, vie with each other for the distinction of being the most perfect "urban donut": Black in the deindustrialized center, lily-white on the job-rich rim.

The Suburban Majority

The age of the edge city, then, is the culmination of a racial sorting-out process. This has had two epochal political consequences. First, the semantic identity of race and urbanity within US political discourse is now virtually complete. Just as during the ethno-religious *kulturkampf* of the early twentieth century when "big city" was a euphemism for the "teeming Papist masses," so today it equates with a Black-Latino "underclass." Contemporary debates about the city—as about drugs and crime—are invariably really about race. Conversely, as Jesse Jackson always underlines, the fate of the urban public sector has become central to the survival agenda of Black America.

Second, 1992 was the watershed year when suburban voters and their representatives became the political majority in the United States (they had already been a majority of the white electorate since at least 1980). The politics of suburbia, notes Fred Siegel in a recent article in *Dissent,* are "not so much Republican as anti-urban ... [and] even more anti-Black than anti-urban."[32] Racial polarization, of course, has been going on for generations across the white-picket-fence border between the suburb and the city. But the dramatic suburbanization of economic growth over the last decade, and the increasing prevalence of strictly rim-to-rim commutes between job and home, have given these "bourgeois utopias" (as Fishman calls them) unprecedented political autonomy from the crisis of the core cities.[33] And vice-versa, "the ascendance of the suburban electorate to virtual majority status, [has] empower[ed] [them] ... to address basic social service needs ... through local suburban government and through locally generated revenues, and to further sever already weak ties to increasingly Black urban constituencies."[34] This, in turn, has greatly simplified the geography of partisan politics: Republican Party affiliation is now a direct function of distance away from urban centers.[35]

Core cities, for their part, have helplessly watched the reapportionment of their once-decisive political clout in national politics. Since Jimmy Carter, their representation in Congress has declined from one-quarter to one-fifth of House seats. In presidential politics, the highwater mark of big-city power was undoubtedly the election of 1960 when the Daley machine resurrected the dead to provide John Kennedy's winning margin over Richard Nixon. In those days Chicago mobilized 40 percent of Illinois votes; today it turns out only 25 percent. Likewise, the capture of decisive majorities in the twenty largest cities was once tantamount to owning the White House. But, as Carter, Mondale, and Dukakis each demonstrated, it was possible to sweep the urban cores and be crushed in the suburbs by the defection of so-called "Reagan Democrats," a stratum largely consisting of blue-collar and lower-middle-class white refugees from the cities.

The Clinton campaign was, of course, the culmination of a decade-long battle by suburban and Southern Democrats to wrest control of the Democratic Party away from labor unions, big city mayors, and civil rights groups. In the aftermath of the Mondale debacle in 1984, Clinton joined with Bruce Babbitt, Charles Robb, and other sunbelt governors to establish the Democratic Leadership Council (DLC) as a competing power center to the Democratic National Committee (DNC). The DLC's principal goals were to marginalize Jesse Jackson (the champion of the urban poor), rollback intra-party reforms, take control of the DNC, and nominate a candidate who could challenge Reaganism in its own crabgrass heartland.

Clinton's genius has been his skill at pandering to the DLC's stereotype of the Reagan Democrat. From his electrocution of a brain-damaged Black convict on the eve of the New York primary to his sudden speech impediment faced with the word "city," Clinton was programmed to reassure white suburbanites at every opportunity that he was not soft on crime, friendly with the underclass, or tolerant of big-city welfare expenditures. This implicitly anti-Black, anti-urban theme music was played in continuous refrain to his promises to reinvest in middle-class economic and educational mobility while continuing to defend George Bush's New World Order.

Despite the clarity of the essential Clinton message, his victory has spawned strange hopes and misguided interpretations. Like the pathetic paupers in Port-au-Prince who were reported to have organized a cargo cult around Clinton in the mistaken belief that he would open America's golden door to the Haitian masses, a jubilant gaggle of rustbelt mayors, community developers, and members of the Congressional Black Caucus have cheered the Bush defeat as the dawning of a New Deal. Some, perhaps, have become intoxicated with Arthur Schlesinger's oft-repeated assurance that the great wheel of American politics was again turning, inexorably, from right to left. Others may have been hallucinating on the even stranger idea, sprouted in DSA circles, that Clinton is actually a "stealth social-democrat" committed to a huge Keynesian expansion of education and health-care entitlements.[36]

In any event, there is virtually no evidence that President Clinton is the "Manchurian candidate" of a largely invisible American social democracy. Nor, for that matter, is there evidence that the 1992 election has actually moved the country back toward anything resembling pro-city New Deal liberalism. As MIT's Walter Dean Burnham has frequently pointed out, Schlesinger's mythical wheel of US politics no longer moves at all, but is stuck, semi-permanently, in a center-right position that corresponds to our current "post-partisan," suburbanized political system.[37]

Most important, there is no obvious reason why a campaign carefully designed to de-emphasize the cities should deliver a president suddenly fixed on their needs. In the aftermath of the Los Angeles rebellion, neither *Business Week* nor the *National Journal* could locate a significant dividing line between the Clinton and Bush approaches to urban policy.[38] Senior Clinton advisor Will Marshall, president of the DLC's Progressive Policy Institute, conceded that there was "very little difference over the central idea," while his Republican counterpart, Heritage Foundation domestic policy director Stuart Butler, saw "no conceptional difference between Clinton and Bush." On the rare occasions when either candidate dealt with urban issues, each used the same debased rhetoric of "empowerment" to espouse enterprise zones, school vouchers, privatization of public housing, and workfare not welfare.[39]

Nor, in the months since the election, have flowers suddenly blossomed in Cabrini-Green or the South Bronx. Attempting to present New York's woes before the Clinton transition team, Congressman Charles Rangel of Harlem complained that "they listen and they say nothing"—not surprisingly since the transition team's "bible," the Progressive Policy Institute's *Mandate for Change*, omits cities altogether from its fourteen topic chapter headings.[40] For his part, new Housing and Urban Development Secretary Henry Cisneros may have aroused great expectations among the cargo-cult crowd, but so far he has only promised to work from existing funds, refloat federal enterprise zone legislation, and preserve "Weed and Seed," which he described as "an important program."[41] Congress, meanwhile, gives little sign that it will challenge a Bush-Clinton continuum of urban neglect. A postelection Gallup poll of Democratic members of the new House revealed that aid to cities was ranked a miserable thirteenth out of eighteen issues (housing was ranked last).[42]

Let us suppose for the sake of argument, however, that escalating urban unrest, perhaps sparked by another Los Angeles riot, forces Clinton—as it did an equally reluctant Richard Nixon in 1969—to attempt to address some of the underlying urban contradictions. Could he actually mobilize the political and budgetary resources to save the cities? It is hard to see how. The forecast for any resumption of urban reformism is quite bleak as long as federal discretionary spending is weighted in chains by the deficit, Perot voters, and a white-collar recession.

Lemmings in Polyester

The chief legacy of the Reagan-Bush era, of course, is the $2 trillion cost of "winning" the Cold War. A generation's worth of public investment—probably the fiscal equivalent of several New Deals—was transformed into stealth bombers and nuclear armadas, financed by the most regressive means imaginable (huge tax cuts for the rich and rampant offshore borrowing). Bipartisan politics then added another half-trillion dollars to bailout wealthy investors from the savings-and-loan debacle. Spent on cities and human resources, these immense sums would have remade urban America into the Land of Oz instead of the wasteland it has become.

The social burden of servicing this deficit may be measured by comparison to the annual combined budgets of America's fifty largest cities. In 1980 the interest payments on the federal debt were twice as large as the aggregate big-city budgets: today they are six times larger. Alternately, the $300 billion 1990 deficit was simply equal to the annual interest cost on a federal debt soaring toward $5 trillion.[43]

Keynesians, pointing to much higher per capita debts elsewhere in the OECD world, may argue that it is ridiculous to allow the deficit to become an absolute fetter on national growth or urban reinvestment. But the deficit is not merely a figure on a balance sheet, it is also the major strategic weapon of the right. It is the Archimedean lever that the conservative coalition in Congress has used to dismantle the citizenship entitlements of the urban and rural poor, and it is the structural guarantee, via Gramm-Rudman and the 1990 budget treaty, that the Reagan Revolution is irreversible. As Guy Mollineux has eloquently argued, the deficit-warriors' call for "shared sacrifice" is a "truly Orwellian inversion of political language" where spending on cities is "pandering to special interests" and "rough choices" means more austerity for the poor.[44]

The most Orwellian voice in American politics, of course, speaks in a just-plain-folks East Texas drawl. Clinton may continue to snub Ross Perot, but the diminutive billionaire's shadow (magnified by the international bond markets) looms enormous over the new administration. Perot's achievement has been to create an unprecedented populist crusade, 19 million strong, around the thesis that the deficit, not the decline of the cities or the plight of the poor, is the epochal issue facing ordinary Americans. Like lemmings in polyester, millions of his followers vow to walk off the cliffs of a major depression in order to balance the federal checkbook.

Perot is also the gatekeeper to any political realignment. Clinton won the election because Perot stole Bush's vote in the edge cities, retirement communities, and high-tech belts. (See Table 5.) By himself Clinton got a 3 percent smaller share of the popular vote than even Dukakis in 1988. The strategic focus of his administration, therefore, will be winning over the Perot voters in the suburbs, who, surveys have shown, overwhelmingly favor tax cuts and less government spending, especially on the urban poor.[45] Not surprisingly, the Clinton cabinet is top-heavy with deficit hawks and admirers of Reagan's New Federalism. In particular, the combination of Leon Panetta ("time to make sacrifices ... cut, not raise public spending" etc.) and Alice Rivlin in the Office of Management and Budget is the moral equivalent of having Perot himself in the cabinet.[46]

Finally, the hope that Clinton will shower the cities with some of his proposed $220 billion investment budget (infrastructure, technology, and education) is perhaps the cruelest mirage of all. As much a subsidy to huge Wall Street municipal bond merchants like Goldman, Sachs and Company—whose chairman, Robert Rubin, is now Clinton's "economic security" chief—as to localities, the fast shrinking investment budget is primarily targeted on costly rail, optical-fiber, and Interstate highway projects that will benefit Perot voters in the suburbs and the traditional highway lobby of state officials, contractors, and white-dominated construction trades.

Ironically, this is the one arena of domestic expenditure—presumably because it is most dear to the hearts of gridlocked suburbanites—that least needs additional federal investment. Reagan and Bush may have decimated urban housing and job-training funds, but they wisely left the freeways alone. The 1983 Highway Act is still generating major road construction, while the 1991 Intermodal Surface Transportation Efficiency Act has allocated $155 billion over the next six years for rail transit, including Los Angeles's pharaonic subway system.[48]

Table 5

The Perot Factor in the "Edge Cities"[47]
(national Perot vote = 15%)

County	Bush 1988	Bush 1992	Decline	Perot 1988
Orange (Cal.)	68%	44%	−24%	24%
San Bernardino (Cal.)	60%	37%	−23%	23%
Santa Clara (Cal.) (Silicon Valley)	47%	28%	−19%	22%
San Diego (Cal.)	60%	35%	−25%	26%
Clark (Las Vegas)	62%	33%	−23%	25%
Orange (Orlando)	68%	46%	−22%	19%
Gwinett (Georgia)	76%	54%	−23%	16%
Du Page (Illinois)	69%	48%	−21%	21%
Fairfax (Virginia)	61%	44%	−17%	14%

If the big city mayors and the Congressional Black Caucus attempt to divert any of this investment toward urgent inner-city needs (for example, schools, affordable housing, environmental cleanup, and public space), they will face unprecedented battles with the suburbs. The current downturn is the worst white-collar recession since the 1930s. Hundreds of thousands of middle-managers, computer programmers, bookkeepers, and salespeople have tumbled out of their safe nests in bank skyscrapers and corporate front offices. They have been joined by regiments of redundant defense workers, aerospace engineers, and skilled construction workers. For the first time, the new edge cities are feeling some of the pain of the older cities, and the competition for resources has become exceptionally intense.

No one better appreciates the internal logic of these redistributional struggles in the shadow of deficit than Richard Darman, Bush's outgoing budget director. At a press conference called to present Clinton with the unwonted gift of a huge prospective increase in the deficit, an almost gleeful Darman reminded the new administration that it was the prisoner of the Reagan-Bush past. It was impossible for Clinton, he emphasized, to simultaneously deal with the deficit and implement his investment program without taking the politically suicidal course of taxing the middle class or reducing their Social Security and Medicare entitlements. Thus the Clinton campaign promises were so much rubbish and the only electorally safe option for the Democrats, as for the Republicans, was to continue to blast away at the big cities and urban poor.

> The political system has accepted the reforms that affect the poor ... but it has not accepted the reforms that affect the rich. Nor, more importantly, has it accepted reforms, by and large, that affect the broad middle and that is half the budget. Where you've got sixty million adults who are beneficiaries of broad middle-class entitlement programs, that's a lot of voters.[49]

Poor Law States

In the dark days of the early Reagan Administration, many big cities looked toward the new light they thought they saw shining from their statehouses. The federal retrenchment in domestic policy (which, as we have seen, actually began with Carter in 1978) opened the way for state governments to assume a more dynamic role in urban finance and local economic development. The California legislature, for example, organized a major fiscal rescue for cities, counties, and school districts threatened by the combined disasters of Proposition 13 (the Jarvis-Gann tax amendment) and federal cutbacks. Michigan and Massachusetts compensated for the absence of a national industrial strategy by enrolling their stricken urban areas into ambitious state-level development programs, while other states assumed higher-profile roles in funding local education.[50] Aggregate state expenditures, only 60 percent of the federal budget during Lyndon Johnson's presidency, drew almost equal to the Bush budget in 1990: $1 trillion versus $1.1 trillion.[51]

By the late 1980s the big Washington and New York policy institutes, from Brookings to the Committee for Economic Development, were abuzz with talk of this extraordinary "state renaissance."[52] Conservative advocates of states' rights complained bitterly about the powers that had been left to liberals in statehouses, while progressives speculated optimistically about the future of "Keynesianism in one (two, three, many?) state[s]." But the illusion that the worst of Reaganism could be halted at the state line, or that the states could replace Washington as the saviors of the city, was sustained only by the relative fiscal autonomy of the richer states amid the "bicoastal" boom of the mid 1980s. (The "new economic role" of the poor states had been confined in most instances, like Clinton's Arkansas, to becoming better salesmen of tax advantages and cheap nonunion labor.)

The onset of a new recession in 1990 kicked out the props from underneath this over-hyped "renaissance" of the states and exposed the real underlying damage done by more than a decade of federal cuts. With state-funded Medicaid and unemployment costs soaring, the Bush Administration poached on states' fiscal capacity by raising federal excise taxes on gas, tobacco, and alcohol. Other traditional state tax resources were put off bounds by Proposition 13 and its progeny across the country. Meanwhile, the War on Drugs was becoming literally a "domestic Vietnam" as out-of-control prison budgets sucked up larger portions of states' operating funds.[53] With no one left to bail them out, the statehouses now followed the city halls into the fiscal black hole excavated by the Republicans in Washington.

The result—according to the principle of "suburbs first in the lifeboat, cities and poor last"—has been the dramatic reduction, even elimination, of cash and medical assistance to the urban poor. The welfare systems of an entire stratum of traditionally progressive, industrial states whose names still resound like a roll-call vote for FDR—Illinois, Michigan, Massachusetts, Maryland, and Minnesota (as well as Ohio and Oregon)—have been leveled downward to the meanness of Mississippi and Arkansas. Nominally Democratic legislatures have radically reduced medical coverage, slashed cash payments, and tightened the eligibility and duration of benefits.

In the most extreme case, Michigan general assistance has been abolished, and unemployed single adults and childless couples are left without any income or medical safety net whatsoever. Maryland has also purged its relief rolls of everyone but the disabled and the very elderly, while Ohio, Minnesota, and Illinois have time-limited assistance payment regardless of hardship or economic climate. Massachusetts, meanwhile, has reduced eligibility for the disabled, and Oregon has excluded hospitalization.[54] Similar measures are close to passage in New York, New Jersey, and, as we shall see, in California. One study suggests that at least forty states are currently weighing a reduction in welfare benefits to children.[55] Like serial murders, the example of one state cutting benefits has spurred others to emulate the same foul deed.

Meanwhile, the current debate in most statehouses is fully up-to-date with the 1830s Poor Law Reform and the Reverend Malthus. In the face of bipartisan abuse against the "welfare underclass," advocates of the poor have tried to point out the relentless attrition of income maintenance standards. Both the minimum wage and the median state welfare (AFDC) benefit have lost 40 percent of their real value (in inflation-adjusted dollars) since 1970, while the median welfare benefit for a family of three now barely equals one-third of the poverty threshold.[56] And, in contrast to the demonology of a welfare system rife with cheats and layabouts, more than 28 percent of the population living below the poverty line receives no public aid at all.[57]

But such statistics make little headway these days in Lansing, Columbus, or Sacramento. In an important article, John Begala and Carol Bethel argue that the current legislative attack on the poor is driven by the same force alluded to earlier: the competitive pressure of anxious middle-income voters, including displaced workers in manufacturing and laid-off second-income wage-earners.[58] The struggle is not about the moral economy of welfare, but about the political precedence of suburbs and the entitlements of the middle class. In Michigan, for example, this has taken the form of a cruel war of Detroit's white suburbs against the unemployed population in the Black core city. In a typical exchange, one suburban legislator suggested that if Detroit's jobless were unhappy with the abolition of general assistance they could "move to sunny California, to stylish New York; or, if they like winter sports, to Minnesota."[59]

Although "the relationship between state and local governments has deteriorated to maybe the lowest level anyone can remember," statehouses have been able to legislate this new, Dickensian immiseration without facing mass revolt in the cities.[60] Shrewd governors and legislative majority leaders have learned to cut Faustian bargains with city, and especially county, authorities. In exchange for acceding to state welfare cuts and tax-poaching, for example, the localities are legislatively released from their mandates to provide certain essential services like relief and medical care for the indigent. The "hit"—as legislators and other hired assassins like to say—is passed directly on to the street, and inner-city property owners are conscripted to common cause with the suburbs.

A Bonfire of Rights

The ultimate casualty of this current wave of state-legislative attacks on the urban poor may be the tenability of belief in common citizenship itself. Whether in the name of the budget or the War on Drugs, social and economic rights that were won through generations of hard-fought struggle are now routinely abridged or even abolished. Not since the end of Reconstruction have so many Americans faced such a drastic devaluation of their citizenship as do urban communities of color today. And no recent sequence of government actions has set this bonfire of rights in more stark relief than the events in Sacramento since the Rodney King verdict.

While the cinders of South Central were still warm, Art Torres, the liberal state senator from East Los Angeles, submitted two bills to the California legislature for urgent consideration. One bill simply funded emergency relief for Los Angeles with the same temporary sales tax increase that had been used in 1989 to aid the Bay Area after the Loma Prieta earthquake. The other bill took a small step toward acknowledging the existence of the police brutality that had sparked the riots by establishing a standardized citizen complaint process and a statewide databank. Neither bill was envisioned as controversial.

To Torres's consternation, however, both bills were quickly incinerated in a suburban anti-L.A. backlash, orchestrated by the powerful law-enforcement lobby. Equal treatment for riot and earthquake victims was dismissed out of hand by Senate Republican leader Ken Maddy (Fresno), who snidely observed to Torres that "there was not the same kind of outpouring of sentiment for Los Angeles."[61] Meanwhile, Torres's modest proposal for state invigilation of police abuse—a barometer of the Capitol's attitude toward Rodney King's near-lynching—was killed and replaced by four criminal bills authored by the Senate Majority leader, David Roberti (Hollywood). Roberti, who over the course of the summer would emerge as Republican Governor Pete Wilson's fifth column in the Democratic Party, proposed an alternative message to the inner city. He wanted to ban probation for convicted looters, increase the sentence for fire-bombing from seven to nine years, extend the deadlines for arraignment, and offer state rewards for the arrest of looters. His bills passed handily.

While the Senate was venting its spleen, the entire legislature was embroiled in an epic debate on the future of California that came to overshadow the riots. Caught between Proposition 13 and the worst recession since 1938, the state budget was $6 billion in the red with the prospect of even larger deficits in the future. The Democrats, under the leadership of Roberti and House Speaker Willie Brown (a Black corporate lawyer from San Francisco), initially proposed to increase taxes for millionaires, close some egregious loopholes, and roll over the rest of the deficit until the economy recovered.

Republican Governor Wilson, on the other hand, blamed the recession on labor and the poor and wanted to make deep, permanent cuts in family assistance, medical care, and higher education. In exchange for releasing county governments from their health and welfare mandates, he also proposed to end Sacramento's Proposition 13 bailout of local government. In a state whose postwar prosperity had been generated by its traditionally high levels of investment in education and public services, Wilson advocated a draconian, Michigan-style retrenchment.

By the beginning of the summer, the Democrats had capitulated almost totally. After a behind-the-scenes blitz by the Chamber of Commerce and the oil and real-estate lobbies, the sons of the people led by Willie Brown abandoned their feeble attempts to raise taxes on the rich and close corporate tax loopholes. Ignoring a report on the state's children that showed youth unemployment and homicide rates soaring in tandem, the House Democrats instead unveiled their own budget plan, which the *Los Angeles Times* described as offering "even deeper cuts in state services than Wilson proposed."[62] A prominent Democratic assemblyman told his Republican colleagues, "Why don't you just declare victory and go home?"[63]

In retrospect, it is hard to say what was more astonishing: the Democrats' phony war and abject surrender, or Wilson's subsequent refusal "to declare victory and go home." Ostensibly the budget crisis dragged on through the summer, forcing the state to pay its bills with IOUs, because the governor dogmatically continued to insist on deep education cuts, which the Democrats, heavily financed by teachers' organizations, could not afford to accept. Speaker Brown, as local officials from Los Angeles seeking riot relief discovered to their horror, wanted to sacrifice aid to the cities instead.

In fact, the two sides were playing different games for unequal stakes. The Democrats, dominated by a neoliberal, lobbyist-fed majority, simply wanted to deflect as much pain as possible away from their core suburban constituencies, whose major concerns were taxes, transportation, and education, not welfare or urban development. They consoled their consciences by proposing "trippers" that would restore cut programs and reduce the suffering of the poor once the recession ended and pork-barrel days returned.

The governor, on the other hand, was playing hardball—that is to say, strategic politics—against the Democrats' softball. Under siege from the aptly named "cavemen" of his own right wing, Wilson ("no more mister nice guy") had decided to abandon bipartisan compromise for ideological confrontation. Like Reagan in 1980 he aimed to permanently shrink the welfare role of the state and fragment the traditional Democratic coalition. He was intransigent about the budget because he was determined to force the Democrats to betray their education allies and concede the structural permanence of the cutbacks. Moreover, when Speaker Brown evoked the transience of the recession, the governor talked about the inevitability of "demographics."

In the summer-long budget battle (which finally ended with the brunt of the cuts, as Speaker Brown wanted, shifted from education to local government), Wilson repeatedly quoted from two official bibles. One was the report of the Governor's Commission on Competitiveness, chaired by Peter Ueberroth and released on the eve of the riots, that blamed California's economic malaise and the "flight of capital" on over-regulation and excessive taxing of business. The other was a 1991 Department of Finance report, *California's Growing Taxpayer Squeeze*, which warned that immigrants and welfare mothers were multiplying more rapidly than taxpayers. Wilson's intellectual originality was to synthesize the two reports into a single demonic vision of white middle-class breadwinners and entrepreneurs under siege by armies of welfare leeches and illegal immigrants, aided and abetted by public-sector unions and Sacramento Democrats. The Los Angeles riots made the images more vivid and colored in the faces of the enemy.

Actually, California's ongoing "structural deficit" is nothing more than the bill finally come due for Proposition 13, which in 1978 cut back and froze the property tax rolls. According to figures in a recent study by the Advisory Commission on Intergovernmental Relations (ACIR), the deficit would disappear if California (29th in national "tax effort") simply taxed property owners at the average national rate. With 6 percent more fiscal capacity per capita, Pete Wilson's golden state demands 38 percent less taxation per capita than Mario Cuomo's New York.[64]

Wilson repeatedly evoked this "demographic" scenario, with its racist and nativist undertones, to justify his radical surgery on the state's public sector. Regulations, entitlements, taxes, and public employment had to shrink permanently, while the parasitic welfare class had to be driven off the dole (the governor drafted a ballot initiative to slash payments and caseloads). In fact, Wilson was building—with the Democratic majority's complicity—an economic atomic bomb to drop on the state's poorer communities, above all the barrios and ghettos of Los Angeles, Oakland, and the Central Valley cities.

Although neither the governor nor the Democrats dwelt on the fact, the deficit bomb was primarily designed to hurt children—who, after all, comprise fully two-thirds of the welfare underclass and half of the immigrants. And like an actual nuclear device, it will continue to inflict damage on them for generations, since it entails a permanent reduction of education, health, and welfare entitlements. The kids of the new immigrants and people of color (now a majority in the state's primary schools) will not be allowed the same opportunities and privileges enjoyed by previous generations of Californians. Citizenship is being downsized.

In the course of the most ignoble summer in modern California history, when budget deficits were used to justify every manner of inhumanity, one veteran legislator confessed his despair to a reporter: "Is state government turning its back on the poor? Yes. Is the Democratic Party turning its back on the poor? Yes. I don't like it, but the fact is that most people up here don't share my values. If poor people starve on the street, they don't care. Any budget we pass is going to wreak havoc on the poor."[65]

1992

Notes

1. Quoted in Burt Solomon. "Bush and Clinton's Urban Fervor...," *National Journal*, 16 May 1992, p. 1196
2. Quoted in the *Los Angeles Times*, 17 May 1992.
3. Cf. Rochelle Stanfield. "Battle Zones," *National Journal*, 6 June 1992, p. 1349; and Kirk Victor, "Fiscal Fire Sale," ibid., 27 June 1992, p. 1514.
4. See Jack Germond and Jules Witcover, "Clinton's at Risk After Riots in L.A.," *National Journal*, 9 May 1992, p. 1137.
5. *Los Angeles Times*, 4 and 5 November 1992.
6. Ibid., 2 June 1992.
7. See criticisms of FEMA by local and state officials in *Los Angeles Times*, 11 January 1993.

8. See Neal Pierce, "A Riot–Ravaged City Is Still on Hold," *National Journal*, 10 October 1992, p. 2325.

9. The local press largely missed this extraordinary story. See Victor, op. cit., pp. 1512–16.

10. It is important to contrast the different urban strategies of the Nixon-Ford and the Reagan-Bush regimes. Nixon's "New Federalism" did not so much seek to dismantle the Great Society as to reallocate its benefits to the "new Republican majority" in Sunbelt cities and suburbs. He and Ford expanded urban grants-in-aid but rerouted them away from the Democratic big-city heartland in the Northeast toward the urban South and West. Nixon also ended the "maximum feasible participation" era of the War on Poverty and returned administrative control over federal grants to traditional city hall elites. Thus in urban policy as in foreign policy, the Reagan revolution was as much oriented against the legacies of Nixon and Ford as against those of Johnson and Kennedy.

11. American Enterprise Institute budget expert Allen Schick quoted in the *Wall Street Journal*, 4 February 1985, p. 4.

12. For an extended discussion, see my essay, "The Lesser Evil? The Left, the Democrats and 1984" in *Prisoners of the American Dream*, London 1986, pp. 267–70.

13. *President's Commission for a National Agenda for the Eighties, Urban America in the Eighties: Perspectives and Prospects*, and *A National Agenda for the Eighties*, Washington, D.C. 1980; quoted in Timothy Barnekov, Robin Boyle, and Daniel Rich, *Privatism and Urban Policy in Britain and the United States*, Oxford 1989, pp. 101–5.

14. Quoted in ibid., pp. 105–7. Barnekov, Byle, and Rich's *Privatism and Urban Policy* is an indispensable analysis of neoconservative urban policy.

15. See Timothy Conlan, *New Federalism: Intergovernmental Reform from Nixon to Reagan*, Washington, D.C. 1988, p. 233. Previously Ted Kennedy had in 1982 joined hands with Dan Quayle to kill tens of thousands of local public sector jobs supported by the Comprehensive Employment Training Act (CETA) of 1973; see pp. 175–76.

16. For Democratic indifference to the seven-point plan proposed by the mayors in January 1992, see Rochelle Stanfield, "Cast Adrift, Many Cities Are Sinking," *National Journal*, 9 May 1992, p. 1122.

17. Demetrios Caraley, "Washington Abandons the City," *Political Science Quarterly* 107, no. 1 (1992), pp. 8 and 11. I have amended Caraley's estimate of the total percentage reduction in federal urban assistance with the figures from "The Economic Crisis of Urban America," *Business Week*, 18 May 1992.

18. Ester Fuchs, *Mayors and Money: Fiscal Policy in New York and Chicago*, Chicago 1992, p. 288.

19. US Bureau of Census, *City Government Finances*, 1977–78 and 1984–85; and Preston Niblack and Peter Stan, "Financing Public Services in L.A.," in James Steinberg, David Lyon, and Mary Vaiana, *Urban America: Policy Choices for Los Angeles and the Nation*, Santa Monica, Calif. 1992, p. 267.

20. See James Fossett, "The Politics of Dependence: Federal Aid to Big Cities," in Lawrence Brown, James Fossett, and Kenneth Palmer, eds., *The Changing Politics of Federal*

Lawrence Brown, James Fossett, and Kenneth Palmer, eds., *The Changing Politics of Federal Grants*, Washington, D.C. 1984, pp. 121–24 and 48.

21. Cf. *Washington Post National Weekly Edition*, 11–17 May 1992; and Caraley, ibid., p. 9.

22. See James Hosek and Jacob Klerman, "Military Service: A Closing Door of Opportunity for Youth," in Steinberg et al., ibid., pp. 165–67.

23. *Los Angeles Times*, 4 January 1993.

24. See Joan Petersilia, "Crime and Punishment in California," in Steinberg et al., ibid.

25. The shortfall of employment in most core cities is further multiplied by the large percentage of high-wage and salaried jobs held by suburban commuters.

26. Cf. Caraley, pp. 5–6, and Fred Siegel, "Waiting for Lefty," *Dissent* (Spring 1991), p. 177.

27. National League of Cities 1992; and Caraley, ibid.

28. See Joel Garreau, *Edge City: Life on the New Frontier*, New York 1991, p. 3.

29. US Bureau of the Census, Population, 1970 and 1990.

30. For the ethnic transformation of Los Angeles's industrial heartland, see my chapter "The Empty Quarter," in David Reid, ed., *Sex, Death and God in L.A.*, New York 1991.

31. Figures calculated from California's Economic Development Department, *Statistical Abstracts*; and the 1990 Census. The "edge-city rim," as I define it, includes Ventura and Orange Counties, as well as northern Los Angeles County (basically Santa Clarita and the Antelope Valley), the I-5 corridor of western San Bernardino and Riverside Counties, and suburban San Diego County. It excludes part of the suburban Inland Empire.

32. Siegel, ibid., pp. 177–79.

33. Robert Fishman, *Bourgeois Utopias: The Rise and Fall of Suburbia*, New York 1987.

34. Thomas Edsall with Mary Edsall, *Chain Reaction: The Impact of Race, Rights, and Taxes on American Politics*, New York 1991, p. 217.

35. See James Barnes, "Tainted Triumph," *National Journal*, 7 November 1992, p. 2541.

36. For the argument that Clinton "has overturned neo-liberal politics," brought an end to the conservative era, and is really a "stealth social-democrat," see Harold Meyerson, "The Election: Impending Realignment?" *Dissent* (Fall 1992), pp. 421–24.

37. Walter Dean Burnham, "Critical Realignment: Dead or Alive?" in Byron Shafter, ed., *The End of Realignment?* Madison, Wis. 1991, pp. 125–27.

38. "Clinton's approach sounds even more Republican ... enterprise zones, government-financed organizations to lend money and give advice to budding entrepreneurs—the sort of public-private cooperation the Bush Administration will be pushing as well" (*Business Week*, 18 May 1992). "[This] increasingly odd campaign ... with its spectacle of Bush and Clinton ... saying much the same things about the agony of the inner cities" (*National Journal*, 16 May 1992, p. 1996 passim).

39. Quoted in ibid., p. 1197.

40. And when Clintonians do talk about cities, they never acknowledge the special circumstances of Blacks or Latinos. Andrew Hacker points out, for example, that Clinton and Gore's own book, *Putting People First*, "hardly ever mentions race, even obliquely. A chapter entitled 'cities' neither uses the term 'inner-city' nor mentions residential or school

segregation" ("The Blacks and Clinton," *New York Review of Books*, 28 January 1993, p. 14).

41. Quoted in the *Los Angeles Times*, 25 January 1993.

42. *Washington Post National Weekly Edition*, 14–20 December 1992.

43. Caraley, ibid., 25.

44. See his opinion piece, *Los Angeles Times*, 5 November 1992.

45. Cf. Rhodes Cook, "Republicans Suffer a Knockout That Leaves Clinton Standing," *National Journal*, 12 December 1992, p. 3810; and James Barnes, op. cit., p. 2541.

46. In her book *Reviving the American Dream*, Rivlin resurrects Reagan's federalism initiative of 1982 that would have returned grant programs to the states and terminated the federal role in welfare in exchange for the nationalization of health-care financing for the poor. Other advisors at the new White House are equally keen on continuing the Reagan revolution, including David Osborne, author of *Reinventing Government* (1992), whose catchphrase "entrepreneurial government" regularly shows up in Clinton speeches. See Rochelle Stanfield, "Rethinking Federalism," *National Journal*, 3 October 1992, pp. 2255–57.

47. Figures from *Congressional Quarterly*, 12 December 1992, pp. 3815–20.

48. Kirk Victor, "A Capital Idea?" *National Journal*, 28 November 1992.

49. Richard Darman quoted in ibid., 7 January 1993.

50. For a comparison of seven states sponsored by the corporatist Committee for Economic Development, see R. Scott Fosler, ed., *The New Economic Role of American States*, New York 1988.

51. *National Journal*, 3 October 1992, p. 2256.

52. The phrase used by Brookings' Timothy Conlan; see *New Federalism*, p. 228.

53. For a succinct overview of states' growing structural deficits, see Penelope Lemov, "The Decade of Red Ink," *Governing* (August 1992).

54. Cf. J. Michael Kennedy, "Cutbacks Push Poor to the Edge," *Governing* (April 1992); and John Begala and Carol Bethel, "A Transformation within the Welfare State," *The Journal of State Government*, 1992.

55. From a 1991 study by the Center on Budget and Policy Priorities (Washington, D.C.) and Center for the Study of the States (Albany, N.Y.).

56. Cf. House Ways and Means Committee, *Green Book*, Washington, D.C. 1991; and Center on Law and Poverty, cited in *Los Angeles Times*, 18 June 1992.

57. Kennedy, ibid.

58. Begala and Bethel, ibid.

59. Kennedy, ibid.

60. State representative, quoted in Rochelle Stanfield, "Rethinking Federalism," op. cit., p. 257.

61. Quoted in *Los Angeles Times*, 15 June 1992.

62. Ibid., 2 July 1992.

63. Children Now study quoted in ibid., 25 June 1992; and Representative Phil Isenberg (D–Sacramento) quoted in Linda Paulson and Richard Zeiger, "Blundering Toward a

Budget," *California Journal* (September 1992), p. 426.

64. Raising California's property tax effort to the national standard would generate approximately $170 per capita in additional taxes. Multiplying this by 30 million disposes of most of the 1992–93 deficit. See ACI, *State Fiscal Capacity and Effort–1988*, Washington, D.C. August 1990, pp. 75, 103, 132, and 133.

65. John Vasconcellos (D–Santa Barbara), quoted in *L.A. Weekly,* 10 July 1992.

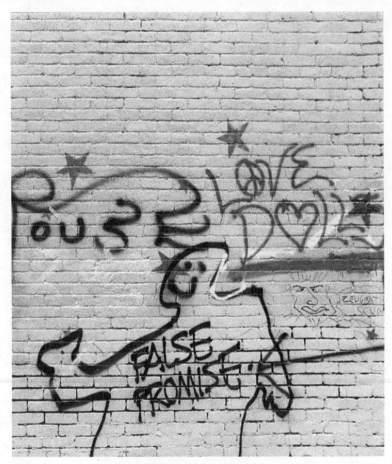

Epitaph for Compton

14

Fear and Loathing in Compton

"This sad-ass town has been dissed to death. Everybody wants to stab Compton in the back." Ricky Miller and I are cruising Compton Boulevard. The full moon—slung low over some elderly palm trees—bathes the tired 1950s landscape of America's most notorious city in a cold white light. The neighborhoods to the north of the boulevard are largely Crips territory; those to the south, predominantly Piru (Bloods) land. There has been an unsteady gang truce since 1992. It doesn't look much like the *Boyz N the Hood* phantasmagoria depicted in MTV videos. It is a hot summer evening, and the quiet streets are mostly empty.

Ricky, thirty-four, is a local rap producer and community activist. He was, as he puts it, "raised up heart and soul in Compton." He wears a jagged row of stitches above his left eye, where a cop recently walloped him with a steel flashlight. "Compton Boulevard used to run clean to the beach, until we became known as the gangsta-rap capital of the universe. Now all these little shit-scared cities west of here [Gardena, Hawthorne, Lawndale, and Redondo Beach] have changed the name to 'Marine Avenue.' Nobody can claim Compton now or give it any respect except for us. We might as well be Haiti."

The comparison with Haiti strikes a strange chord, Compton being the only American city ever invaded and occupied by the Marines. This occurred during the riots in April 1992, when forty-three stores were burned and several people died. South Central L.A., of course, was supersaturated with media, but hardly anyone bothered to drive the extra mile south of Watts to Compton. Thus few people outside the "Hub City" understand that this was the first time that the Black poor looted and burned the property of the Black bourgeoisie on any large scale (damage estimates run as high as $100 million).

Compton, the oldest Black majority–governed city west of the Mississippi, also claims the unhappy distinction of having criminalized more of its youth than any other place on the planet. Indeed, Compton Police Department statistics portray a gang membership rate greater than 100 percent. That, at least, is the peculiar quotient that arises from dividing the number of "Compton gang members" kept on the police database (10,435)—in forty Black, Latino, and Samoan sets— by the actual number of 15- to 25-year-old males recorded by the 1990 Census as living in Compton (8558). Perhaps Crips moonlight as Bloods and vice versa?

To complete a cruel litany, Compton—with a population of 91,600—also has a school district singled out by the National Education Association as "horrible," a homicide rate usually higher than Detroit's or Washington's, the highest school dropout rate in California, and Los Angeles County's highest property tax and unemployment rates. Moreover, during the past eighteen months, Compton has experienced something akin to a complete meltdown of its governing institutions.

Early last year its former chief of police pleaded no contest to stealing from a fund for undercover drug buys that was kept locked in his office. A bit later the president of Compton Community College was fired after an audit revealed that nearly $500,000 earmarked for low-income students had been improperly spent on clothing, office furniture, carpeting, and nepotistic stipends for administrators' relatives. That July the bankrupt and scandal-ridden Compton Unified School District was placed under receivership by the state for failing to maintain minimum education standards.

Then on 29 July of this year, a local television station broadcast an amateur videotape of a slightly built Latino teenager being batoned and stomped by a furious African-American Compton cop. Immediately compared to the Rodney King

King case, the beating of seventeen-year-old Felipe Soltero has become a lightning rod for accumulated Latino grievances against Compton's Black political elite. "Have the oppressed now decided to become the oppressors?" demanded a prominent Latino activist in a city that demographers estimate has a 51 percent Spanish surname majority.

Two weeks later, an even larger piece of the sky fell on Compton Boulevard. Freshman Congressman Walter Tucker III, the talented and handsome scion of a political dynasty known locally as the "Kennedys of Compton," was indicted by a federal attorney for extortion and tax evasion. While mayor in 1991–92, Tucker is alleged to have accepted $30,000 in bribes and demanded $250,000 more from the promoters of a huge solid-waste incinerator. (The project was eventually defeated by militant community opposition.)

The FBI's two-year investigation of City Hall is expected to yield indictments of other locally prominent politicians and developers. No one yet has any idea how many skeletons in the closet the feds may have discovered. According to *The Sentinel*, L.A. County's major Black newspaper, Council member Bernice Woods supposedly told a federal grand jury that "everybody in the city of Compton was a crook, but me."

The Color Blue

For Ricky Miller this "growing burden of negativity ... so much scandal and corruption" was all the more reason to be proud that he lived on the same street with Clarence "D.J. Train" Lars. Only twenty-three years old, he was a nationally known scratch deejay who had appeared with rock stars throughout Europe. Moreover, unlike some of the other bright young rappers and musicians who have come "straight outta Compton" in the last generation, Train refused to leave his Southside neighborhood. Living in his grandparents' duplex—with his mom, his three sisters and his gold records—he brought dignity and respect to the whole street. He and Ricky were planning a multicultural peace concert in October to bring Black and Latino youth closer together.

Ricky now finds it difficult to acknowledge that Train is dead, or that he died so needlessly. "It was about half past one in the morning [25 July] when my sister woke me, screaming about a fire. I ran outside and saw flames shooting from Train's kitchen windows. Several of us fought the fire with garden hoses, but we could hear someone inside, disoriented, pounding frantically at the walls. It was impossible to get into the house from the kitchen side, so I ran around to the front. The Compton police had already arrived, but they were just standing there with their arms folded, doing nothing at all. Although Train's door was heavily barred, the living room window was accessible. When I approached the window, a young cop yelled at me to get back. I explained that there was someone trapped inside. I asked why the police weren't doing anything. That seemed to piss him off. He tried to put me in a chokehold, but I just stepped out of his reach. Then he pulled a long flashlight from his belt. Swinging it two-handed, like a batter, he caught me twice in the face. Pop! Pop! It was a home run: My head exploded. Blood splattered everywhere.

"Then other cops came and they threw me across the hood of Train's car. As they handcuffed me, all the neighbors were screaming: 'Leave him alone. What did he do?' I begged them to save Train, but one cop said, 'Let the nigger burn.' They locked me in a squad car for another fifteen or twenty minutes. I never saw the firefighters arrive, but I am told they found Train near the living-room window. Later at Martin Luther King [hospital], I got a glimpse of him after they stitched my eye. He was naked on a gurney, his skin the color of raw meat, all bloated and blistered. He died a day later in the USC burn ward. I was charged with 'battery on an officer' and released on $1000 bail."

When I dumbly asked Ricky what color the cop who hit him was, he looked right through me. "He was Latino, and that is neither more nor less important than the fact that the cop who beat Soltero was Black. Thing you got to understand, partner, is that all cops are colored blue."

The White Embargo

A week after Train's death, Ricky led a large crowd in a chant of "No justice! No peace!" in front of the Compton Police Department. Dozens of Black residents added their voices to those of their Latino neighbors protesting the savage beating of Felipe Soltero. Many argued, like Ricky, that police brutality in Compton, while certainly victimizing many young Latinos, is essentially a class not a race issue. One person carried a sign in memory of two young unarmed Samoans shot nineteen times in the back by a Compton police officer in 1991. Others handed out fliers demanding the establishment of a civilian review board to investigate police misconduct.

Theresa Allison is a supporter of the Soltero family. The founder of Mothers Reclaiming Our Children, she works to unite Black and Latina women whose kids are lost in the labyrinth of the state prison system. She is saddened and puzzled by Compton's decline: "Who now remembers the pride and bright hope this city once represented? Who could have predicted our own people would come to treat us this way?"

Indeed, in 1969, when Douglas Dollarhide took over as Compton's first Black mayor, after a heroic fifteen-year struggle for civil rights, the city was widely seen as a flagship for Black power. But what the *Los Angeles Times* patronizingly labeled "an experiment in Negro self-government" was structurally disabled from the beginning. Old-timers still recall how the "plantation regime" of Col. Clifton Smith—the city's long-time political boss and publisher of the *Compton Herald-American*—had earlier fought each Black advance with scorched earth. It had stopped building schools and parks, floating bonds, or soliciting new industrial investment. It had scorned federal aid and ignored the deterioration of the downtown retail district, which had supplied almost all the city's sales tax revenue.

Despite the quiet that reigned in Compton's Black neighborhoods during August 1965, the nearby Watts rebellion was the signal for a veritable white Dunkirk. In headlong flight toward the enclaves of South Gate and Lakewood, whites abandoned almost two thousand homes and stripped the city of most of its retail capital. Compton Boulevard was left without a single restaurant, hardware store, movie house or pharmacy. The city budget, meanwhile, was woefully insufficient to deal with the needs of a much younger Black population. The only sector that the old regime left intact—in fact, expanded—was the Police Department, and for many years the Black majority had to live with the mocking anomaly of a white-dominated police force.

By the late 1960s, when Dollarhide came in, it was clear that Compton's fiscal survival depended on an aggressive campaign to annex industrial tracts within its unincorporated periphery. But the all-white County Local Area Formation Commission, which then dispensed permission for annexation, systematically discriminated against Compton. Although surrounded by new industry and port-related warehousing, Compton had to watch its potential tax base stolen instead by white-majority cities like Carson, Torrance and Long Beach. In the microcosm of Los Angeles County, this was the structural equivalent of a white embargo.

Without an expanded retail or corporate tax base, Compton had to increase the fiscal pressure on its homeowners relentlessly. At the same time, the major banks cut off most home and small-business financing. Residents were thus unable to rehabilitate old and substandard housing, while new home construction virtually ground to a halt. High property taxes, redlining and residential dereliction, in turn, were major causes of the exodus of the Black middle class from the city during the 1970s (many of them headed, ironically, a few miles farther south to white, tax-rich Carson).

Thus, the racial transition of the 1960s was followed by a class recomposition in the 1970s. Home ownership declined and median family incomes plummeted (from 92 percent of the county average in 1970 to 62 percent in 1990), while poverty, overcrowding, welfare dependence, illiteracy, and unemployment soared. Absentee landlordism began to spread over the city like a fungal infection. Increasingly, Black public-sector professionals, together with white cops and Latino store-owners, commuted to work in Compton but lived elsewhere. Community activists began to use the term "neocolonialism."

This vicious circle had two final, decisive arcs: First, expenditure on the Police Department virtually swallowed up the rest of the city budget after the mid-1970s. Compton today devotes an incredible 70 percent of its general fund revenue to "public safety"—possibly the highest percentage in the country. The most dramatic escalation occurred between 1985 and 1990, when the police budget increased by a staggering 195 percent while funding for parks and recreation collapsed (minus 97 percent). Local government had been reduced to an iron heel.

Second, federal aid had been Compton's financial last resort, and no city suffered more brutally from the Reagan/Bush counterrevolution in urban policy.

Yet, as the city's 1991 General Plan frankly acknowledges, federal funds were largely squandered in a series of redevelopment chimeras that have left Compton with a ghostly, two-thirds-empty auto mall, an insolvent hotel-and-convention center kept afloat by bingo, and supersubsidized shopping centers that return almost no revenue to the city.

The major winner in Compton's redevelopment fiasco has been developer Danny Bakewell, who has leveraged fabulous profits from city land discounts and direct subsidies. A flamboyant figure often compared to the Rev. Al Sharpton, Bakewell directs California's leading Black charity, the Brotherhood Crusade, and has been prominent in sensationalized fights with the Latino and Korean communities. Not surprisingly, he has also been one of the most generous campaign contributors to members of the Compton City Council, who also sit as commissioners of the redevelopment agency.

While the FBI continues to investigate possible corruption in the city's management of redevelopment funds, the City Council majority has been busy worshiping a new golden calf: a proposed card casino in the defunct auto mall. Instead of putting gambling to a citywide vote, and thereby risking the mobilized opposition of Compton's powerful churches, the Council simply smuggled the decision through one of its agendas. Critics charge that the projected revenues from the casino—expected to open in 1995—are wild overestimates that ignore competitive conditions in an already overcrowded local gaming market.

A New Plantation?

The excesses of redevelopment have also deepened the chasm between City Hall and Compton's burgeoning Latino community. For years tax revenue has been drained from north Compton's Latino neighborhoods with little visible return investment. Some Latino leaders worry that the city is being so thoroughly strip-mined by outsiders that soon there will be nothing left to meet the needs of their kids. They bitterly contrast the favoritism shown to politically connected developers like Bakewell with the string of ordinances—prohibiting outdoor sales, outlawing street vending and so on—that they say discriminate against Latinos.

For more than a decade Latinos have been trying to negotiate a power-sharing arrangement with Compton's Black political elite. Constituting 11 percent of the city workforce and only 5 percent of the school district teaching staff, the emergent Latino majority sees itself effectively locked out of public employment. In 1988–89, both City Hall and the school board rejected affirmative-action proposals to open opportunities for Latinos as older Black employees move or retire. "We do not need affirmative action; the majority of employees are minority," argued school board president John Steward.

At the same time, Latinos were struggling to increase their representation in elected office beyond the single school board seat that they had occupied since 1963. According to City Attorney Legrand Clegg II, the active electorate is still 80 percent African American and "largely composed of senior citizens who have resided in the city for thirty years or more." Latino candidates, therefore, have sought to build inclusionary "good government" coalitions that appeal to Black voters.

In 1991, Pedro Pallan, the popular owner of Compton's major Mexican bakery, captured nearly 1000 Black votes to force Omar Bradley into a runoff for the City Council. Bradley won, but two years later he recruited Pallan as his ally in a bruising battle for the mayoralty, vacated when Walter Tucker was elected to Congress. Pallan agreed to swing Latino and Samoan voters behind Bradley and his slate in return for appointment to Bradley's empty seat on the Council. Latinos flocked to the polls to cast what they believed was a dual vote for Bradley and Pallan. Every Latino vote counted as Bradley edged out feisty Councilwoman Joan Moore by a mere 349 votes. Then, before anyone could say "multiculturalism," Bradley reneged on his public promise to Pallan and appointed a relatively unknown African American to the Council.

When I interviewed Pallan last year, a few weeks after the election, he suggested gloomily that Compton's Black elite seemed intent on rebuilding the same kind of political plantation based on a disfranchised majority that Dollarhide and the NAACP had overthrown in the 1960s. "What irony," he observed, "if we Latinos have to ask the Justice Department to intervene in Compton to enforce our civil rights." Late last month, they did just that.

Despite a brave "Multicultural Unity Summit" last spring, co-sponsored by the

Mexican-American Legal Defense and Educational Fund and the NAACP, ethnic tempers remain raw. After viewing the video of the Soltero beating, Mayor Bradley evinced little respect for the feelings of Latinos when he icily commented, "No reaction." At the other extreme, Arnulfo Adatorre Jr., one of the leaders of the new Latinos United Coalition, enraged African-Americans with his silly allegation that "Latinos are treated worse by [Compton] city officials than the Blacks were treated by whites in South Africa."

Meanwhile, Ricky Miller, Compton patriot, remains optimistic. He thinks the Soltero beating has galvanized unity "where it really matters—not on top but on the bottom, between the activists, the rappers, the homeboys and *chollos* in the hoods." Honoring his promise to D.J. Train, he is organizing that concert for Black and Latino youth in the fall.

1994

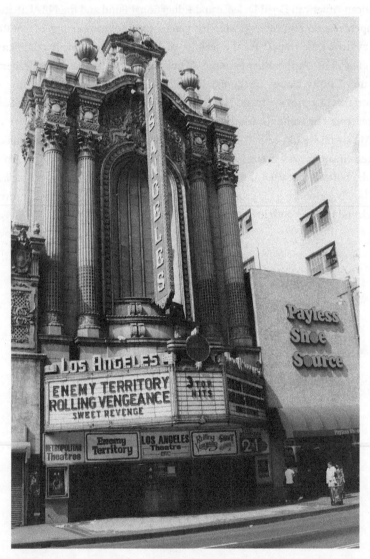

It's not just a movie…

15

Dante's Choice

A Double Funeral

Blondie's grief has hushed the overcrowded funeral chapel. All coughing and rustling have ceased as the mourners focus on the tense sentry in front of the altar. For fifteen minutes Blondie has stood motionless next to Oscar's coffin, scrutinizing every detail of the waxen death-mask as if searching in vain for some hint of a familiar wry grin. In life Oscar always had a handsome sparkle in his eyes, but today the lids are shut and his mouth is drawn preternaturally tight.

Finally Blondie bends over the open coffin. With great tenderness he lifts Oscar's head and gently emplaces a homemade wreath of white carnations. In the folded hands he inserts a small silver crucifix. Then, shivering, he kisses Oscar on the cheek. As he turns to leave, the stricken expression on his face is beyond pain. Someone in the back of the chapel begins to sob.

Four days earlier, 22-year-old Oscar Trevizo had been drinking beer with a Black friend on a bus bench at the edge of Los Angeles's Koreatown. Although Oscar's family had moved out of the neighborhood (Hobart Boulevard between Venice and Washington Boulevards) several years before, he frequently stopped by to "kick it" with Blondie and other guys on the block. According to an eyewitness, he smiled as two young African-American men (later identified by the police as members of Avenue 20s Bloods) approached the bus bench. One of them quietly ordered Oscar's companion to move aside. Then, without further warning, the other emptied a .22-automatic into Oscar's chest and abdomen.

Now a large crowd (the funeral director estimates 400) is lined up behind Oscar's hearse. Blondie leads the pallbearers, followed by the immediate, then the extended families. There are almost too many cousins (from Texas, New Mexico, Chihuahua, and Sonora) to count. The *compañeros* from Mr. Trevizo's workplace—big, mustachioed men in Sears workshirts—walk side by side with the eldest daughter's colleagues from UCLA in their twill-tweed suits.

But the biggest contingent has come from the 1700 block of Hobart Boulevard, past and present. Black, Mexican, Japanese, Irish, Hawaiian, Korean, Guatemalan, and Salvadoran—they represent two generations of neighborhood unity and interdependence. And, from the elderly Nisei widow to the young Salvadoran homeboy, they have come not just to bury Oscar—the innocent victim of an incipient, citywide gang war between Blacks and Latinos—but also to say goodbye to some ineffable part of themselves and their life together. It is a double funeral for a boy and a neighborhood.

Ultimate Urban Hell

Oscar was buried in Inglewood Cemetery, across the street from the Lakers' homecourt in the Forum. Like other inner-city graveyards it is overcrowded with too many victims of the childhood disease called gang violence. Nearly 10,000 infants, kids, teenagers, and young adults have been slaughtered in Los Angeles County since Oscar entered kindergarten in 1976. Most have died in the streets, but others have been murdered in their homes, schools, playgrounds, and even in their mothers' wombs. It is chilling to realize that there are neighborhoods in Southern California where children are more likely to visit the coroner's morgue by age eighteen than Disneyland.

Yet for all this appalling carnage, there has also been the grim solace that gang warfare largely has been confined within specific ethnic communities. For reasons too vast or capricious to fully understand, Black and Latino communities (and sometimes Asian-Pacific communities as well) have imploded in murderous shards of intra-ethnic rage and self-hatred. Crips have concentrated on killing Bloods, and vice versa, while Eastside homeboys have decimated each other over intricate territorial boundaries inscrutable to the outside world. Miraculously, in a metropolis transformed by immigration and dramatic spatial and demographic shifts, ethnicity or race per se have seldom been the *casus belli* of gang violence. A surprising attitude of "peaceful coexistence" (the more cynical would call it "mutual deterrence") along the major cultural divides has kept alive—amid the bloodshed—the dream of a rainbow city.

That dream may now be dying. If the thousand fires of April 1992 illuminated the angry chasm between Blacks and Koreans, the last year has witnessed an ominous escalation and randomization of violence between Latinos and Blacks. Riots between the two groups have become brutal weekly occurrences (for example, fifty-five in 1993 alone) in Los Angeles County's huge, overcrowded jail system. Less lethal, but equally virulent disturbances between Black and Brown youngsters have brought police in riot gear to a dozen high schools from Inglewood to Palm Springs. Chilling examples of "ethnic cleansing" against Black residents have been reported from the heavily Latino suburbs of Paramount, Norwalk, Azusa, and Hawaiian Gardens. And a few blocks from the famed Venice boardwalk, local Black and Latino gangs are locked in a deadly vendetta that has taken seventeen lives and threatens to ignite a citywide explosion.

Oscar's murder on 28 June 1993—part of a continuing cycle of interethnic shootings in the Midcity area—coincided with this dramatic increase in Black–Latino tensions and thus forms a rough baseline for narrating some of the disturbing events of recent years. But before retracing the latest steps in the descent toward the ultimate urban hell, it is necessary to briefly tell the story of what was, and in many places still is, the everyday reality of inter-group relations in Los Angeles. The current trend toward deadly, self-perpetuating ethnic violence must be measured against the remarkable success of many inner-city neighborhoods in peacefully managing unprecedented cultural and economic change. Consider, for example, the 1000 block of Hobart Avenue.

An Extended Family

Massive stone pillars, a block away from the house where Oscar grew up, are reminders that Hobart Boulevard was originally part of an exclusive subdivision—home to dentists, real estate dealers, and wealthy widows—known as Westmoreland Heights. In 1910 this was the very edge of the city—further west there were only miles of eucalyptus-lined dirt roads and scattered farms until you reached the National Soldiers' Home in Sawtelle. A pioneering 1909 zoning ordinance, the first of its kind in North America, had excluded noxious industries west of Downtown and determined, once and for all, that middle-class Los Angeles would grow toward the sunset.

Over the next three generations this bourgeois "Westside" repeatedly shifted further west and north, closer to the ocean and hills, but leaving behind a few fossil enclaves of old money like Fremont Place and Hancock Park. In the 1920s, a handful of wealthy Black families managed to circumvent restrictive covenants and buy homes in the West Adams area, adjacent to Westmoreland Heights. By 1940 this was Los Angeles's own "Sugar Hill," with an African American elite that included insurance pioneer William Nickerson and film stars Stephan Fetchit and Louise Beavers. Despite the protests of civil rights groups, the Santa Monica freeway (now I-10) was bulldozed through the heart of Sugar Hill in the early 1960s, displacing hundreds of families. To the embittered survivors the freeway would always represent the Black counterpart to the infamous demolition of the Chavez Ravine barrio to make way for the Dodgers in 1958.

Oscar's parents, of course, were scarcely aware of this history or its scars when they moved into the neighborhood in 1969. To them Hobart Boulevard was simply a welcome urban oasis. In a city notorious for its shortage of housing for large families, the huge Craftsman homes were a wonderful bargain. The Trevizo family with its four kids was, in fact, one of the smallest on the block: the nearby Diaz and Pinela clans had eighteen kids between them. On Saturday mornings in the early 1970s, Hobart Boulevard was a pandemonium of kids riding bikes, kicking balls, throwing water balloons, playing hide and seek—all watched over by Mrs. Hagio, a widowed Nisei woman, who more than once had saved a child's life from speeding cars. In the fond recollection of Oscar's sisters, everyone—Mexican, Black, Anglo, and Asian—blended together into a single extended family.

Next door, for example, were the Taylors, an Anglo family who were very active in St. Thomas, the local Catholic church, as were the large Diaz clan down the street. Oscar's three sisters usually played with the Pinela kids, whose father was an RTD foreman, or with the Berrios daughters, whose dad, from Puerto Rico, was a chef at a famous French restaurant in Beverly Hills. From the age of four Oscar had been inseparable from the Garcias' son, Gilberto (or, as he became known in junior high school because of his *huero* complexion, "Blondie"). The street's hero, meanwhile, was undoubtedly Steven Beamon. A high school basketball star whose mother was a veteran volunteer for Tom Bradley, he was tall, handsome, and incorrigibly kind. When not showing the older Mexican kids how to shoot baskets, he was buying ice cream for their younger siblings. Not surprisingly, he was also the heartthrob of every teenage girl on the block.

Despite creeping decay and a tightening perimeter of gang violence, Hobart Boulevard—almost exclusively composed of homeowning blue-collar and civil-service families—prospered through the late 1970s. The older kids enjoyed extraordinary academic success. The Trevizos, for example, marvel at the strict and industrious Pinelas, who sent all eight of their children to colleges that included Harvard, Yale, USC, and Berkeley. Oscar's older sister Dolores, meanwhile, won a scholarship to Occidental and later UCLA, while his middle sister Lupe attended Cal State Los Angeles.

Lost Tribes

In the early 1980s, however, Hobart Boulevard began a disturbing metamorphosis. First, after their son was killed in the street by a hit-and-run driver, the Nisei couple who owned the apartment building on the corner of Pico sold out to an anonymous slumlord who packed it with poor Central Americans and refused to make any repairs. Then, as the older Mexican families started to move out to the new Latino suburbs (like Whittier, Fontana, and Huntington Park), their huge, Edwardian homes were also converted into apartments for Salvadoran or Guatemalan families.

In the wake of Proposition 13 the local schools were stripped of the resources they needed to deal with a second, larger wave of Spanish-speaking kids. Class sizes soared, teachers quit, and drop-out rates increased. Local teenagers, meanwhile, could no longer find jobs. Oscar's older sister Dolores bitterly remembers making more than forty employment applications throughout the Koreatown area in 1981–82. Korean merchants as a rule refused to hire Black or Latino youngsters, except in garment sweatshops. Rejected by teachers and employers, these "throw-away kids"—as they often called themselves—found surrogate families in the streets.

As a result, the 1700 block of Hobart Boulevard gradually divided into two worlds. At the north end children still played in the street while their parents mowed lawns and painted garages. The south end, however, was taken over by an outlaw Central American street gang—the Crazy Riders—and nearby apartments began to fill up with *los zombies* (crackheads). Growing like weeds in a suddenly derelict garden, other gangs proliferated on the edges of Hobart Boulevard: the Playboys, Clanton, the Midcity Stoners, and a local franchise of the nation's largest gang, Eighteenth Street (which claims an estimated 10,000 to 20,000 members in Southern California). Black residents, whose roots in the neighborhood reached back to the early 1920s, were especially alarmed by the rise of the Latino street gangs. Their worst fears were confirmed in 1984. Oscar's sisters dolefully recall the day that Stevie Beamon was stabbed in Normandie Park by a Latino gang member who stole his beautiful ten-speed bike. In the hospital he developed serious complications that staunched his dream of winning a major athletic scholarship. He became increasingly morose and withdrawn. Then, a few weeks after the death of his father, he killed himself. The old-timers on Hobart Boulevard have never gotten over his suicide.

Little brother Oscar, meanwhile, like thousands of other inner-city kids, was being shunted between schools dozens of miles apart. After spending his elementary years in the majority-Black 21st Street School, he was bussed (at 6 a.m. each morning) to a white section of the Valley for junior high school, then transferred back to high school in South Central Los Angeles. Manual Arts High was ethnically polarized between the notorious Rollin' 60s Crips and 18th Street. Although

Oscar spoke Black English and relished Rap culture, he sought the protection of the 18th Street homeboys after his best friend, Blondie, was jumped and beaten by some Crips. As Oscar later explained it, ethnic solidarity had become a matter of sheer survival.

Yet it had dangerous repercussions back on Hobart Boulevard, where Latino gang rivalries—frequently reflecting tensions between Mexicans and Salvadorans—were also growing more violent. The Midcity Stoners, for example, originally had been little more than a gaggle of boys who hung around Bishop Conaty High in the afternoons to flirt with the girls. They then slowly evolved into a heavy metal clique—common enough in Los Angeles during the mid 1980s—before falling under the proselytizing spell of charismatic *veteranos* who persuaded them to adopt the style and aggressiveness of a traditional Chicano street gang. Thus *"chollo-*ized," the renamed Midcity Locos mounted a violent offensive that drove the Crazy Riders off Hobart Boulevard. Oscar suddenly found himself the odd man out.

At this point Oscar's parents providentially moved him to Whittier. His youngest sister, Cesy, and her Nicaraguan husband stayed behind to manage the rental units into which the old family home had been subdivided. They nervously watched as Hobart Boulevard and its environs became increasingly militarized. As yet more kids dropped out of the overcrowded schools, forming a lost tribe with an estimated 50,000 members citywide, gangs like Midcity, which started out with a dozen kids, evolved into heavily armed small armies with over a hundred members. Veteran Black gangs, like the Avenue 20s Bloods from the Gramercy and Washington area, were also seen more frequently.

The second Los Angeles riot in April 1992 was yet another watershed in the violent transfiguration of Hobart Boulevard. The nearby intersection of Western and Venice became a major node of arson and looting. Cinders from a fiercely burning McDonald's ignited palm trees in the backyards of two Westmoreland Heights mansions that had been lovingly restored to their Art Nouveau glory by a tiny colony of gay professionals. Oscar's fair-skinned middle sister Lupe (a social worker for the Inglewood School District) laughingly remembers being accosted in front of her old house by some gangbangers who warned: "Hey, white bitch, better get off the street. This is a Black and Latino thing!" Despite such protestations of unity, however, many local Black-owned as well as Korean-owned businesses were heavily damaged or destroyed.

Fourteen months later, on a warm Saturday evening (24 June 1993) a Midcity Loco known as "White Boy" shot an Avenue 20s Blood who was eating dinner at TNT Tacos at the corner of Western and Venice. A few hours later there was more shooting at Venice and Hobart. The next day LAPD CRASH (anti-gang) officers came through the neighborhood warning Latino gang members of the inevitable retaliation. Oscar, home in Whittier, knew nothing of these events.

On Monday he drove to Hobart Boulevard to finish some yardwork around his parents' apartments. Afterwards he hung out with an Anglo neighbor, shared a Wendy's with his little nephew, and then, around nine o'clock—and still unaware of the killing two days earlier—went to the corner for a smoke. Passing by a laundromat he recognized an old friend, who offered him a beer after Oscar volunteered to help fold clothes. Ten minutes later a white Cadillac slowly circled the block, then pulled to a stop a few yards from the bus bench where Oscar was sitting. He smiled as two strangers approached...

His sister Lupe has never been able to get out of her mind the conversation she had with Oscar a few days before his murder: "Stay off Hobart Boulevard," she warned, "Everything has changed. You don't know the block anymore. You'll get killed." Oscar just laughed softly. "No, *mija*, not me. I like everyone. Why would they want to hurt me?"

The Hate Factory

Two days after the funeral Oscar's *Novenaria* began: nine evenings of rosary for the soul of the deceased. "Dios me salve Maria..." An old-fashioned Catholic custom, the same in the Latino neighborhoods of Los Angeles as in Palermo or Galway, it is a vigil kept by women, praying to another woman whose son was murdered two thousand years ago. It is also an opportunity for Oscar's three sisters to begin talking through their grief. They repeatedly focus on the terrible question of why Blacks and Latinos are suddenly killing one another.

Dolores, the eldest, recalls how years before some Black families were driven out of the Pico-Union neighborhood by a Salvadoran gang. Lupe talks about the racial conflict she has increasingly encountered in the Inglewood and Hawthorne schools. Cesy vaguely remembers a bad incident during one of the semesters she spent at Los Angeles High. But all these examples are outside the neighborhood; they do not explain why Oscar was murdered or, for that matter, why one of his closest Black friends was immediately targeted for retribution (he was severely wounded, but survived). As Lupe emphasizes, "Hobart Boulevard has always been different. We have always been united."

"Homegirl," a seventeen-year-old neighbor who sometimes hangs out with a Salvadoran street gang, has overheard the conversation. She has a better explanation. "No, man," she drawls, "it comes from jail. They make the hate in County. Up in Wayside especially. The homies bring it back to the street when they get out. Things are really fucked up in Wayside. Like a race war. Check it out."

A few exits beyond Six Flags Magic Mountain amusement park in northern Los Angeles County, "Wayside," or the Peter J. Pitchess Honor Rancho as it is officially known, is a fallen utopia. It was established in the late 1930s by Sheriff Eugene Biscailuz as a "revolutionary experiment" in rehabilitating minor offenders. Biscailuz, a legendary outdoorsman with "an inborn dislike of confinement," ran Wayside as a working ranch where prisoners could experience the rugged and rehabilitative life of cowboys in "the unbounded hills." Cattle still graze in the meadows at Wayside, but the benign "honor rancho" has evolved into a monstrous version of the overcrowded jails that Biscailuz wanted to reform: 9000 inmates—90 percent Black and Latino—are shoehorned into facilities designed for fewer than 6000. Only a handful of trusties still enjoy Biscailuz's outdoor life; everyone else is confined 16 to 23 hours a day in claustrophobic dormitories. Formerly Wayside raised alfalfa; now—as "Homegirl" accurately explained—its main product is hate.

Dewayne Holmes has risked his life to fight this hate. A year older than Oscar, he is a veteran PJ Crip from the Imperial Courts housing project in Watts. After the police killed his cousin in 1991, Dewayne organized a ceasefire between Watts street gangs that grew into the permanent truce between the Crips and Bloods. Recognized as a "community hero" by Congresswoman Maxine Waters and former governor Jerry Brown, Dewayne has just been sentenced to seven years in prison for a $10 robbery he claims he did not commit. (A dozen eyewitnesses in Imperial Courts support his story.)

Awaiting final sentencing, Dewayne has spent six months in Wayside's ultimate human pressure-cooker: "Super Max." He laments the violent breakdown in Black–Latino relations: "When I first arrived, I met this old partner from Watts—one of my Mexican homeboys. We came up together and have always been very tight—almost like best friends. But this time he looked worried, very cautious. He whispered to me: 'Hey, Sniper [Dewayne's nickname], I love you, man, but things are different now. When shit goes down, it goes down. Everybody has to fall in line. You know what I mean? Watch your back.' We didn't speak again."

Dewayne has been ambushed twice by Latino inmates. In one encounter he nearly lost an ear ("Almost like that Van Goth dude," he jokes.) Now he is vigilant even in his sleep. He explains the simple, implacable logic of hate-making: "You see, the Crips and Bloods are still the majority downtown [at Men's Central Jail in Bauchet Street], but the Latinos outnumber us here two-to-one. When brothers whip the Mexicans downtown, somebody gets on the phone and a little while later we catch hell here. It's like this infernal system that keeps the havoc going round and round. Nobody feels they can control it or stop it. And pretty soon people are gonna start dying."

In fact the Los Angeles County Sheriff's Department, responsible for the daily management of an estimated 22,000 prisoners (47 percent Latino, 33 percent Black), admits that hundreds of violent confrontations have occurred since 1991. Ordinary fist-fights easily escalate into full-fledged riots. A typical sequence began on Monday, 3 January 1993, when twenty Black and Latino inmates in an overcrowded holding cell in Downtown's Criminal Courts Building engaged in a brief brawl. The sheriff's deputies apparently made no effort to separate the combatant sides, and the battle resumed the next day with sixty prisoners involved. Three Blacks were stabbed in the neck with a six-inch shank (homemade knife) while eight other inmates suffered minor injuries.

When word of the Downtown stabbings reached Wayside it immediately ignited a chain reaction of attacks and counterattacks. On Wednesday a fight broke out in the east facility, followed the next day by a vicious battle in a medical dorm that injured fifteen inmates. Meanwhile, on a Sheriff's Department bus en route from Wayside to San Fernando Superior Court, a half-dozen shackled white and Latino prisoners were severely beaten by two Blacks who managed to break free of their handcuffs. On Friday there were more casualties during a melee in the north facility. All this simply set the stage for a huge eruption on Sunday, 9 January.

With military precision, Latinos, armed with shanks and jagged broom handles, simultanously ambushed Blacks in all twenty maximum-security dormitories at 3:55 that afternoon. Desperate hand-to-hand combat, involving as many as a thousand inmates, raged for hours until deputies, firing rubber bullets point-blank, finally restored control. Incredibly no one was killed, although eighty were injured, including twenty-four with serious stab wounds or broken bones.

Sheriff's Commander Robert Spierer promised a crash program of reforms to curtail further outbreaks, but another major riot broke out at the beginning of the summer (14 June). An estimated eight hundred Black and Latino inmates, again wielding broom handles, knives, and rock-filled socks, battled for almost half an hour until they were quelled by deputies with batons and pepper spray. The Sheriff's Department claims that this virtually incessant violence has been imported into the county's jails from the streets and prisons. Since 1988, when Latinos became a majority in the system—according to the official explanation—there has been a nonstop struggle for internal control of the jails, aggravated by the ambitions of the Mexican Mafia to wrest street narcotic sales away from the dominant Black gangs.

Although many prisoners will concede that there is a grain of truth in this scenario, they primarily fault the inhuman overcrowding within the juvenile detention, county jail, and state prison systems (a situation soon to be made even more barbarous by California's new "three strikes" law). "If we weren't packed into these dorms all day long like slaves on an old slaveship," Dewayne points out, "relations between Black and Brown might not be so explosive." Furthermore, inmate-rights advocates like the Coalition Against Police Abuse and Mothers

Reclaiming Our Children have accused deputies of deliberately inflaming antagonisms, even setting up prisoners for attack. Yet Sheriff Sherman Block, the most unassailable and perhaps the most powerful politician in Los Angeles County, ignores such criticism. Despite daily evidence that interethnic violence within the jail system is poisoning the entire city, no public official has yet had the courage to propose a public investigation of the Sheriff's Department.

Small Infernos

Jordan Downs is a low-rise public-housing community of 2500 people (80 percent African American, 20 percent Latino) a few blocks from the famous Watts Towers. Early in the morning of Saturday, 15 June 1992, several figures—later identified as notorious crack-cocaine dealers—were seen dousing gasoline on the outside of the apartment occupied by the Zuniga family, recent arrivals from Mexico City. The ensuing inferno consumed the lives of two adults and three small children. (The Los Angeles Times gave a poignant account of the heroic but unsuccessful efforts of the 78-year-old great-grandmother to shield her two-year-old great-grandchild from the 1200-degree flames.) A Black neighbor, rushing to help save the family, was mistakenly shot and paralyzed by the panic-stricken grandfather.

Although investigation revealed that the Zunigas were targeted because of their repeated protests against drug sales outside their door and not necessarily because of their ethnicity, the tragedy was broadly interpreted by the press as the consequence of long-simmering Black-versus-Latino tensions in the L.A. public-housing system. Some terrified Latino residents of Jordan Downs urgently demanded to be rehoused in a "safer," segregated complex. Thanks to arduous efforts by project leaders and the organizers of the Watts Gang Truce (whose unofficial slogan is "Crips plus Bloods plus Mexicans—Unite!"), a wholesale exodus of Latino families was avoided and intercommunal relations gradually improved.

But—just as the nightmare in tenant relations appeared to have passed—there were new firebombings at the end of summer, this time against Black households in Boyle Heights's Ramona Gardens project. Often described by the LAPD as "the cradle of the Mexican Mafia," Ramona Gardens was once substantially integrated, but Black families fled after a series of fatal shootings in the late 1960s. In the 1980s, however, a half-dozen Black families moved back into the project and were well treated by most of their neighbors. The 30 August arson, which destroyed two apartments but amazingly caused no injuries, was rumored to be the initiative of younger gang members envenomed by ethnic warfare in juvenile halls and jails.

Although a tense quiet has returned to Los Angeles public housing (in Ramona Gardens' case because almost all the Black tenants moved out), isolated African American families have continued to be attacked in predominantly Latino suburbs. Hawaiian Gardens, for instance, is an obscure one-square-mile city—without significant industry or commerce—wedged between the San Gabriel River Freeway (605) and the Orange County line (Coyote Creek) just north of Long Beach. Originally one of the two distinctively "Okie" suburbs in Los Angeles County (Bell Gardens was the other), it is now overwhelmingly Latino, albeit with rapidly growing Asian (8 percent) and African American (4 percent) minorities. The 620 Black residents, however, have been subjected to a virtual reign of terror by local Latino gangs.

In March 1993, for example, sheriff's deputies and the FBI were called into Hawaiian Gardens after a molotov cocktail was thrown into the kitchen of a Black family who had earlier been the target of racial slurs. Graffiti along the side of another Black family's home warned: "We gonna shoot up your house and we don't care if you got kids inside." Two weeks before, Latino gang members had ambushed and severely beaten a young Black man in a nearby shopping center. Black children were temporarily pulled out of a local junior high school after they were repeatedly taunted and attacked.

A few months later, African-American families in Azusa—a small Latino-majority city in the San Gabriel Valley—charged that they had also been the victims of systematic gang-led intimidation. According to a suit filed against the Azusa Police Department (which was accused of inaction), two families were forced to move after shots were fired into their homes and their children attacked in a nearby park. Azusa police conceded that five classified hate crimes had been logged since the beginning of the year, all with "gang implications."

The same grim pattern repeated itself in February 1995 in a Latino area of Norwalk in southeastern Los Angeles County. After persistent gang attacks on their children and homes had driven several Black families out of his Walnut Street neighborhood, Robert Lee Johnson vowed to stay. On 10 February a sniper wounded him in the leg and foot as he was watching a video in his living room. Then, two weeks later, a molotov cocktail exploded on his front porch and burned his house down. Despite crutches and his injured leg, Johnson managed to rescue his wife and seven sleeping children and grandchildren (ages two to twenty-two). As he explained wearily to the *Times:* "It's a lot of racial stuff flaring up, Hispanics taunting Blacks. My family was all in the house, in bed. They didn't care what happened. They didn't care who they killed in there."

Los Angeles County's aging, blue-collar suburbs—the major destinations for both Blacks and Latinos leaving the inner city—have also witnessed ugly racial outbreaks in their overcrowded high schools. The typical scenario involves a change in the ethnic balance of power (usually an emergent Latino majority) combined with a marked deterioration in the educational environment. In 1991, for example, riot-equipped police were repeatedly called onto campuses in the contiguous southwest Los Angeles County communities of Lennox, Lawndale, Hawthorne, Inglewood, and Gardena to break up fighting between Black and Latino students who were sometimes armed with pipes and other weapons.

In fall 1993 almost identical disturbances—again requiring the intervention of police in full riot gear—broke out in a belt of South Bay campuses (Compton, Centennial, Dominguez, and Long Beach's Jordan High Schools) as well as at Pomona High School in the San Gabriel Valley. Eighteen months later, tensions between Paramount High School's 500 Black and 2300 Latino students erupted in a major melee that led the Sheriff's Department to clear the campus. Paramount, a poor blue-collar city east of Compton, has also been the scene of several gang-related racial murders.

It would be misleading to suggest that these micropogroms and schoolyard riots have an inexorable, cumulative momentum. In some instances, community workers and school officials have been surprisingly successful in mending group relations, although the current poverty of public resources makes this everywhere more difficult. Across Los Angeles County, however, from Midcity to Pomona, and wherever Blacks are at the losing end of the demographic transition, there is an ominous trend toward greater inter-ethnic violence. This logic has been played out most completely, and appallingly, in the Oakwood area of Venice by the sea.

Oakwood Dispossessed

Like Hobart Boulevard, the Oakwood story begs respect for a dream that once thrived. Elderly Black people in Southern California can still recall the old adage: "Jim Crow lives at the beach." Indeed, in the 1920s many seaside towns, including Redondo Beach and Manhattan Beach, were fervent outposts of the KKK's Invisible Empire. The coastal strip was the most rigorously segregated real estate in California, with a single famous exception: the one-square-mile Oakwood tract in Venice, which had been established as a "servants' tract." Over the years it evolved into a proud enclave of Black homeownership, tidy bungalow streets a few blocks from the Boardwalk and Pacific Ocean Park.

In the late 1960s, however, as large developers began to "Miamize" the shoreline between Ocean Park and Venice Pier, speculative shockwaves undermined Oakwood. In a classic example of block-busting-in-reverse, real estate operators cajoled hundreds of traditional homeowners to sell out. When Miami refused to cross Ocean Boulevard, developers instead exploited the lucrative tax advantages of the federal Section 8 housing program to plant fifteen cheaply built stucco tenements in the heart of Oakwood. Other landlords discovered a profitable rental market among counterculture whites who were being pushed out of Venice's beachfront walks and courtyards by soaring property values. A precarious new ecology emerged between the older Black homeowners, the poor Black and Latino Section 8 tenants, the young white bohemians, and a handful of wealthier, established artists and architects (principally on the area's western edge).

The simultaneous arrival of cocaine and gentrification in the late 1970s and early 1980s exploded this taut neighborhood equilibrium. Oakwood, unfortunately, was perfectly sited to become a 24-hour narcotic supermarket for upscale consumers in the white beach communities. Jobless, bitter youth in the Section 8 tenements were easily recruited as a low-wage and expendable sales force. Meanwhile, gated condos and designer homes began to displace ragtime bungalows and railroad shacks. By 1988 the neighborhood was officially deemed "one of the Westside's hottest real-estate markets." Even "stealth mansions" appeared, including Dennis Hopper's notorious pleasuredome disguised as a concrete-block bunker (a white picket fence deliberately underlines the bad joke). Oakwood may be the only place in urban America where movie stars live fifty yards from Crips-dominated tenements.

As a result, class and ethnic tensions have been close to a boil for most of the 1980s. A final ingredient was the rapid growth of an immigrant Latino population in overcrowded, rack-rented cottages and apartments. By the 1990 Census, Oakwood's 9200 residents were half Latino, one-quarter white, and only one-fifth Black. Not surprisingly, many Black residents have become embittered against what they perceive as a deliberate campaign to push them out of their own historic neighborhood. The area's two major gangs—the (Black) Venice Shoreline Crips and the (Latino) Venice-13s—first came to blows in the late 1970s, in a brief but savage street war that left four dead and a dozen wounded. Community activists, however, were successful in brokering a truce, and for the next decade, as friction rose between white gentrifiers and the Black poor, the Crips and V-13s managed to stay out of each other's way.

During the citywide rioting in April 1992, Black youth—identified by the police as Shoreline Crips—attacked the most visible symbols of gentrification: parked Mercedes-Benzs, expensive new condos, and the houses of prominent members of the anti-gang Neighborhood Watch. A white bicyclist was beaten unconscious, and several homes were ransacked. Later the LAPD charged five young Blacks, including the son of a leading opponent of gentrification, with attempted murder and street terrorism. This so-called "Denny West" case, however, quickly collapsed after the key prosecution witness—Oakwood's most vocal anti-gang activist—admitted that he had initially identified the wrong suspects. (Shown a photographic lineup, he selected two pictures: one of a dead man, the other of a prison inmate.)

Despite the dropping of the main felony charges against the "Venice Five," the neighborhood remained in turmoil. There were further arson attacks against the homes of alleged police informers, and some of the most outspoken white residents fled Oakwood. Despite an outcry from national Jewish organizations, desperate tenants hired the Fruit of Islam—the security wing of the Nation of Islam—to clamp down on drug sales in the Section 8 projects. The defiant Shoreline Crips, however, soon forced the unarmed followers of Louis Farrakhan to retreat from Oakwood. At about the same time, a V-13 *veterano*—32-year-old Mark Herrera—was stabbed to death by a Black woman named Diane Calhoun, who successfully pleaded self-defense. A few weeks later she was killed in what was widely assumed to be a retaliatory drive-by.

The 1992 murders of Herrera and Calhoun festered under the surface but did not provoke an immediate cycle of vengeance. Then, on 27 September 1994, 41-year-old Benjamin Ochoa—whom police identified as another V-13 *veterano*—was shot by an unidentified Black in an alley off trendy Rose Avenue. Oakwood exploded. In the next three weeks there were ten serious shooting incidents, including two attacks on the LAPD. In separate ambushes on 10 October, V-13 gunmen killed two well-known Shoreline Crips. Eleven days later—following a racial brawl at Venice High School—a Latino student with a large V-13 tattoo was shot and wounded a block from campus. Over the following six weeks *chollos* and Crips brutally traded tit-for-tat in an escalation of violence that claimed eight more victims (five Black, three Latino) and wounded thirty others. At least one of the dead—24-year-old UCLA nursing assistant Shawn Patterson (shot on 16 November)—had no gang affiliation whatsoever and seems to have been targeted simply because he was Black.

Nevertheless, in various interviews in the *Los Angeles Times* and the (Santa Monica) *Outlook*, neighborhood activists argued that Oakwood was engulfed in a gang, not an ethnic, war. In their view, the violence had arisen almost inadvertently—albeit in tinderbox conditions—from the killing of Ochoa. Older Black residents, especially, discerned a *Chinatown*-like meta-plot behind the bloodshed. As one of them told the *Times*, "The big developers are sitting there twiddling their thumbs, saying as soon as they kill each other, it's [Oakwood] going to be Marina Venice."

The LAPD, meanwhile, continued to elaborate the master theory of an EME-backed takeover of the street narcotics trade. Specifically, they charged that the V-13s had united with their former enemies, the (Latino) Culver City Boys, in a bold invasion of some Shoreline Crips' drug turf in the Mar Vista projects, a mile east of Oakwood. The Crips, now outnumbered and outgunned, were desperately fighting back to protect their livelihood.

As Christmas approached, however, the vendetta seemed to lose momentum. Although there were several nonfatal shootings in January (including Jimmie Powell, a popular gang mediator wounded by Latino teenagers), which may have been related to the New Year's riots in Wayside, the winter passed without further deaths. Some community leaders spoke hopefully of "the storm that has passed." But neighborhood gunmen were merely biding their time. In early March shootings resumed—not only in Oakwood but, increasingly, throughout the rest of Venice and Mar Vista as well. On the 20th the V-13s ambushed a prominent Shoreline Crip O.G. ("original gangster"). Five days later Anselmo Cruz, a 30-year-old nursing home cook with no gang affiliation, was murdered while driving his two small daughters and a friend to school. The children were injured by flying glass. Two weeks later a young Latino was killed in the Mar Vista projects. V-13 or Culver City Boys, in turn, wounded actor Byron Keith Minns, who had played a gang leader in the movie *Southcentral*. At the end of May they shot a Black Oakwood resident on busy Lincoln Boulevard—the major thoroughfare between Santa Monica and LAX.

As school graduation approached, teenagers in the Venice area began to apprehend that they were all potential targets in a street war that no longer made fine distinctions between combatants and their ethnic group. On 7 June, for example, six Latino youth were wounded when gunmen in a truck opened fire on a high school graduation party near the Penmar Recreation Center in Venice. Three days later, two young Latinos were killed and two wounded when their car was ambushed near Venice High School by three heavily armed Black youths. The seventeen- and eighteen-year-old victims, who came from the Mid-Wilshire area not far from Hobart Boulevard, were scheduled to graduate the next week.

Already by the beginning of the spring carnage, Marilyn Martinez—the *Out-*

look reporter who has been the principal chronicler of the Oakwood wars—was already pointing out "signs that the 1.1-square-mile neighborhood is starting to unravel under the pressure of the violence." Like Sarajevans, the residents of Oakwood have had to learn how to avoid sniper fire by living in the back rooms of their houses, not venturing outside after dark, and constantly changing their daytime routes to school and work. First-graders nervously converse about bodies they have seen in the streets and the gunfire and police helicopters that keep them awake all night. Their parents discuss security fencing, plunging home values, and the impotence of the police to stop the killings. Hundreds have simply moved out.

Postscript

In the three years since Oscar's death, the inter-ethnic conflict in the county jail and state prison systems has only escalated. Savage cutbacks in county services, public employment, and federal aid have increased the violence of everyday life in Los Angeles's inner-city neighborhoods and aging blue-collar suburbs. But the street war in Oakwood has ended.

On the brink of their conflict becoming a citywide Armageddon between Black and Latino gangs, the Shoreline Crips and the V-13s accepted last-minute mediation, and with the help of two veteran youth-probation officers negotiated a truce that has so far weathered the inevitable provocations. As one of the mediators recently told me, "Like the gang truce in Watts, this is a small social miracle that flies in the face of hardened hearts and cynical minds. But the gentrification juggernaut is reving up again in Oakwood, and the key issue for these kids remains JOBS. Make sure you spell that with capital letters."

1995

PART IV

EXTREME SCIENCE

Impacts: Earth's existential truth?

16

Cosmic Dancers on History's Stage?

> Upon the streets the people stand and gaze
> Transfixed at Monstrous portents in the sky.
> Where creeping fiery-snouted comets blaze,
> Threatening the jagged towers as they fly.
>
> *Stefan Heym*

Early on the morning of 1 February 1994, President Clinton, Vice President Gore, the Joint Chiefs of Staff, and the members of the National Security Council were awakened from their sleep by Pentagon officials.* A military surveillance satellite had detected the brilliant flash of a nuclear explosion over the western Pacific. There was intense concern that the strategic warheads aboard a Russian or Chinese missile submarine might have detonated accidentally. US military aircraft failed to detect any unusual radiation in the indicated ocean sector, however, and Defense Department intelligence experts soon concluded that the satellite had actually witnessed the explosion of an asteroid fragment, which they later estimated to have been the equivalent of a 200-kiloton nuclear blast. The President went back to bed.[1]

Five months later, beginning on 16 July, hundreds of millions watched in awe as the Hubble space telescope transmitted images of Comet Shoemaker-Levy 9's fiery death in the dense atmosphere of Jupiter. For nearly a week, the plummeting trail of cometary fragments produced a succession of huge fireballs—equivalent to many million megatons of explosive energy—that left dark, temporary scars on the giant planet. Then, on 9 December 1994, an object comparable to one of Shoemaker-Levy 9's fragments—the asteroid 1994XMI—approached within 105,000 kilometers of the Earth, a record close call in the brief annals of monitoring so-called Near-Earth Objects (NEOs).[2]

These events made 1994 something of a watershed in public awareness of the Earth's vulnerability to comet and asteroid bombardment.[3] Indeed, the spectators at Shoemaker-Levy 9's immolation were the first generation of humans to observe a major planetary impact since medieval monks recorded the collision of an asteroid with the moon, forming the crater Giordano Bruno, in 1178.[4] Congress was sufficiently impressed to fast-track a major study of NEO-detection technology and a probe to the asteroid Eros—launched on 16 February.[5] Meanwhile, the friends of Star Wars, including H-bomb father Edward Teller, lobbied for an orbital anti-asteroid defense of super-lasers and thermonuclear weapons. (Both of which, as Carl Sagan and others immediately pointed out, could be turned against Saddam Husseins on Earth as easily as NEOs.)[6]

Beyond the predictable media hyperbole about exterminators from outer space—so reminiscent of "comet hysteria" throughout human history—the events of 1994 were also an incomparable "teach-in" on the Earth's citizenship in the solar system. February's giant fireball over the Pacific, July's fusillade against Jupiter, and December's breathtaking near-miss—were all cram sessions in the new Earth science being shaped by comparative planetology and the neo-catastrophist reinterpretation of the stratigraphic record. It is a lesson, of course, that many geologists, as well as geographers and historians, have great difficulty accepting. Even more than plate tectonics, an "open system" view of the Earth that recognizes the continuum between terrestrial and extraterrestrial dynamics threatens the Victorian foundations of classical geology. To cite only one example, a single impact event can compress into minutes, even seconds, the equivalent of a million years or more of "uniformitarian" process.

A Revolutionary Science?

But this is not a mere family feud. The "golden age" of Cold War space explora-tion, now drawn to a close, has seeded the fields of philosophy with discoveries every bit as strange and revelatory as those of Magellan and Galileo—the names, appropriately enough, of our most recent planetary galleons. I must confess that as an aging socialist who spent the glory years of the Apollo program protesting the genocidal bombing of Indochina, it has taken me half a lifetime to warm to a scientific culture incubated within Cold War militarism and technological triumphalism. Yet it is also the contemporary home of luminous and, dare I say, revolutionary attempts to rethink the Earth and evolution within the context of other planetary histories.

While postmodernism has defoliated the humanities and turned textualism into a prison-house of the soul, the natural sciences—which now include plan-etology, exobiology, and biogeochemistry[7]—have once again, as in the time of Darwin, Wallace, Huxley, and Marx, become the sites of extraordinary debates that resonate at the deepest levels of human culture. In this chapter, I explore how one debate—over the role of asteroid and comet impacts in mass-extinction events—has opened a door to a new vision of the Earth and, perhaps, even of human history.

I begin with a polemical question: If postwar oceanography produced a revolution known as "plate tectonics," what has the geological exploration of the solar system produced? This is a ploy to discuss the "axiomatic" deep struc-ture of traditional Earth science, surprisingly undisturbed by plate tectonics but mortally threatened by the post-Newtonian perspective of comparative planetol-ogy. A review of the debate over impact tectonics and "coherent catastrophism" then introduces three case studies. Herbert Shaw's *Craters, Cosmos, and Chronicles* (1994) is a disconcerting work—of Rabelaisian energy and squalor—which uses nonlinear dynamical systems theory (a.k.a. chaos theory) to rethink Earth his-tory as the "coevolution" of mantle dynamics and asteroid bombardment.[8] Stuart Ross Taylor's *Solar System Evolution: A New Perspective* (1992) provides a dignified funeral to "Grand Unified Theories" in the tradition of Kant and Laplace. Taylor offers instead an intellectually breathtaking tour of a radically contingent and historical solar system, which leads, in turn, to a brief rendezvous with Vladimir Vernadsky, Stephen Jay Gould, and the fierce god Shiva. Finally, the cometary astronomers Victor Clube (Oxford) and William Napier (Edinburgh) have developed in the course of several dozen articles and books the case for a "microstructure of terrestrial catastrophism" that includes devastating meteoroid storms every few thousand years. Where is the archaeological and geological evidence for the role of their "Taurid Demons" in human history?

The Dragon and the Comet

> As dwellers on the land, we inhabit only about a fourth part of
> the surface; and that portion is almost exclusively a theatre of
> decay, and not of the reproduction.
> *Charles Lyell*[9]

Imagine a scientific expedition to a distant world that, after nearly 175 years of intensive exploration, utterly failed to discover that planet's most spectacular surface feature: a volcanic mountain chain, stunningly rifted along its spine and nearly 60,000 kilometers long. Given such an improbable case, we would likely consider the expedition's leaders, if not literally blind or mad, to be captive to some lethal epistemological conceit.

Yet the planet is the Earth; and the expedition, modern geology before 1956. The serpentine mountain chain is the Ocean Ridge system—or, as one famous geophysicist likes to call it, the "Dragon."[10] A segment of it, the Mid-Atlantic Ridge, was discovered during efforts to lay trans-Atlantic cable in the 1870s, but it was not reconnoitered until the German Meteor expedition in 1925–27, whose soundings also revealed a prominent median rift. In the early 1930s, the British John Murray expedition confirmed a similar ridge and valley topography in the Indian Ocean.

Finally in 1956, after pioneering explorations of the submarine Pacific Basin, Bruce Heezen, Maurice Ewing, and Mary Tharp were able to demonstrate that the ridges-with-rifts—which they recognized coincided with the distribution of mid-ocean earthquake foci—constituted a continuous global belt, rising an average 2.8 kilometers above the ocean floor. Heezen and Tharp's map, published in the *New York Times*, was the most dramatic addition to human knowledge of the Earth's rock-surface since Columbus. As Menard points out in his memoir of the period, it also "provided a target that unified global exploration during the International Geophysical Year"—the Great Leap Forward of Earth science.[11]

The blinding conceit of traditional geology, of course, was its faith in the uniformity of continental and oceanic crustal processes. Until the 1950s, it was generally accepted that the bulk composition of the Earth's crust was horizontally homogenous, and that continental geology could be extrapolated to the ocean floors. The ocean crust, although putatively more ancient and devoid of relief, was conceived to be similar to the thick granitic crust of the continents. (Some even argued that large parts of the ocean floor consisted of Atlantis-like floundered continents.) Geological exploration of the continental shelves—which indeed proved to be "continental" in lithology—seemed to ratify the orthodox model. It was considered very unlikely that future exploration of deep ocean basins would unveil any unfamiliar tectonic features.

In the event, the frenetic Cold War efforts to map the ocean floors revealed a radically different reality. The ocean crust was basaltic not granitic; thin, not thick; and young, not Archaean. Instead of the predicted abyssal plain, marine geologists were shocked to discover startling families of new landforms: thousands of hillocks and sunken island mounts (guyots) as well as the globe-girdling immensity of the mid-ocean ridge belt.[12] They also found scores of inexplicably long escarpments—fracture zones—offsetting the ridge axes like the cross-ties of Neptune's railroad.

Novel structures, moreover, indicated novel processes, as it gradually became apparent that the rifted ridges were magma factories—"spreading centers"—producing new crust that was eventually swallowed (subducted) at the deep ocean-margin trenches. The great engines of crustal reproduction, in other words, were hidden under the seas.

The Conservative Revolution of Plate Tectonics

These fundamental revelations—produced by the shift of perspective from Lyell's "theatre of decay" to Menard's "ocean of truth"—made a new view of the Earth inevitable. Indeed, plate tectonics is now conventionally recognized as one of the classic, Kuhnian "revolutions" in the history of science. Yet, precisely, what was "revolutionary" about the new theory? To purloin a distinction from social history, was it truly a "radical" revolution that rebuilt the Earth sciences on new foundations, or was it a "conservative" revolution that saved the foundations while erecting a reformed structure of explanation?

Certainly plate tectonics swept away the last remnants of a late-Victorian geophysics that envisioned mountains as the crumpled expressions of internal cooling.[13] In place of "contraction theory"—and its alternative, thermal expansion—it offered a new view of the Earth as a restless chemical factory, where the ocean plates act as conveyor belts between the mantle and crust. As the plunging plates are digested, they produce belches of magma that become volcanoes or the plutonic roots of mountains that eventually erode back into sea-floor sediment to be subducted all over again. Some of the revolution's leaders, like the great Canadian geophysicist J. Tuzo Wilson, even perceived a majestic mega-cycle, many geological epochs in duration, of oceans opening and closing to create or dismantle a Pangaean supercontinent.[14]

Geology's old guard—especially the Soviets—bunkered down in their theoretical winter palaces, desperately clinging to their geosynclines in the face of the sweeping challenge of the new paradigm. But when the sound and fury had died away, it became clear that plate tectonics, for all of its demolition work on the superstructure, nonetheless had left intact the core doctrines underpinning Earth science. "At the time, the new global tectonics appeared to be an extremely radical departure from classical geology. In retrospect, however, we can see that plate tectonics, as envisioned today, is fully consistent with the uniformitarian concepts inherited from Hutton and Lyell: the plates gradually split, slide, and suture, driven by forces intrinsic to the globe."[15]

It is commonplace, of course, to equate foundational doctrine with Hutton and Lyell's uniformitarianism per se. I would argue, however, that Lyellian geology borrowed a decisive, if unspoken premise from Newton (the independence of Earth's process from any astronomical context), and, in turn, passed on an all-important axiom to Darwin (evolution as a gradualistic reform of natural design). Each of these root principles can be represented as simple syllogisms, and together they constitute an axiomatic framework that was rarely questioned by geologists before the 1980s.

Table 1

Earth as a Closed System (Old Axiomatic Framework)

1) *Newton's Guarantee* (he expelled chaos from the solar system)

   a) The solar system is a precision orrery—a well-regulated mechanical system.

   b) Earth's astronomical context, consequently, is unchanging except in the largest time-frames (earliest Earth, oldest Earth).

   c) The Earth, therefore, can be studied as closed system.

2) *Lyell's Principle* (he expelled catastrophe from Earth process)

   a) Tectonic change is gradual over vast periods. Catastrophism, biblical or secular, is a misreading of the geological record.

   b) From a whole-Earth perspective, there is a steady state. Any cross-section of geological time reveals the same processes and landforms.

   c) The present, therefore, is an analogue for the past.

3) *Darwin's Corollary* (he expelled saltation from evolution)

   a) Biological evolution follows Lyell's gradualistic pace of environmental evolution. "Nature never progresses by leaps."

   b) Extinction and speciation, as a result, are uniformitarian in scale and rate. Natural selection fine-tunes adaptation.

   c) Evolution, therefore, has a subtle, progressive logic.

Lyell's vision of a uniformitarian Earth, whose surface is sculpted by the continuous action of small causes over great intervals of deep time, effectively dehistoricized natural history by excluding the unique events—catastrophes—that gave it narrative directionality. At the most fundamental level, moreover, Lyell asserted that geology was immune to astronomical chaos. This guarantee of "cosmic security," which Clube and Napier call the "lynch-pin" of modern science, was provided by Newton's celestial mechanics "in which the Earth moves untroubled by cosmic forces."[16] Although the actual seventeenth-century author of the *Principia* was a fervent astrologer, the Newton deified by the Enlightenment was seen to have expelled from the heavens the apocalyptic Comet—symbol of extraterrestrial influence over Earth's history—that had excited medieval imaginations.[17]

Geology would later concede the role of planetesimal bombardment in the history of the early Earth, but this Hadean eon, with its magma oceans and infernal visitors from space, was conceptually partitioned off from the rest of "normal, closed-system" geological time. According to E. G. Nisbet, for example, "in the Hadean, the Earth was in its formative stage, subject to external influence, while in the Archaean, it was evolving through its own internal constraints as a closed system."[18]

Lyell's anti-catastrophism, moreover, wound Darwin's evolutionary clock, which, in turn, kept perfect Newtonian time:

> Lyell is the source of Darwin's assumption that, viewed from the proper perspective, organic change, both within species and across species boundaries, moves at rates that may speed up here and there but are nonetheless gradual. Darwin also adopted Lyell's perspectivalism about evidence. If we think that geological and biological history are punctuated by discrete, dramatic, catastrophic changes, that is only because, with all our scratching and digging at the earth, we come up only with isolated pieces of data that we falsely aggregate into sudden large changes.[19]

Nothing in plate tectonics directly challenged this closed-system model of gradual Earth evolution or the sacrosanct boundary between geology and astronomy. After 1960, to be sure, some anxiety was aroused by the growing evidence for episodic mass extinctions, as well as the discovery of an unexpected number of probable impact craters. But the ancient foundations of Earth science repelled all assaults until the 1980s. Then, appropriately enough, the blow came from outer space.

Coherent Catastrophism?

> Large-body impacts are not *deus ex machina* explanations, they
> are inevitabilities.
>
> *Walter Alvarez et al.* [20]

The shock of recognition, of course, was the discovery, near Gubbio, Italy, of an indelible extraterrestrial signature—an improbable concentration of iridium—in a pencil-thin layer of clay corresponding to the Cretaceous/Tertiary (K/T) boundary, 65 million years ago. In their famous 1980 article interpreting the significance of the anomaly, physicist Luis Alvarez and his Berkeley-based team—including his geologist son Walter—boldly claimed to have found the smoking gun responsible for the extinction of the dinosaurs.[21] A subsequent cascade of corroborative research, including the identification of a worldwide pattern of iridium anomalies and the 1990 confirmation of the gigantic Chicxulub impact crater at the tip of the Yucatan Peninsula, established beyond reasonable doubt that the Earth had been struck by a bolide—a comet nucleus or asteroid—at least ten kilometers in diameter and with an explosive power equivalent to five billion Hiroshima-sized atomic bombs. Its deadliness was redoubled by the coincidence that the vaporized target rock, which included thick layers of sulfur-rich anhydrite, produced an estimated 600 billion tons of sulfuric acid aerosol that fell as a hellish acid rain and temporarily turned the seas into a "Strangelove ocean."[22]

Despite a long history of speculation about the catastrophic origin of the K/T extinction, the Alvarez-Berkeley group—as William Glen has emphasized—were the first to propose a genuinely "testable," that is, falsifiable, hypothesis.[23] In this sense, their iridium anomaly was comparable to the famous magnetic anomaly—the Vine-Matthews hypothesis—that had cinched the case for plate tectonics. Yet, just as Cold War oceanography had created the larger context of discovery that made plate tectonics necessary, so too has the emergence of comparative planetology out of the space race radicalized the conceptual landscape of the K/T impact debate.

Indeed, many geophysicists, at least, now talk about "two parallel revolutions in the Earth sciences": one that has revealed the unexpected landforms of the ocean floor, and another that has "changed the Moon and planets from astronomical into geological objects."[24] Just as a whole-Earth geology was impossible before the discovery of the mid-ocean ridges and trenches, so, likewise, there was "no way to develop a decent comprehension of the original and evolutionary history of a single, highly evolved, complex planet (Earth) by studying it in isolation from the class of objects of which it is but one member."[25]

The study of comparative planetary history, moreover, requires new understandings of the dynamic processes that organize planetary bodies, large and small, into a single complex system of interaction. Thus, satellite reconnaissance of the Earth, by deciphering the outlines of dozens of astroblemes in remote areas, has confirmed impact cratering as a fundamental geological process. According to one research team, the Earth has experienced at least 200,000 impacts equal to or larger than the famous Meteor Crater in Arizona since the beginning of the Devonian period, 408 million years ago.[26]

The Earth as an Open System

A handful of astrophyscists and planetary geologists, to be sure, had long advocated the likely role of extra-terrestrial agents, comets and asteroids, in Phanerozoic Earth history. But they were voices crying in the wilderness until the 1980s, when mainstream geology finally had to confront powerful new evidence arising both from planetary exploration and the K/T debate. With the advantage of hindsight, it is now clear that an "open-system" view of Earth history, struggling against the hegemony of the old axiomatic framework, has had to surmount four major scientific hurdles.

First, it was necessary to establish unequivocally the extra-terrestrial credentials of proposed impact craters on Earth. Grove Karl Gilbert, the founder of modern geomorphology, opened the debate with his research on Arizona's so-called "Meteor Crater" in the early 1890s. Although Gilbert was the leading advocate of an impact, rather than volcanic, origin for the Moon's craters, he was baffled by the implausible geometry of the Arizona structure and the absence of any large meteoritic mass.[27] The physics of the impact—including the

vaporization of the sixty-meter bolide—was finally clarified by the distinguished
astrophysicist Forest Ray Moulton in 1929, but general acceptance of the impact
hypothesis among geologists had to await Eugene Shoemaker's groundbreaking
studies on shock metamorphism in the 1950s and early 1960s.[28] Later, satellite
and space-shuttle photography revolutionized the search for impact footprints,
while the K/T controversy dramatically raised the scientific stakes. Currently, 145
craters have been authenticated, ranging in diameter from a few hundred meters
to 300 kilometers, including two megastructures—Vredefort in South Africa and
Sudbury in Ontario—that may be the terrestrial equivalents of lunar *maria*.[29]

Second, it was essential to identify a plausible reservoir of potential Earth
impacters in unstable orbits. Here the key event was the 1932 discovery of the
first asteroid—1862 Apollo—in an orbit that intersected the Earth's. By the late
1940s a whole family of Earth-crossing asteroids—known as Apollos—had been
identified, and some astronomers, like Fletcher Watson and Ralph Baldwin, were
warning that major collisions might occur every million years or so.[30] Indeed, a
previous Soviet expedition to Siberia in 1927 had confirmed the meteoroid origin
of a huge 1908 air blast—now estimated to have been equivalent to a twenty-
megaton hydrogen bomb—that leveled nearly 2000 square kilometers of taiga in
the remote watershed of the Tunguska River.[31]

Since Shoemaker, using widefield Schmidt telescopes, launched the Planet-
Crossing Asteroid Survey in 1972, the detection of NEOs has increased almost
exponentially. Currently about 163 Earth-crossing asteroids—many of them
extinct short-period comets—with diameters greater than one kilometer have
been identified: less than 10 percent of the estimated total population.[32] Meanwhile, the Spacewatch program at Kitt Peak Observatory in Arizona has discovered a previously unsuspected "Near Earth Asteroid Belt" composed of smaller
NEOs, called Arjunas, in the range of tens of meters. At least fifty of these
Tunguska-sized mini-asteroids pass between the Earth and the Moon each day.[33]
NASA's NEO Survey Working Group recently estimated that there are between
500,000 and 1.5 million Arjunas in this near-Earth swarm.[34]

It is now understood, moreover, that the asteroids and short-period comets
of the inner solar system have very unpredictable orbits. Indeed chaos ultimately

rules the entire solar system,[35] but on radically different timescales for different classes of planetary objects. As Carlisle points out: "Below a certain size, which depends both on the distance from the Sun and the proximity to other planets, an orbit is stable for no more than about ten million years before it decays into chaos, while a larger object in the same orbit may be stable for ten billion years."[36] The orbits of NEOs, in particular, evolve so chaotically that "they cannot be computed far enough into the future to determine reliably the risk of planetary impact."[37] In addition to chaotic objects, moreover, the solar system also contains numerous chaotic zones, including the unstable 3:1 "Kirkwood Gap" in resonance with Jupiter that some experts believe is a primary source for Earth-crossing asteroids and meteoroid debris.[38]

Third, using the cratering records of the Moon and the other terrestrial planets as a comparative archive, it was necessary to establish some general parameters for impact size and frequency. Within geology, at least, a die-hard volcanist contingent rejected the asteroidal origin of the Moon's craters until the Apollo missions brought back incontrovertible "ground truth" in the form of impact breccias rather than basalts from the lunar highlands.[39] Although most planetary theorists had long regarded the Moon as a fossil mirror-image of the Earth's early impact history, direct exploration has provided measurements of the cratering rate since the end of the so-called "late heavy bombardment" 3.8 billion years ago. NASA's Magellan probe, meanwhile, has produced similar estimates for Venus, where the crater population, because of the relative youth of the planetary surface—approximately 500 million years old—exclusively consists of events within Phanerozoic time.[40]

Planetologists originally had hoped to create a master impact stratigraphy for the entire solar system. However, as planetary scientist Stuart Ross Taylor has pointed out, "cratering fluxes appear to vary widely in different parts of the system, and there does not appear to be that prerequisite, a uniform solar-system-wide flux of impactors."[41] Still, extrapolating from the lunar and Venutian cases, and allowing for differences in gravity and atmospheric density, it has been possible to estimate terrestrial impact frequencies, which can, in turn, be double-checked against the age and size distribution of the 140-plus known cra-

The pioneering calculations of the Shoemakers are reproduced in Table 2.[42] (Atmospheric bolide explosions, of course, are far more common. The current estimate is that a Hiroshima-size event occurs annually, while a megaton airburst is expected once or twice per century.)[43]

Table 2

Estimated Production of Impact Craters on Earth During the Last 100 Million Years

Impacting object	Minimum crater diameter (km)						
	10	20	30	50	60	100	150
Asteroids	820	180	73	10	4.5	0.3	0
Comets	(270)	60	24	8	5.3	1.7	1
All objects	(1090)	240	97	18	10	2	1

Finally, to put all the pieces in place, it was crucial to show a correlation between impact catastrophes and significant watersheds in the history of life. The K/T controversy, as Raup wryly points out, is "box office" and has focused unprecedented interdisciplinary resources on the twin problems of mass extinction and bolide impacts.[44] One result was a dramatic renewal of interest in the approximate thirty-million-year cycle first identified by the eminent Scots geophysicist Arthur Holmes during the 1920s in the course of his research on large-scale sea-level fluctuations. In 1984 geologists tunneling their way through data on the ages of impact craters unexpectedly ran into palaeontologists digging through extinction records. The two chronologies—cratering and mass extinction—coincided (within a rather generous margin of error) around the Holmsian wavelength of 26–33 million years.[45] Writing a few months later about this serendipitous convergence, Stephen Jay Gould proposed naming the cycle of impact and death after Shiva, the dancing Hindu god of destruction and rebirth.[46]

The Cosmic Carousel

Although many geologists immediately disputed the correlation, others detected Holmsian periodicities in an astounding range of tectonic phenomena. Michael Rampino of the Godard Institute for Space Studies, for example, discerns a causal linkage between major impact events, flood-basalt volcanism, and extinctions.

Using lunar analogies, he argues that the largest impacts can sufficiently disturb the crust and mantle to produce flood-basalt eruptions. Seismological models, as well as counterpart structures on Mercury and the Moon, suggest that the shear waves are most likely to be focused by the mantle upon the spot that is antipodal to the impact crater itself. Intriguingly, plate-tectonic reconstructions of the late Cretaceous indicate that the Deccan Traps—a vast flood-basalt province in India—was directly opposite the Chicxulub crater.[47] Rampino's colleague Richard Stothers has further amplified this tectonic connection with a proposed synchronism between lunar cratering history (as a surrogate of Earth's) and the six major episodes of terrestrial mountain-building.[48]

The increasingly sweeping claims of impact theorists—including the Rampino-Stother hypothesis that global tectonic upheavals are periodically driven by collisional energy—soon produced a partisan realignment within the K/T debate. The original battle-line between volcanists and impacters was supplanted by what the British astrophyscist William Napier has characterized as "simple giant impact theory" versus "coherent catastrophism." Advocates of the former position, according to Napier, accept the Chixulub impact but reject terrestrial catastrophism; the K/T event, in their view, was a unique exception to the otherwise general rule of uniformitarian processes.[49] The case for "coherent catastrophism," on the other hand, was most forcefully argued by Harvard astrophysicist Ursula Marvin at a 1988 meeting on the K/T controversy. "It is time to recognize," she told the Second Snowbird Conference, "that bolide impact is a geologic process of major importance which by its very nature, demolishes uniformitarianism itself as the basic principle of geology." "Once the full implications of bolide impact are clearly understood," she added, "geologists will realize that this violent force carries with it a far more revolutionary departure from classical geology than did plate tectonics."[50]

Unlike the plate-tectonicists of the 1960s, contemporary neo-catastrophists like Marvin accept the need to reconstruct the very foundations of Earth science. In their struggle against what Victor Clube calls the "terrestrial chauvinism" of mainstream geology, they repeal Newton's guarantee of Earth's immunity from malign cosmic forces.[51] They instead propose a new view of the Earth (summa-

rized in Table 3) as an "open system" integrated into the solar system's complex
and unpredictable ecology of impacts and chemical exchanges. In an important
sense, they are finally completing the Copernican revolution.

Table 3

Earth as an Open System (New Axiomatic Framework)

1) *Halley's Comet* (chaos theory reveals deep structures of singularity)

   a) The solar system is fundamentally historical: a *bricolage* of unique events and
      assemblages, governed by deterministic chaos and open to galactic perturba-
      tions.

   b) Earth's astronomical environment forms a dynamic continuum with geophys-
      ics and plate tectonics.

   c) Only comparative planetology—a *historical science*—can establish the real
      specificity of solid Earth and biosphere evolution.

2) *Cuvier's Revenge* (catastrophe organizes geology as history)

   a) Process regimes—from earthquakes to supercontinent cycles—are
      (re)structured by unique events (catastrophes); periodicities actually unfold as
      nonlinear histories.

   b) Catastrophic and uniformitarian processes are interwoven at all temporal
      scales.

   c) The past is only a partial analogue for the future.

3) *Vernadsky's Legacy* (Gaia dances with Shiva)

   a) The biosphere is adapted, via the evolution of biological cooperation, to cha-
      otic crises of its planetary environment. Nature usually proceeds by leaps.

   b) Mass extinction events are non-Darwinian factories of natural selection. At
      its extremes, evolution is a punctuated equilibrium between autonomous
      dynamics of environmental and genetic change.

   c) Natural history, like planetary history, is characterized by its irreversible and
      unpredictable contingency.

The new axiomatic framework, not surprisingly, has been most easily
embraced by astronomers and planetologists like Marvin at Harvard, Clube and
Napier at Oxford, Rampino and Stothers at the Goddard Institute, the Shoemakers at Lowell Observatory, R.A.F. Grieve at the Geological Survey of Canada,
Jay Melosh and John Lewis at the University of Arizona, and Stuart Ross Taylor
and Duncan Steel in Australia. Within palaeontology and evolutionary biology,

its major supporters—like David Raup and John Sepkoski at the University of Chicago—tend to be allies of the "punctuated equilibrium" camp of Stephen Jay Gould and Niles Eldridge. Intriguingly, the most radical advocates of neo-catastrophism within academic geology are Asian, like Kenneth Hsu in China (and Switzerland) and Mineo Kumazawa at Nagoya University; a fact that may be related to the cultural specificity of the uniformitarian tradition. Chinese philosophical traditions, in particular, privilege the role of astronomical events in Earth's history. Thus Chinese geologists embraced the Alvarez hypothesis with alacrity and immediately launched a remarkable nationwide hunt for extraterrestrial isotopic anomalies at other extinction horizons, which were later confirmed at the base of the Cambrian and the end of the Permian. They also have been the boldest in reconceptualizing stratigraphic boundaries: "It is obvious that the subdivision of the major stages of geological history should not be dependent solely on the Earth's evolution, but chiefly on the occurrence of astrogeological events."[52]

Aside from their united front on the impact origin of the K/T extinction, many of the neo-catastrophists have also organized their research around the same hypothetical game of planetary billiards. They conjecture that the Shiva cycle of impacts is driven by a galactic tide, probably the Sun's vertical oscillation in the plane of the Milky Way Galaxy.[53] Each time that the solar system, like a carousel horse, rises or falls through the galactic plane, the gravitational attraction of stars or, more rarely, interstellar molecular clouds—the most massive objects in the galaxy—loosen giant comets from the Oort Cloud. As these comets eventually arrive within the planetary belt, Jupiter's gravity eventually ejects most of them out of the solar system, but a few are shunted into short-period orbits close to the Sun.[54] Solar energy disintegrates the cores of the giant comets, which become long trails of smaller comets, asteroid fragments, and zodiacal dust. In Earth-crossing orbits, they produce regular episodes of relatively intense bombardment and extinction, as well as meteor-induced changes in climatic regimes.[55]

The evolution of the Earth, in other words, is galactically controlled through a Rube Goldberg–like chain of gravitational accidents: a hypothesis that is as exhilarating to most astronomers as it is preposterous to most geologists. It remains, however, simply a grand hypothesis, lacking proof in all the crucial details. It is not yet the holy grail of "coherent catastrophism" that Clube, Napier, Steel, and others so ardently seek. Indeed, there has been much confusion about whether a new synthesis, as theoretically unified and testable as plate tectonics, was even possible. Then, in 1994, Herbert Shaw published his magnum opus: a treatise of nearly 700 pages—*Craters, Cosmos, and Chronicles*—which is bravely subtitled "A New Theory of Earth."

Chaotic Couplings

> Meteoroid impacts are not simply cosmic events independently imposed *on* Earth. They are events that have evolved *with* the Earth...
>
> *Herbert Shaw*[56]

The impact debate is sometimes portrayed as the resurrection of the epic nineteenth-century struggle between uniformitarianism and catastrophism. Palaeontologist David Raup, for one, sees the ghosts of Cuvier ("the father of catastrophe") and Lyell ("the father of gradualism") hovering over the K/T battlefield.[57] Yet, as I have tried to show, these traditional labels inadequately capture the even deeper epistemic conflict between closed and open-system views of the Earth's interaction with a stable—or chaotic—solar system. The biggest step for the Earth sciences has not been the admission of an occasional catastrophe or two, but rather the acceptance that terrestrial events, at a variety of timescales, form a meaningful continuum with extraterrestrial processes.

Where most geologists fear to tread, of course, has long been familiar territory to astrophysicists and geophysicists. Space plasma physics, for example, studies a single Sun–Earth system: "a chain of intimately coupled regions extending from the Sun's surface to Earth—the solar photosphere, the solar corona, the solar wind, Earth's magnetosphere, Earth's ionosphere, and Earth's atmosphere."[58] Herbert Shaw, a research geologist with the USGS in Menlo Park, has boldly extended these couplings into a limitless chain of "resonances" linking terrestrial microcosms to galactic macrocosms. Although the crux of Shaw's theory concerns complex feedbacks between the Earth's thermodynamic regime (its core-mantle system) and its impact history, his concept of "Earth system," like Hegel's concept of History, is dialectically all-encompassing:

> It seems necessary to conclude that we are faced with a revolution in concepts of natural history. This revolution is more profound than suggested by correlations of volcanic events with meteoroid impact events ... the most mind-bending implication is the possibility that synchronicity may extend between phenomena as widely separated in space and time as biochemical genetics and intergalactic dynamics. Every type and scale of scientific discipline has been thrown together in ways that violate (or enlighten) long established traditions of autonomy.[59]

Shaw is a tireless missionary within the Earth sciences for the viewpoint of non-linear dynamical systems theory, which, as we shall see, he has used to revise the Alvarez-Berkeley hypothesis in a startling new direction. Like other scientists bored with orthodoxy who lived in the Bay Area during the intellectually restless 1970s, he first saw the Burning Bush of Chaos on the road to Santa Cruz. In his prologue he fondly recalls UC Santa Cruz's legendary "Dynamical Systems Collective" and their pioneering analyses of dripping faucets and other chaotic phenomena. As a rheologist with an extensive background in magma research, a notoriously nonlinear subject, who became fascinated during the 1970s with the role of tidal deformations in the thermal history of the Earth, he was predisposed to relish the "conceptual resonances"—especially between astrophysics and geophysics—that chaos theory seemed to reveal.[60]

In common with other champions of nonlinearity, Herbert Shaw sometimes seems to speak in fractal tongues. *Craters, Cosmos, and Chronicles* is a daunting rainforest of tightly coiled allusion, luxuriant digression, and dense polymathy. Yet it is also infused with the almost drunken energy of theory working at the edge. Two things, I suspect, will arbitrate the reception of this strange but visionary work, even among hardcore neo-catastrophists. First, it will be fascinating to see whether other geophysicists are willing to endorse his "Celestial Reference Frame" theory of ballistic guidance and bolide/mantle-anomaly coupling—to be discussed in a moment—as a viable hypothesis. Second, there is the larger question of chaos theory's credibility, under tough logical interrogation, as a subsuming worldview and revolutionary epistemology.

Critics may be able to treat these two issues separately, but for Shaw himself the validity of his impact scenario is obviously indissociable from his radical anti-reductionism. Consider, for example, his provocative definitions of "causality" and periodicity. Shaw rejects linear cause and effect—for example, "planetary billiard games"—in favor of a structural causality arising from "strange loops" of complex feedback. "The universe of experience," he adds, "is ruled by processes that rarely, and in detail never, boil down to direct proportional parts—parts that can simply be summed to describe the net effects of those processes."[61]

Causality, Complex and Tangled

The statement "impacts cause extinctions," therefore, is nothing but careless shorthand. No single subsystem, however significant, can be causative by itself, since "behaviour exists only by virtue of the non-linear synchronization of the whole." Moreover, "because igneous, solar system and galactic periodicities are founded in common non-linear phenomena, the separation into endogenous and exogenous effects is lost as a basis for classifying mechanisms." "To a non-linear dynamicist," Shaw adds, "the fact that the solar system is imbedded in the galactic system indicates that the former is a special substructure that exists because of resonant interactions with the system as a whole." ("Resonance" by the way, is more than a good vibration; in Shaw's writing it frequently has the same connotation—of multiple, entangled causality—as "overdetermination" in Freud's theory of dreams or "structural causality" in Althusser's theory of modes of production.) Therefore the rigorously correct formulation—although try explaining this to a dinosaur—is that impacts and extinctions are "coupled oscillators."[62]

Similarly, because there are no perfectly independent processes, there are no perfect Newtonian clocks. Abstract, linearly calibrated time—like its negation, pure stochasticity—is an illusion, reinforced by our own species-specific dimensional scaling and self-referentiality. "Only the behaviour of the natural system itself (its properties of non-linear recursion under specified conditions of observation) identifies the contextual meaning of time."[63] Like the geological column, nonlinear time is a fractal order of periodic but non-uniform punctuation (for instance, Mandelbrot's notorious "devil's staircase"), typically scaled at irrational-number ("quasi-periodic") frequencies. As a result, it has intervals of both sharp and fuzzy calibration.[64] In his own reading of Shaw, the British geographer and historian of catastrophism, Richard Huggett, sees a "fundamental question" at stake: "how periodicity develops in systems replete with rate-dependent interactions and complex forms of feedback with other processes. A tentative answer is that fundamental periodicities in any system arise from the variability of non-linear coupling, that they emerge as sets of interacting resonances in the course of an irreversible evolution."[65]

The "interacting resonances" at the core of Shaw's reformulation of the Impact Hypothesis ally the evolution of the Earth's interior with the configuration of near-Earth orbits intermittently occupied by bolide impacters. He brazenly wagers the family farm on the proposition that the major impact events over the last half-billion years have not been random, but ordered; not only in their temporal pattern—as many neo-catastrophists believe—but also in their spatial pattern. Astonishingly, he finds that most known impacts cluster around three "Phanerozoic cratering nodes" (PCNs) connected by a single great circle (or "Phanerozoic swathe"). The PCNs, Shaw claims, are locked in a stable angular relationship with the Earth's axis of rotation that is more fundamental and invariant than the "hot spot reference frame" normally used in palaeogeographical reconstructions. He consequently proposes a "celestial reference frame" (CRF), created by the celestial-sphere projection of the Phanerozoic cratering pattern, as a new framework for interpreting the surface history of the Earth.[66]

But what established this remarkable cratering pattern in the first place? Again, Shaw's answer is breathtaking: the Earth has a "meridional geodetic keel" (MGK)—a longitudinal mass anomaly frozen deep in the mantle (and corresponding to observed primary-wave seismic anomalies)—that is the legacy of the Moon's violent birth 4.4 billion years ago. Through the MGK, the Earth's past controls its future. When "chaotic crises"—including the thirty-million-year galactic tide—swell the population of near-Earth objects (NEOs), the gravitational resonances of the MGK, perhaps assisted by the Moon, recruit some as temporary natural satellites of the Earth. As their orbits decay, the MGK then provides "flight control" that guides the bolides through the CRF to ground zero within the PCN, where their impact reinforces the ancient patterning of the MGK. (A virtuoso example, to be sure, of circular positive feedback!)[67]

Table 4 rudely converts Shaw's "multifractal" argumentation into a linear summary or cartoon. But *Craters, Cosmos, and Chronicles* digresses so frequently into a teach-in on nonlinear dynamical systems theory that it sometimes fails to adequately spotlight or elaborate some of its most daring theses. The most important of these, I think, is the primacy that Shaw assigns to impact cratering as a geological agency. For most geologists, of course, it has been difficult enough to accept extraterrestrial bombardment as an ongoing Earth process. But Shaw goes much further.

Warmed by Bombardment

One of the major loose ends in current geophysical theory, for example, is resolving the ultimate nature of the Earth's thermal regime. Many theorists now doubt that core heat loss is by itself sufficient to produce the mantle plumes that supposedly blowtorch rifts in continents and drive plate motion. At the same time, there is little agreement about the quantitative contribution of radiogenic heating in the lithosphere. At worst, geophysicists see a mysterious discrepancy. In Shaw's view, however, this is an unnecessary conundrum rising out of the underestimation of the cumulative input of energy from impacts. Correctly calculated, impacts and other external energy sources, including tidal friction, balance the books.[68]

Shaw estimates, for instance, that the extraterrestrial donation to the Earth's energy budget over the last billion years has been at least 10^{33} ergs—the equivalent of one billion earthquakes as powerful as the largest historical quake—in Chile in 1960, of 9.5 magnitude.[69] Whereas most geologists, and even some neo-catastrophists, seem to visualize impact cratering as a separate and independent process from plate tectonics, Shaw asserts that they have been indivisibly coupled since the formation of the first cratonic proto-continents in the Early Archean era. He clearly implies that the plate-tectonic "revolution" will only be completed when it integrates the role of impacts and other external energy sources.[70]

Table 4

A Linear Summary of Shaw's Nonlinear Theory of Impact Tectonism

---

1) Collision between Earth and Mars-sized proto-planet (4.4 billion ybp) creates Moon; leaves immense scar basin/magma ocean, probably ancestral Pacific Basin (pp. 110–12, 271–77).

2) Secondary bombardment by orbiting debris reinforces longitudinal mass anomaly (meridonal geodetic keel, or MGK) in deep mantle that evolves from root system of magma ocean (pp. 110–12).

3) This MGK—which Shaw identifies with the present-day seismic (fast P-wave) anomaly around the Pacific Ocean—exerts gravitational influence (spin-orbit resonance) on temporary natural satellites of the Earth (pp. 271–80).

4) A geocentric (orbital) reservoir intermittently fills with small (10 km or less) asteroids and comet debris; chaotic crises—perhaps caused by the periodic galactic tide—produce cascade of impacts on Earth (pp. 3, 11–12).

5) The MGK exercises flight control over impacters, whose trajectories, therefore, are not random walks, but "focused barrages" governed by an invariant celestial reference frame (CRF) locked to the long-term motion of the mean axis of rotation (pp. 165–202, 559).

6) Impacts, therefore, are not random but patterned in space and time. Where swathes of bombardment intersect, phanerozoic cratering nodes (PCN) can be identified. Shaw also accepts the nonlinear (devil's staircase) version of the Holmsian 30-million-year cycle for the peak power of the impacts (pp. 30–41).

7) Thus, there is a circular positive feedback (MGK—CRF—PCN—MGK…) between the Earth's internal mass anomalies and the history of impacts. Shaw refers to this as a kind of "orbital ballistic telegraphy" or "nonlinear laser printer" (pp. 3, 10, 35).

8) Patterned impacts make an important contribution to the Earth's energy budget; there is persistent feedback between the cumulative energy of impacts and net plate-tectonic motion (pp. 245–59).

9) MGK also guides plate motion; principal subduction zones correlate crudely with deep mass anomalies. In a nontrivial sense, biostratigraphy is ultimately structured by the solar system (pp. 277–83, 312).

10) This system is specific to Earth: other planets have not "experienced the same degree of coupling between their internal and external non-linear-dynamical resonance states versus the chaotic evolution of the solar system" (p. 305).

From one daring hypothesis to another, *Craters, Cosmos, and Chronicles* tickles the geological imagination with new and unorthodox interpretations of Earth history in the light of its cosmic context. In a strictly formal sense, it provides an anticipatory model of the kinds of complex feedback and resonance relationships that any comprehensive open-system theory would have to explain. Yet Shaw's extreme antireductionism also acts as a fail-safe mechanism against the potential testing—that is to say, falsification—of his individual hypotheses. For example, when discussing the hypothetical correspondence between cratering patterns, deep mantle anomalies, and geomagnetism, Shaw concludes that they form a "global system of multiple time-delayed coupled oscillators."[71]

"Time-delayed coupled oscillators?" What Shaw actually seems to be saying is that while there is no straightforward relationship in time between these phenomena, they can be correlated at some deeper "time-delayed" or fractal level. In other papers, he evokes "rhythmic patterns of orchestral proportions."[72] This, of course, follows almost tautologically from Shaw's premise of a holistic Earth-cosmos system where phenomena are time-connected by definition rather than by empirical linkages. After all, what can't be coupled in oscillation at some "time delay"? Indeed, Shaw calmly acknowledges that "this is not good news from the standpoint of sorting out specific causes and effects, or from the standpoint of assigning unique frequencies to specific mechanisms."[73] In effect, his resonances, in their "indifferent diversity," threaten to dissipate themselves in the same "bad infinity" that Hegel cautioned was the graveyard of all nondialectical Absolutes (read, global systems).[74]

Elephants on Mars

> Do Shiva and Gaia represent a coupled and coevolved
> system—the stability of one somehow dependent on dis-
> turbances caused by the other?
> *Michael Rampino*[75]

In December 1998, after two decades of exasperating delay and disaster, the Galileo spacecraft finally reached its rendezvous with Jupiter and launched a probe into the hydrogen-and-ammonia maelstrom of the Jovian atmosphere. In the hour before NASA's artificial meteor was incinerated, it transmitted data "so off-

Most of the predicted atmospheric water was missing: a shortfall that invalidates existing models of Jupiter's energy budget and chemistry. The *New Scientist* predicted that "planetary scientists may now have to completely rethink their theories of the giant planet's structure."[77]

This was no more than business as usual. Since 1959 when the USSR's Luna 3 documented the absence of *maria* on the Moon's far side, surprise has been the standard ration of planetary reconnaissance. The geological survey of the solar system has revealed new realities as completely unexpected as those discovered by Cold War oceanography during the 1950s. As one team of researchers pointed out, "The sense of novelty would probably not have been greater if we had explored a different solar system."[78] In essence, theory has been unable to predict planetary composition or dynamics in advance of exploration. The solar system is distinguished by the conspicuous absence of "normal planets." Each instead is an eccentric individual with its own unique chemical and tectonic identity. Moreover, the same rule applies to miniature as well as major worlds, since "every satellite has turned out to differ in some significant feature from its neighbour." Singularity, in other words, seems to have a fractal distribution across scales.[79]

The distinguished Australian cosmochemist Stuart Ross Taylor has drawn important epistemological conclusions from this planetary exceptionalism. "The complexity of the solar system," he argues, "is not in accord with theories that start from some simple initial condition." In his view, planetary diversity confounds the classical Kant-Laplacean project of discovering "some uniform principles, analogous to the Periodic Table or Darwinian evolution, from which one might construct clones of our solar system." In particular he disputes elegant, "top-down" theories of the system's origin like the "equilibrium condensation model," so popular in the 1970s, with its postulates of a chemically zoned nebula and an orderly process of planetary accretion.[80]

In contrast, Taylor views accretion as inherently messy and event-driven. Instead of a "grand unified theory," he proposes a "bottom-up" narrative in which planetary formation is the outcome of a kind of deterministic chaos. "If large impacts of planetesimals are a characteristic feature of the final stages of planetary accretion, then the details of the individual impacts become to some extent free parameters."[81] Even in its most general features, then, the present solar system cannot be theoretically "deduced" from the equations of the state of the original solar nebula. Singular impact events, unpredictable in any Laplacean model, have produced some of the characteristic anomalies itemized in Table 5.

Table 5

Historical (Chaotic) Features of the Solar System
(summary from Taylor)

1) atmospheres
2) obliquities
3) irregular satellites (in retrograde and nonequatorial orbits)
4) density of Mercury
5) slow retrograde rotation of Venus
6) Earth/Moon system
7) Asteroid belt
8) crustal dichotomy of Mars
9) rings of giant planets
10) Pluto/Charon system

Taylor's solar system, in a word, is radically historical—which is to say, chaotic—and impact cratering is its existential moment. Classical celestial mechanics, of course, allowed no role for collisions, or for intrinsically unpredictable or irreversible outcomes. Taylor's conception of the solar system as *bricolage*, however, prescribes innumerable possible evolutionary paths out of the same initial conditions. The major planetary features "are the result of events that might readily have taken a different turn."[82] Although "other planetary systems doubtless exist," the duplication of the singular sequence that has produced the present solar system is as likely as "finding an elephant on Mars."[83]

Life on Earth
This is a view broadly shared by other planetary scientists, who, in recent years, have given more precise definition to the external preconditions for the existence of the Earth's biosphere.[84] There seem to be four paramount contingencies. First

of all, computer modeling indicates that a climate conducive to life on Earth depends upon the extraordinarily narrow orbital parameters that define a "continuously habitable zone" (CHZ) where water can exist in a liquid state. This is sometimes called the "Goldilocks problem," since Venus is too hot, Mars too cold and only the Earth "just right" for life.[85]

> If the Earth's orbit were only 5 percent smaller than it actually is, during the early stages of Earth's history there would have been a "runaway greenhouse effect" [like Venus], and temperatures would have gone up until the oceans boiled away entirely!... [On the other hand] if the Earth–Sun distance were as little as 1 per cent larger, there would have been runaway glaciation on Earth about 2 billion years ago. The Earth's oceans would have frozen over entirely, and would have remained so ever since, with a mean global temperature less than –50 degrees F [like Mars].[86]

(Even if Mars, as some exobiologists speculate, has preserved a "stealth biosphere" of primitive anaerobic bacteria in the interstices of its Archean-aged crust, its engine of evolution, for all intents and purposes, is turned off.)[87]

Second, Jupiter plays an essential role as the Earth's big brother and protector. Its vast mass prevents most Sun-bound comets from penetrating the inner solar system. Without this Jovian shield, it has been estimated that the Earth would have experienced a flux of comet-sized impactors *a thousand times larger* than actually recorded during geological time. K/T-sized catastrophes, in other words, would have taken place at 100,000-year rather 100-million-year intervals. If the Earth's surface were not actually sterilized by such a bombardment, it is hard to imagine the survival of taxa beyond the most primitive levels of evolution. (Nils Holm and Eva Andersson suggest deep-sea hydrothermal systems as the only possible refuges from heavy bombardment.)[88] This suggests, as a minimum precondition, that only planetary systems that contain both terrestrial planets and gas giants are capable of sustaining complex life forms.[89]

Third, the gravitational shield of the giant planets, while highly efficient, must occasionally fail to protect the Earth. One of the central paradoxes of planetary science—the so-called "temperature-volatiles conundrum"—is that the temperatures for the existence of liquid water exist only in the inner solar system, while the key building blocks of life, including water itself, occur primarily beyond the asteroid belt. Indeed, the Earth "probably formed almost entirely devoid of the biogenic elements." Thus some modulated frequency of cometary impacts has been necessary to convey oceans of water, as well as carbon and nitrogen, from the volatile-rich regions of the solar system to Earth. The evolution of the biosphere, in other words, has been dependent upon a subtle cometary trade surplus in imported volatiles that stops short of the impact magnitudes that would erode the atmosphere or vaporize the oceans.[90]

Fourth, the Earth's unique and massive satellite, the Moon, plays a crucial role in stabilizing the obliquity of the Earth's rotational axis. Locked into spin-orbit resonance with the Earth—that is, the lunar day is equivalent to the lunar month—the Moon with its high angular momentum keeps the Earth tilted within one degree (plus or minus) of 24.4° relative to its plane of revolution. Obliquity, of course, is what creates the terrestrial seasonality so important to the evolution and diversity of life. Mars, in contrast, has a wildly oscillating tilt and chaotic seasonality, while Venus, rotating slowly backward, has virtually no seasonality at all. It may be impossible for a "Gaian-type" biosphere, with its complex network of self-regulating biogeochemical cycles, to evolve under such conditions, regardless of the presence of water or not.[91]

Extraterrestrials: Where Are They?

The "historical" interpretation of the solar system, therefore, would seem to reinforce the existentialist pessimism of Jacques Monod—Resistance veteran and Nobel-winning biologist—who in *Chance and Necessity* (1970) argued that "the universe was not pregnant with life, not the biosphere with man."[92] Alternately, in the words of Zuckerman and Hart, "the US program of planetary exploration, while highly successful from a technical and scientific standpoint, has failed to produce even a hint of an extraterrestrial biology."[93] The recent identification of planetary objects outside the solar system, and the likelihood that many more will soon be discovered, does not necessarily increase the chances of finding alien life-forms.[94] Indeed, another paradox implicit in our new understanding of the solar system is that while planetary systems may be common, life-sustaining planets are probably exceedingly rare. Planetary exploration, in other words, has drawn conclusions about the prevalence of life—or, at least, its environmental preconditions—that are antipodal to contemporary molecular biology which tends to see life as everywhere emergent, robust, even a "cosmic imperative." (In *Vital Dust*, for example, Nobel laureate Christian de Duve confidently asserts that "the universe is a hotbed of life" and that "trillions of biospheres coast through space on trillions of planets."[95]) Likewise, the new view of Earth history afforded by solar-system exploration and the K/T debate confounds the ultra-Darwinist dogma that gene-centered natural selection is the dominant, if not exclusive, evolutionary process. A planetary perspective demands a reconceptualization of the fossil record, not only in terms of impact catastrophism and mass extinctions, but also in terms of the largely noncompetitive interactions that weave local ecologies into an evolving global biosphere.

Indeed, one of the major intellectual fruits of the NASA era has been the rekindling of interest in the "biosphere" concept proposed in 1926 by the Soviet mineralogist Vladimir Vernadsky (1863–1945). Vernadsky attempted to break down the artificial barriers between biology and geology that had arisen during the late nineteenth century. He argued that life is the true architect of the Earth's surface, and that evolution necessarily reorganizes the chemistry of the atmosphere, oceans, and lithosphere. The biosphere, therefore, encompasses all the biogeochemical networks—"transformers" in Vernadsky's terminology—by which life increasingly commands geology and accelerates the "the biogenic migration of atoms."[96] (Although life, at any one time, may seem only an insignificant scrim on the face of the Earth, the total mass of all organisms that have ever lived has been estimated as 1000 or even 10,000 times the mass of the Earth itself!)[97] Moreover, as Alexej Ghilarov has emphasized, Vernadsky believed in a "deep natural connection between all organisms and the planetary scale." As founder of the USSR'S Committee on Meteorites, he was also fascinated by the possible chemical interchanges between biosphere and cosmos.[98]

Vernadsky's ideas exercised a subtle influence on the emergence of the "ecosystem" paradigm in North American ecology—largely through the friendship of his son with the Yale biologist C. Evelyn Hutchinson[99]—but biospheric thinking had little prestige in non-Soviet Earth science until the late 1960s when NASA began investigating techniques for discovering life on other planets. The debate about life on Mars, in particular, prompted a re-examination of life's global imprint on Earth processes, especially the maintenance of an oxygen (21 percent) atmosphere in radical chemical disequilibrium. James Lovelock, one of the NASA researchers, more or less reinvented Vernadsky with his bold "Gaia" hypothesis that, for organisms to survive, they must "occupy their planet extensively and evolve with it as a single system. ... The evolution of the species and the evolution of their environment are tightly coupled together as a single and inseparable process."[100]

Although most scientists remain skeptical of Lovelock's claims that Gaia is a literal "superorganism" with the ability to regulate environmental conditions for its own survival, the Gaia controversy, like the K/T debate, has refreshed the Earth sciences by stimulating interdisciplinary research on global biogeochemical cycles and the evolutionary implications of plate-tectonic theory.[101] The "Vernadskian worldview," as Lovelock calls it, has also been congenial to those biologists who see symbiosis and mutualism as major mechanisms in evolution.[102] As Lynn Margulis and others have argued, the Darwinian arena of competitive natural selection assumes a pre-existing biosphere largely constructed by cooperative guilds of bacteria, protozoa, fungi, and plants.[103] Moreover, the ultra-Darwinist camp has failed to recognize that a competitive organism at one level of analysis is somebody else's ecology at another. ("The 'individuals' handled as unities in the population equations are themselves symbiotic complexes involving uncounted numbers of live entities integrated in diverse ways in an unstudied fashion."[104])

The greatest challenge to Darwinist gradualism, however, remains the role of periodic mass extinctions. On the one hand, intransigent geneticists and population ecologists, led by Richard Dawkins, continue to define evolution primarily in terms of "background extinction" rates and the gradual turnover of species. On the other hand, geologists and palaeontologists are focused on the irrefutable evidence of catastrophic cascades of extinction that have decimated higher taxa, including entire orders and even classes. Can the event at the end of the Permian era, for example, that extinguished 96 percent of the Earth's species really be compared to Darwin's thin wedge and the fine-tuning of natural selection?

The neo-catastrophist response is scathing. Stephen Jay Gould and Niles Eldredge are cruel but honest when they insist that mass extinction really means "evolution by lottery" and "survival of the luckiest." Extinction events ruthlessly reset all ecological clocks.[105] Species interactions play little role in determining survival in the aftermath of great bolide impacts, and the major watersheds in the evolution of life do not work by orthodox natural selection. (For example, the decisive "adaptive advantage" of mammals during the K/T catastrophe simply may have been their concentration in circumpolar regions least affected by the low-latitude Chicxulub impact.) Hence Darwin, like Laplace, must submit to revision by chaos and historical contingency. Discussing the radical evolutionary implications of the Alvarez hypothesis, Eldredge clearly echoes Taylor and Monod: "Evolutionary history ... is deeply and richly contingent. Gone are the last vestiges of the idea that evolution inevitably and inexorably replaces the old and comparatively inferior with superior new models. Evolution, at least on a grand scale, is not forever tinkering, trying to come up with a better mousetrap."[106] Evolution by catastrophe, Michael Rampino adds, also entails speciation through a different process than the classic gradualist mechanisms of geographic isolation and adaptive change. Catastrophe replaces the linear temporal creep of microevolution with nonlinear bursts of macroevolution. Comet showers accelerate evolutionary change by injecting huge pulses of sudden energy into biogeochemical circuits. Nutrient recycling is stimulated and bolides add new stocks of organic molecules.[107] (According to the comet astronomer Duncan Steel, the Earth accumulates an average of 200,000 tons of cosmic debris per year.)[108]

Most importantly, catastrophes break up static ecosystems and clear adaptive space for the explosive radiation of new taxa—like mammals after the K/T horizon. Rampino, awed by this dialectic of creative destruction, openly wonders if impact catastrophe is not the real driving force behind the movement toward greater biological diversity, and if Gaia has not evolved in intricate choreography with Shiva—the ultimate form of macroevolution?[109]

Rampino is, of course, not the first to wonder about the Earth's strange waltz with apocalyptic comets. Indeed, the postmodern hypothesis that life on a planeetary scale is periodically renewed by extraterrestrial cataclysm uncannily echoes the baroque cosmology of Restoration England. In a paper read to the Royal Society in 1694 Edmund Halley—after whom, of course, the famous comet is named—argued that the Earth over time inevitably becomes sterile and infertile. Great comets annihilate existing populations, but also renew the Earth's fertility and prepare the way for new creations. "Deadly in the short term but healthful in the long run, cometary collisions allowed for the succession of worlds."[110]

Taurid Demons

> Even within the short history of *Homo sapiens*, the most violent events on Earth have been extraterrestrial impacts.
> *John Lewis*[111]

Like the supporters of Gaia, the advocates of the Shiva hypothesis are a small but seminal scientific minority. In light of the overwhelming evidence that impact cratering has remained a significant geological process throughout the Phanerozoic aeon, they have built a respectable case for its episodic role in detouring evolution down new and unpredictable pathways. And, together with other neo-catastrophists, they have added impressive scaffolding to the Gould-Eldredge theory of punctuated equilibrium and chaotic Earth history. The Impact Hypothesis, in other words, now has a firm purchase within geological time (10^7 to 10^9 years), and an important beachhead, established by the K/T debate, within evolutionary time (10^6 to 10^8 years). But what about ecological time (10^4 to 10^6 years) and cultural time (10^2 to 10^4 years)? Have impact events left their catastrophic imprints within human history? Few questions in Earth science are more controversial.

In 1993, for example, two Austrian scientists published a book in which they claimed to solve the mystery of "the darkest chapter in human history": the deluge catastrophe chronicled in the Gilgamesh Epic, the Old Testament, the Vedas, and scores of oral traditions all over the world. Edith Kristan-Tollmann and Alexander Tollmann marshaled anthropological and geological evidence to support their thesis that seven large cometary fragments had struck the ocean nearly ten millennia ago, causing terrible tsunamis now recalled as the Flood. On the one hand, they cited numerous ancient accounts of "seven invading stars," ranging from the "great burning mountains" of the Jewish prophet Henoch to the "fiery sons of Muspels" in Icelandic saga, which they interpreted as contemporaneous with flood legends. On the other hand, they presented "geological proof" in the form of tektites (glassy impact ejecta) "with an age of nearly ten thousand years" from Australia and Vietnam, as well as the small Kofels impact crater in the Austrian Tyrol—caused, they said, by a "splinter" of the Noachian comet.[112]

Although the Tollmans created a predictable stir in the popular media, they were punctually massacred in the scientific press. In one review, a team of eminent meteoriticists, including R.A.F. Grieve, dismissed their "evidence" as "sheer fantasy" and characterized their approach as "pseudo-science in the tradition of Donnelly and Velikovsky." The critics systematically demolished their tektite dating as well as their exaggerated claims for the Kofels structure.[113]

Of course, this is hardly the first case where the self-proclaimed confirmation for biblical or mythical events has turned into a fiasco. The whole intellectual terrain of archaeological and historical catastrophism has been polluted by far too many bizarre hypotheses and spurious discoveries. Rare or unique astronomical phenomena have become the staple diet of a burgeoning genre of fringe-science and mega-disaster books.[114] Yet, at the risk of ridicule, cometary astronomers—led by Victor Clube at Oxford and William Napier at Edinburgh—have persevered in arguing a scientific case for cosmic intervention in human history. They claim, in fact, that some ancient societies almost surely experienced the shattering equivalent of nuclear warfare.

In an important restatement of the theory that they have been developing over the last twenty years, Clube and Napier—together with David Asher and Duncan Steel from the Anglo-Australian Observatory—contrast two different interpretations of impact tectonics. "Stochastic catastrophism," as they call it, is concerned with the extraterrestrial influence upon the geological *longue durée*. It relies on averaged cratering rates, derived from known terrestrial structures and from the impact records of the Moon and inner planets. The history of small, but more frequent impactors (less than 1 kilometer in diameter) is discriminated against in this approach because they do not individually produce global consequences, and because terrestrial erosion more quickly erases their footprints.[115] Moreover, the data set is too coarse-grained to resolve temporal heterogeneities—clustered events, for instance—within frequencies of less than 1 million years. As a result, it cannot differentiate what Clube and Napier call the "microstructure of terrestrial catastrophism" within the time periods relevant to human evolution.[116]

"Coherent catastrophism," on the other hand, contends that "the overall effect of giant comets on terrestrial evolution is far more complex than that of single giant impacts." Impact events operate on all timescales, and punctuational crises are "hierarchically nested in the overall manner of glacial-interglacials." To visualize this entire spectrum of phenomena, especially the influence of small-body impacts, Clube and Napier have augmented cratering data and near-Earth-object censuses with a wealth of historical data, including medieval European records and Chinese astronomical archives.[117]

While other researchers have been absorbed in the search for catastrophic celebrities, like billion-megaton exterminator bolides, Clube and Napier have been focused on the study of the more prosaic population of comminution products—small Apollo asteroids, meteoroidal swarms, zodiacal dust—resulting from the breakup of giant comets. Although the comets arrive in Earth-crossing orbits only at intervals of 100,000 years or so, their debris "interacts catastrophically with the Earth on relatively short time-scales: 10^2–10^5 years." Clube, in fact, has argued that because of the frequency of small-body impacts, terrestrial catastrophism may be "uniformitarian" at all timescales greater than a millennium. Thus, a "new world view, embracing the effects of the full range of 'small bodies' in the Solar System … has become one of the outstanding imperatives of our time."[118]

Astronomy Meets History

The foremost concerns of Clube and Napier are the organized swarms of 10- to 300-meter objects—Arjunas and small Apollos—that they claim bombard the Earth every few thousand years.[119] These bodies are responsible for the "low-level catastrophism on 'biblical time-scales'" that was "the subject of interest to Newton and his contemporaries as well as to the early-nineteenth-century catastrophists." "We believe," they add, "that there is at least one such cluster of material currently existing, which over the past twenty thousand years has produced episodes of atmospheric detonations with significant consequences for the terrestrial environment, and for mankind." The "major danger to mankind" that Clube and Napier have in mind are the Taurids.[120]

The Taurid complex is a large circumsolar stream of cometary and asteroidal debris in sub-Jovian orbit that Clube and Napier believe resulted from the hierarchical breakup of a large comet over the last 20,000 years. In addition to the remnant Comet P/Encke, the complex includes a half-dozen Apollo-type asteroids more than 1 kilometer in diameter, four major meteor streams—responsible for the daytime and nocturnal Taurid showers—a zodiacal dust cloud, and, most menacingly, hundreds of thousands of boulder-sized Arjunas that, like so many orbiting hydrogen bombs, are capable of multimegaton atmospheric blasts or surface impacts on Earth. (As we saw earlier, recent Spacewatch data suggests that the general population of these mini-asteroids may be forty times greater than previously believed.)[121]

The Taurids, according to Clube and Napier, pose a maximum hazard every few thousand years when the precession of the stream's orbit produces an intersection with the Earth's. These "comet ages," which may last several centuries, are characterized by clusters of Tunguska-sized impacts mostly coordinated with the daytime beta-Taurid meteor showers (24 June to 6 July), as well as unusual concentrations of meteoric dust. Duncan Steel confirms a Taurid signature in a variety of geophysical phenomena, including unusual iridium and nickel anomalies within the Greenland ice cap. Within the late Holocene, Clube and Napier, supported by Steel and Asher, identify three periods of intense bombardment: in the centuries around 3000 BC, 1200 BC, and 500 AD.[122]

Where, then, are the ensuing Taurid catastrophes in the human record? Staring us in the face, according to Clube and Napier, in the form of the religious and secular texts from an extraordinary diversity of cultures. They have devoted three books—The Cosmic Serpent (1982), The Origin of Comets (1989), and The Cosmic Winter (1990)—to an exhaustive elaboration of possible correlations between human accounts of cometary crises and Taurid storms as predicted by orbital mechanics. To take two examples: around 3000 BC, they find a universality of malevolent sky gods and the political dominance of astronomer priests in an era which may have been environmentally disrupted by the breakup of Comet P/ Encke's huge progenitor. Then, nearly 3500 years later, when the Taurids are again crossing the Earth's orbit, Chinese astronomers warn of a "strange comet" in the same year that a monk describes scenes—still historically unexplained—of extraordinary devastation and social collapse in Britain."[123]

But Clube and Napier do not confine themselves to mere correlation and anecdote. In addition to providing a new framework for understanding ecological upheavals, they also make sweeping claims about the history of human attitudes toward nature. Virtually all major cultures have experienced alternating periods of cosmic optimism and cosmic despair. At different times, the heavens have been seen as providential or as menacing. The Taurid storms, they propose, have been the secret prime movers of this eschatological cycle. The two cometary crises of the pre-Roman world, for example, were "fundamental turning points in human history and they fathered doom-laden beliefs in the end of the world which were probably realistic and which have never since entirely disappeared." Again, in the comet-vexed seventeenth century, Cromwell rose to power on a millenarian wave fueled by "supernatural illumination."[124]

Clube and Napier make the intriguing suggestion that materialism in the classical world arose as a reaction to the tyranny of prehistoric sky gods and their malign cosmos. Indeed, the mainstream of European philosophy from Aristotle to Kant has been obsessed with exorcising the dread of comets and providing rational explanations for their appearance. The "critique of comets," therefore, has been an essential precondition for establishing a view of history as open to infinite possibility within a benevolent or at least neutral cosmos. Conversely, philosophical materialism has always been opposed by millenarian ideologies, including neo-Platonism, that see history as bounded by cataclysm, even human extinction. In this tradition comets have always been recognized as potent agents of destruction and/or rebirth. Clube and Napier challenge historians to look for an actual environmental cycle—like the Taurid precessions—that may lie behind this alternation of optimistic and catastrophist philosophical systems.[125]

Yet the Clube and Napier thesis remains more fascinating than convincing. Although they bring new tools—especially Clube's study of the Chinese fireball records—to the interpretation of the transcultural "apocalyptic record," the correlations with natural history are hardly more than circumstantial. What really can be deduced, after all, from such "evidence" as the fact that the Hebrew word for iron ore *(necoshet)* literally means "snake shit" or that friezes of Quetzlocoatl sometimes make the plumed god look like a fiery comet?[126] Clube and Napier, unfortunately, are rarely able to corroborate exegesis from myth or religion with material evidence from archaeology—as, for instance, in the recent excavation in Holland of 1700-year-old pit configurations in the shape of the constellation Taurus.[127]

Bombardment or Eruption?

More egregiously, they have bent events out of all recognition to support their contention—derived from orbital calculations—that Taurid bombardments produced the crisis of civilization in the eastern Mediterranean around 1450 BC. The Minoan world did collapse catastrophically, but the explanation does not require the resolution of any extraterrestrial mysteries. Indeed, the modern tourist to the island of Santorini (Thera) can gaze straight into a huge sea-filled caldera that was produced by four major volcanic events between 1628 and 1450 BC. The final eruption of Thera, which blew the entire volcanic cone into the atmosphere, has been described as "a natural catastrophe unparalleled in all of history," and its indelible signature can be found all over the world, in California tree rings as well as Greenland ice cores. With ten times the explosive power of Krakatau, Thera did terrible damage to Crete and produced a tsunami that engulfed the coasts of Palestine and Egypt. Such geomyths as the biblical plagues and the exodus event may be descriptions of the eruptions and their aftermaths.[128]

Nor is even the Taurid Complex itself established as an indisputable fact. In a recent review of orbital calculations, one team of researchers concluded that the "hypothesis of genetic relationship among some or all of these bodies in [comet] Encke-like orbits is not supported [nor was it refuted]." They also disputed Clube and Napier's other key hypothesis that Encke had been formed from the breakup of a giant cometary nucleus over the last 20,000 years. And, finally, they predicted that most objects in Taurid orbits would fall into the Sun, not collide with the inner planets.[129] Clube and Napier's task, it seems, is like trying to capture an elephant with a butterfly net. Textual evidence alone, even where reinforced with Chinese and Japanese astronomical records, is obviously insufficient to establish strong correlations between predicted Taurid storms and the dark ages of human history. Like other impact theorists, Clube and Napier are monocausalists who minimize the role of diverse catastrophic agencies, especially volcanism, in Earth history. Yet, unlike the Tollmans, the Oxford team has marshaled considerable evidence—from studies of cratering fluxes and NEO populations—to support their core hypothesis that major bombardment episodes are likely to have intersected human history.[130]

One missing link, of course, is archaeological evidence of otherwise unexplained destruction; the other is geological documentation of impacts or airbursts. It should not be supposed that the Holocene cratering record is in any sense complete. For example, the recently discovered Iturralde impact basin (eight kilometers in diameter) in the Bolivian Amazon—the product of a 500-megaton impact around 8000 BC—has not yet been inspected from the ground.[131] Nor have geologists yet explored the Curuca River region, near the Brazil–Peru border, where Landsat photographs suggest a crater swathe produced by the "Brazilian Tunguska event" of 13 August 1930 that has attracted the interest of researchers trying to calculate the impact frequency of Arjunas.[132] Even more spectacular, and still incompletely studied, is the 30-kilometer chain of craters near Rio Cuarto, Argentina, which was not discovered until 1989. The best radiocarbon dating so far is around 2900 BC: well within the time-window of Clube and Napier's first Taurid catastrophe. Moreover, the impact by the 300-meter bolide packed a punch (1000 megatons) equivalent to Thera, ten Krakataus, or 50,000 Hiroshima-sized nuclear weapons.[133]

Yet only one-quarter of the cratering record is preserved on land. As NASA's John Lewis reminds us, "three out of every four impacts on Earth hit water."[134] Since the beginning of the First Dynasty in Egypt, it is estimated that 500 NEOs, equal in size or larger to the Arjuna that created Meteor Crater in Arizona, have struck the earth. Approximately 375 of these should have impacted in the ocean.[135] When a high-velocity object weighing hundreds of thousands or millions of tons strikes the ocean, the resulting hydrodynamics are extraordinary. Large-body impacts, for example, will produce huge, explosive steam bubbles—up to 500 km³ in the case of the Chicxulub bolide—and heat vast areas of the ocean.[136] Some theorists predict that such extreme ocean warming could feed energy to runaway super-hurricanes, known as "hypercanes," which, in turn, might lift vast, climate-changing quantities of water and aerosols into the stratosphere.[137]

Bodies bigger than 100 meters, moreover, will manufacture nightmare tsunamis right out of science fiction. Steel, for example, predicts that a 500-meter-diameter asteroid impact in the Pacific, say 1000 kilometers off the coast of Los Angeles, could produce a tsunami several kilometers in height.[138] Even an impact one-tenth that size would still result in a super-wave nearly twenty stories high, and two cheerful Japanese researchers recently warned of a 1–2 percent possibility that "most artificial construction on the coastline of the Pacific will be destroyed in the next century by asteroidal impact in the Pacific."[139]

Given the threefold greater frequency of marine impacts and the tenfold greater lethality of tsunamis versus ground impacts—where is the smoking gun? So far, there are only a few tantalizing clues. Lewis points to an extraordinary tsunami deposit on Lanai as high as the Empire State Building—but it is more likely due to a landslide on the neighboring island of Hawai'i.[140] Perhaps some of the great waves depicted in Japanese painting and traditionally attributed to earthquakes are actually meteoroidal in origin. But nobody really knows, since "the entire research field of geologic assessments of tsunami produced by impactors is virtually nonexistent and needs to be initiated."[141] In the last instance, Clube and Napier are probably correct to surmise that some outpost of ancient humanity was subjected to a sudden horror almost beyond description. But the where and when of this holocaust remain one of history's greatest enigmas.

Hidden Histories

> Hutton's rigidity is both a boon and a trap. It gave us deep
> time, but we lost history in the process. Any adequate account
> of the earth requires both.
>
> *Stephen Jay Gould*[142]

On 1 March 1996, a team of geologists from the United States, Austria, and South Africa announced a stunning discovery: a 90-kilometer impact crater buried beneath the mouth of Chesapeake Bay. After using seismic reflection to determine the outline of the multiring structure, they drilled scores of boreholes inside and outside the crater rim. They recovered samples of impact melt breccia and other telltale evidence of massive shock metamorphism in the target rock. In addition, the detailed chemistry of the breccia cores coincided with the isotopic composition of tektites previously retrieved by the Deep Sea Drilling Project from the seabed 330 kilometers northeast of Chesapeake Bay. The researchers dated the impact at 35.5 million years ago, near the boundary of the Eocene and Oligocene epochs.[143]

The Chesapeake impact—30 million years after the K/T cataclysm—broadly correlates with the "terminal Eocene event." Unlike the K/T boundary, however, where majority scientific opinion now supports the instantaneity of extinction,[144] the drastic biotic changes at the end of the Eocene probably occurred in "stepwise" fashion over several million years. Indeed, iridium anomalies and tektite deposits indicate serial impacts—an "Uzi burst" in Shaw's terminology—of which the Chesapeake bolide was perhaps the largest. Rampino and Haggerty suggest a complex sequence of feedbacks between individual impact events, atmospheric dust veils, and rapid climate shifts, leading to the formation of the East Antarctic ice cap and a colder global climate in the Oligocene.[145]

The discovery of the Chesapeake structure, at the predicted Holmsian frequency, as well as the recent correlation of a chain of craters—probably cometary in origin—in northern Chad to the great Devonian extinction, reinforces the case for coherent catastrophism.[146] Even the spacing between impacts conforms to the Clube-Napier hypothesis of episodic giant-comet storms unleashed by galactic tides.[147] Yet, as the radical divergence between the K/T and terminal Eocene upheavals demonstrates, it is probably foolish to expect any replication in detail between different catastrophes. As one of the leading researchers on mass extinctions has emphasized, "nonlinearities, thresholds, and elaborate feedbacks often rule out the reconstruction of simple cause-and-effect cascades. The same forcing factor might have radically different effects depending on the state of the system at the time of perturbation, and several alternative forcing factors might produce the same biotic response."[148] What great impact events share in common is their capacity to reorganize the global biosphere on a supra-Darwinian scale. But each major extinction sequence—including those in which impacts may not have played a role—has been a unique historical conjuncture.[149]

Indeed, from a geological standpoint, impacts are the functional equivalents of wars and revolutions in human history. As we have seen, catastrophes are both condensations of temporal process—for instance, a million years of "normal" environmental work condensed into hours, even seconds—and exponential escalations of the energy circulating through the planetary metabolism. In this dual sense, comet bombardments act as superchargers of geological and biological evolution. But catastrophe is equally an eruption of historicity: a literal cascade of events that are too singular, complex, and, perhaps, sensitive to initial conditions to be captured in any exact model or grand unified theory. Whereas Shaw appeals to the philosopher's stone of chaos theory, Clube and Napier evoke the concept of "episodicity" to characterize periodic functions modified by random events. Yet the banned word "narrative" might also be relevant. The permanent revolution in earth science, first and above all, has been an insurrection in the name of Natural History with a capital "H." And the greatest discovery of solar system exploration has been an existential Earth shaped by the creative energies of its catastrophes.

1996

Notes

*I am very grateful to Phil "Pib" Burns (Northwestern University), Andrew P. Ingersoll (Cal Tech), and Herbert Shaw (USGS, Menlo Park) for their generous comments.

1. The mass of the boulder-sized fragment was estimated as 2500 tons. It was the fourth multikiloton meteoric explosion detected by satellites since 1988. See I. Nemtchinov, T. Loseva, and A. Teterev, "Impacts into Oceans and Seas," *Earth, Moon, and Planets* no. 72, (1996), pp. 414–16. Also see Duncan Steel, *Rogue Asteroids and Doomsday Comets*, New York 1995, pp. 203–5.

2. John Lewis, *Rain of Iron and Ice*, Reading, Mass. 1996, pp. 146–49; "Comet Shoemaker-Levy 9," special section in *Science*, no. 267, 3 March 1995, pp. 1277–1323. A 120-yard-diameter asteroid came almost as close (75,000 miles) on 14 June 2002. An object this size caused the Tunguska catastrophe in Siberia in 1908, flattening 800 square miles of forest.

3. The traditional distinction between asteroids and comets has eroded with the recognition that many near-Earth asteroids are actually extinct (degassed) comets. See David Jewett, "From Comets to Asteroids: When Hairy Stars Go Bald," *Earth, Moon, and Planets*, no. 72 (1996), pp. 185–201.

4. This much discussed event was recorded in the Canterbury Chronicle and probably occurred on 25 June in the modern Gregorian calendar. This corresponds to the annual arrival of the Beta Taurid stream discussed later in this chapter. A prominent Muslim astronomer is convinced that the same meteoroid swarm produced an earlier impact on the Moon on 26 or 27 June 617 AD—an event described in the Qur'an as the "splitting of the Moon." J. Hartung, "Was the Formation of a 20-km-Diameter Impact Crater on the Moon Observed on June 18, 1178?" *Meteoritics* 2 (1976), p. 187; and Imad Ahmad, "Did Muhammad Observe the Canterbury Meteoroid Swarm?" *Archaeoastronomy* 11 (1989–93), pp. 95–96.

5. See Steel, *Rogue Asteroids*, ch. 12. Japan, meanwhile, is planning to land on an asteroid in 2002, while NASA hopes to bring back dust samples from Comet Wild–2 in 2003.

6. Ibid.; and D. Morrison and E. Teller, "The Impact Hazard: Issues for the Future," in T. Gehrels, ed., *Hazards Due to Comets and Asteroids*, Tucson, Ariz. 1994, p. 1140.

7. "Biogeochemistry" studies the global transformation and movement of chemical substances whose cycles pass through the biosphere. Exobiology is a comparative science, arising out of solar-system exploration, concerned with the conditions for life on Earth and other planets, solar and extra-solar.

8. It is important to lay my cards on the table from the very beginning. In what follows I understand chaos theory to entail three principal experimental results: 1) most deterministic motion—temporal change—in nature is sensitively dependent upon initial conditions; 2) the fine structure of most "random" phenomena is actually some form of complex order; and 3) the phase transition from one ordered state to another is

"avalanche" of determinate, but unpredictable events organized via feedback relationships. Chaos, however, reveals itself in strikingly different patterns: as an infinite flowering of complexity (Mandelbrot sets); an eternal recurrence of alternating domains of order and disorder (meandering rivers); or as a dialectic of evolution and revolution (natural and human history).

9. Charles Lyell, *Principles of Geology*, vol. 1, London 1872 (12th edn.), p. 97.

10. William Kaula, "The Earth as a Planet," *Geophysical Monograph* 60, American Geophysical Union, Washington, D.C. 1990, p. 18.

11. See H. W. Menard, *The Ocean of Truth: A Personal History of Global Tectonics*, Princeton, N.J. 1986, pp. 94–107. Unfortunately, Mary Tharp's pioneering contribution is ignored in most later histories of the plate-tectonic revolution, including William Glen's *The Road to Jaramillo, Critical Years of the Revolution in Earth Science*, Stanford, Calif. 1982, and H. E. Le Grand's *Drifting Continents and Shifting Theories*, Cambridge 1988.

12. "The most common landform on the face of the Earth was the previously unsuspected abyssal hill. Largely on the basis of Midpac and Capricorn echograms, I concluded in 1956 that 90 percent of the Pacific sea floor is a hilly terrain and that the remaining 10 percent is smooth only because hills have been buried by sediment or fluid lava flows" (Menard, *The Ocean of Truth*, p. 52).

13. For a brilliant intellectual history, see Mott Greene, *Geology in the Nineteenth Century: Changing Views of a Changing World*, Ithaca, N.Y. 1982.

14. The Open University faculty has produced an outstanding introduction to contemporary geology in the light of plate tectonics: F. Brown, C. Hawkesworth, and C. Wilson, eds., *Understanding the Earth: A New Synthesis*, Cambridge 1992.

15. Ursula Marvin, "Impact and Its Revolutionary Implications for Geology," in V. Sharpton and P. Ward, eds., *Global Catastrophes in Earth History: An Interdisciplinary Conference on Impacts, Volcanism, and Mass Mortality*, GSA Special Paper 247, Boulder, Colo. 1990, p. 153.

16. See Stuart Ross Taylor, *Solar System Evolution: A New Perspective*, Cambridge 1992, p. 287; and Victor Clube and William Napier, *The Cosmic Winter*, Oxford 1990, pp. 96, 127.

17. For a fascinating discussion of Newton as astrologer, alchemist, and catastrophist—including his views on cometary portents—see David Kubrin, "'Such an Impertinently Litigious Lady': Hooke's 'Great Pretending' vs. Newton's *Principia* and Newton's and Halley's Theory of Comets," in Norman Thrower, ed., *Standing on the Shoulders of Giants: A Longer View of Newton and Halley*, Berkeley, Calif. 1990.

18. E. G. Nisbet, "Of Clocks and Rocks—The Four Eons of the Earth," *Episodes* 14, no. 2 (1994), p. 326. In another context, however, Nisbet explains that Archaean geologists have always been uncomfortable with orthodox uniformitarianism: "The fabric of interpretation of Archaean rocks," he emphasizes, "must be built up again from first principles" (*The Young Earth: An Introduction to Archaean Geology*, London 1987, pp. 3–6).

19. David Depew and Bruce Weber, *Darwinism Evolving: Systems Dynamics and the Genealogy of Natural Selection*, Cambridge, Mass. 1995, p. 109. The authors stress the

importance of Newtonian dynamics as the ontological ground under Lyell and Darwin's feet (see chapter 4).

20. Walter Alvarez et al., "Uniformitarianism and the Response of Earth Scientists to the Theory of Impact Crises," in V. Clube, ed., *Catastrophes and Evolution: Astronomical Foundations* (1988 BAAS Mason Meeting of Royal Astronomical Society at Oxford), Cambridge 1989, p. 14. The idea of an asteroidal or cometary origin for the K/T extinction, as opposed to the discovery of the decisive iridium anomaly itself, had been previously advanced by several researchers, including Victor Clube and William Napier in a remarkable paper, "A Theory of Terrestrial Catastrophism," *Nature*, no. 282 (1979), p. 455.

21. L. Alvarez, W. Alvarez, F. Asaro, and H. Michel, "Extraterrestrial Cause for the Cretaceous-Tertiary Extinction," *Science*, no. 208 (June 1980). A Pemex geophysicist, Antonio Camargo, was the first to actually propose (in 1980) an impact origin for Chicxulub. It took nearly a decade, however, for other Mexican and US researchers to review the drill core samples and publish the evidence. For a comprehensive description, see Virgil Sharpton et al., "A Model of the Chicxulub Impact Basin Based on Evaluation of Geophysical Data, Well Logs and Drill Core Samples," *Geol. Soc. Am. Spec. Paper* no. 307, 1996.

22. Richard Grieve, "Impact: A Natural Hazard in Planetary Evolution," *Episodes* 17, nos. 1–2 (1995), p. 14. The fortuitous chemical composition of the Chicxulub target rock may also explain why the two other Phanerozoic impacts, Manicouagan in Canada and Popigai in Siberia, which are comparable in magnitude, failed to produce extinction events on the same scale.

23. William Glen, "What the Impact/Volcanism/Mass-Extinction Debates Are About," in William Glen, ed., *The Mass-Extinction Debates: How Science Works in a Crisis*, Palo Alto, Calif. 1994, pp. 7–12.

24. James Head, "Surfaces of the Terrestrial Planets," in J. Kelly Beatty and Andrew Chaikin, eds., *The New Solar System*, Cambridge, Mass. 1990 (3d edn.), p. 77.

25. Noel Hinners, "The Golden Age of Solar-System Exploration," in Beatty and Chaikin, eds., *The New Solar System*, p. 7.

26. G. Neukum and B. Ivanov, "Crater-Size Distributions and Impact Probabilities on Earth from Lunar, Terrestrial-Planet, and Asteroid Cratering Data," in Gehrels, ed., *Hazards Due to Comets*, p. 411.

27. Kathleen Mark, *Meteorite Craters*, Tucson, Ariz. 1987, pp. 25–39.

28. Astronomers, on the other hand, had little difficulty accepting the impact hypothesis. By the 1940s, clear expositions of explosive cratering and matter-of-fact acknowledgements of the impact origin of the Arizona crater could be found in standard textbooks. See Steel, *Rogue Asteroids*, p. 34. For the Shoemaker story, see David Levy, *The Quest for Comets*, New York 1994.

29. R. Grieve and L. Pesonen, "Terrestrial Impact Craters: Their Spatial and Temporal Distribution and Impacting Bodies," *Earth, Moon, and Planets*, no. 72 (1996), pp. 357–76.

30. Steel, *Rogue Asteroids,* pp. 28–22.

31. E. Krinov, *Giant Meteorites,* Oxford 1966.

32. David Morrison, Clark Chapman and Paul Slovic, "The Impact Hazard," in Beatty and Chaikin, eds., *The New Solar System,* p. 61.

33. T. Gehrels and R. Jedicke, "The Population of Near-Earth Objects Discovered by Spacewatch," *Earth, Moon, and Planets,* no. 72 (1996), pp. 233–42; and Richard Kerr, "Earth Gains a Retinue of Mini-Asteroids," *Science,* no. 258 (1992), p. 403.

34. "The recent discovery that we exist in an asteroid swarm has enormous long term consequences, and its historical importance may someday be seen to rank with Columbus's discovery of the New World." NASA Ames Space Science Division (on the Internet), NEO Survey Working Group's 1995 Report, sections IV and XI.

35. "Indeed, the development of chaos theory goes back to the prize question of the Swedish Academy in 1890: 'How stable is our planetary system?' This question was finally answered by the French mathematician Henri Poincaré, who resolved that even the three-body problem is mathematically non-integratable and that even the planetary system has its history. He created the mathematics of bifurcation dynamics which has led to the contemporary concept of deterministic chaos" (Friedrich Cramer, "The Two Modes of Time—of Planets and of Life," *Interdisciplinary Science Reviews* 20, no. 1 [1995], p. 62).

36. David Brez Carlisle, *Dinosaurs, Diamonds, and Things from Outer Space,* Palo Alto, Calif. 1995, pp. 82–83

37. J. Tatum, D. Balam, and G. Aikman, "Astrometric Recovery and Follow-Up of Near-Earth Asteroids," *Planet. Space Sci.* 42, no. 8 (1994), p. 611.

38. See "Band Gaps," chapt. 8 in Ivars Peterson, *Newton's Clock: Chaos in the Solar System,* New York 1993.

39. Ursula Marvin and David Raup have criticized the indifference of mainstream geology to the stunning advances in meteoritics and planetology during the 1960s and 1970s. Claude Allegre, on the other hand, has rebuked Western geologists for ignoring—at least until the mid 1970s—the pioneering Soviet theory of "progressive planetary accretion" that had been developed by Otto Schmidt and his students back in the 1940s. See Marvin, "Impact and Its Revolutionary Implications for Geology," pp. 152–55; David Raup, "The Extinction Debates: A View from the Trenches," in Glen, ed., *The Mass-Extinction Debates,* pp. 146–47; and Claude Allegre, *From Stone to Star: A View of Modern Geology,* Cambridge 1992, pp. 109–15. For an overview of the Soviet contribution, see A. Levin and S. Brush, *The Origin of the Solar System: Soviet Research 1925–1991,* New York 1995.

40. See Henry Cooper, *The Evening Star: Venus Observed,* New York 1994, pp. 238–49.

41. Taylor, *Solar System Evolution,* p. 146.

42. Shoemaker, Eugene, and Carolyn Shoemaker, "The Collision of Solid Bodies," in Beatty and Chaikin, eds., *The New Solar System,* p. 261. "Comet nuclei tend to break up as they pass through the atmosphere, so the rate at which they produce craters smaller than 20 km may have been suppressed." See also Neukum and Ivanov, "Crater Size Distributions and Impact Probabilities," pp. 359–416.

43. Morrison et al., "The Impact Hazard," p. 63.

44. David Raup, *Extinction: Bad Genes or Bad Luck?* New York 1991. See also Philippe Claeys, "When the Sky Fell on Our Heads: Identification and Interpretation of Impact Products in the Sedimentary Record," *US National Report to International Union of Geodesy and Geophysics, 1992–1994,* supplement to *Review of Geophysics* (July 1995).

45. See Walter Alvarez and Richard Muller, "Evidence from Crater Ages for Periodic Impacts on the Earth," *Nature,* no. 308 (1984), pp. 718–21; and David Raup and Jack Sepkoski, "Periodicity of Extinctions in the Geologic Past," *Proc. Natl. Acad. Sci. USA,* no. 81 (1984), pp. 801–5. Critics, on the other hand, claim that the terrestrial cratering record has too many serious biases in age and size, resulting from erosional effects, to statistically support this periodicity. Grieve, in particular, thinks that the controversy will be more likely resolved by a more thorough study of the lunar cratering archive. See R.A.F. Grieve et al., "Detecting a Periodic Signal in the Terrestrial Cratering Record," *Proc. Lunar Plant. Sci. Conf. 18th,* 1988, pp. 375–82; and Alexander Deutsch and Urs Scharer, "Dating Terrestrial Impact Events," *Meteoritics,* no. 29 (1994), p. 317.

46. "[Shiva] holds in one hand the flame of destruction, in another (he has four in all) the *damaru,* a drum that regulates the rhythm of the dance and symbolizes creation. He moves within a ring of fire—the cosmic cycle—maintained by an interaction of destruction and creation, beating out a rhythm as regular as any clockwork of cometary collisions". (Stephen Jay Gould, "The Cosmic Dance of Shiva," *Natural History,* August 1984, p. 14).

47. There is also a proposed correlation between the end Permian extinction, the Siberian Traps, and a fossil crater on the undersea Falkland Plateau. See Michael Rampino, "Impact Cratering and Flood Basalt Volcanism," *Nature,* no. 327 (1987), p. 468; and (with K. Caldiera), "Major Episodes of Geologic Change: Correlations, Time Structure and Possible Causes," *Earth and Planetary Science Letters,* no. 114 (1993), pp. 215–27.

48. Richard Stothers, "Impacts and Tectonism in Earth and Moon History of the Past 3800 Million Years," *Earth, Moon, and Planets,* no. 58 (1992), p. 151.

49. William Napier, "Terrestrial Catastrophism and Galactic Cycles," in Clube, ed., *Catastrophes and Evolution,* pp. 135, 160.

50. Marvin, "Impact and Its Revolutionary Implications for Geology," p. 153. It is interesting, of course, to speculate whether a middle-of-road position—that is, that Earth history alternates between periods of catastrophism (acceleration of rates of change) and uniformitarianism (uniformity of rates of change)—can be made theoretically consistent. See the discussion in Richard Huggett, *Catastrophism: Systems of Earth History,* London 1990, pp. 194–200.

51. Victor Clube, "The Catastrophic Role of Giant Comets," in Clube, ed., *Catastrophes and Evolution,* p. 85.

52. Dao-yi Xu et al., *Astrogeological Events in China,* New York 1989, p. 221.

53. Napier, however, argues that the 30-million-year cycle is actually a harmonic artifact of a more fundamental fifteen-million-year periodicity. See Napier, "Terrestrial Catastrophism and Galactic Cycles," p. 141.

54. For a recent review of the periodic versus stochastic galactic forces acting upon the Oort comet cloud, see John Matese et al., "Periodic Modulation of the Oort Cloud Comet Flux by the Adiabatically Changing Galactic Tide," *Icarus*, no. 116 (1995), pp. 255–68.

55. Ibid., pp. 138–57; also Steel, *Rogue Asteroids*, pp. 97–103. In addition, see the special issue, "Dynamics and Evolution of Minor Bodies with Galactic and Geological Implications," *Celestial Mechanics and Dynamical Astronomy* 54, nos. 1–3 (1992). R. Mullet argues that changes in meteor flux are responsible for the 100,000-year glaciation cycle that has dominated the Quaternary period ("Extraterrestrial Accretion and Glacial Cycles," *New Developments Regarding the KT Event and Other Catastrophes in Earth History*, Houston, Tex. 1994, pp. 85–86).

56. Herbert Shaw, *Craters, Cosmos, and Chronicles: A New Theory of the Earth*, Palo Alto, Calif. 1984, p. xxvii.

57. David Raup, *The Nemesis Affair*, New York 1986.

58. Donald Williams, "Space Plasma Physics," in *Geophysical Monographs* 60, p. 21.

59. Herbert Shaw, *Terrestrial-Cosmological Correlations in Evolutionary Processes*, USGS, Open-File Report 88-43, 1988, p. 5.

60. Shaw, *Craters, Cosmos, and Chronicles*, pp. xxvii–xxxvii. Shaw used the Jovian example—where huge tidal forces produce cataclysmic thermal convection on major moons—to make the subversive argument that "many significant entries to Earth's mass energy budget in addition to solar energy come from the outside." See "The Periodic Structure of the Natural Record, and Nonlinear Dynamics," *Eos* 68, no. 50 (1987), p. 1654.

61. Shaw, *Craters, Cosmos, and Chronicles*, p. 27.

62. Shaw, "The Periodic Structure," pp. 1653, 1665. Shaw thus finds much of the controversy over statistical testing of the Holmsian periodicity as almost beside the point.

63. Shaw, *Craters, Cosmos, and Chronicles*, p. 20; and his "The Liturgy of Science: Chaos, Number, and the Meaning of Evolution," in Glen, ed., *The Mass-Extinction Debates*, p. 171.

64. Shaw emphasizes "that our knowledge of nonlinear periodicity owes much to the pioneering studies by V. I. Arnold, B. V. Chirikov, and others in the Soviet Union" ("The Periodic Structure," p. 1665).

65. Huggett, *Catastrophism*, p. 198. He is referring to "The Periodic Structure."

66. Grieve and Shoemaker also conclude that the "spatial distribution of known craters is not random," but in their eyes this is simply a trivial consequence of differential rates of erosion. "Very few known craters occur outside cratonic areas, which, with their relatively low levels of tectonic and erosional activity, are the most suitable surfaces for the acquisition and preservation of craters in the terrestrial geological environment." See R.A.F. Grieve and E. Shoemaker, "The Record of Past Impacts on Earth," in Gehrels, *Hazards Due to Comets*, p. 419.

67. In a recent article Michael Rampino and Tyler Volk analyze a linear swathe of eight Palaeozoic craters across Kansas, Missouri, and Illinois which they believe were probably produced by objects in natural Earth orbits. They note that this offers support for "Shaw's thesis that nonlinear resonance effects in the inner Solar System might make capture in

Earth orbit more probable" ("Multiple Impact Event in the Palaeozoic: Collision with a String of Comets or Asteroids?," *Geophysical Research Letters* 23, no. 1 [1996], pp.49–52).

68. See Shaw, *Craters, Cosmos, and Chronicles*, pp. 245–59.

69. Ibid. p. 258. (The conversion of earthquake energy is my own.)

70. "The central thesis of my work holds that a system of geodynamic feedback, between the cumulative energy of impacts and the net motions of plate tectonics and continental drift, has persisted throughout geologic history" (ibid. p. 35).

71. Ibid. p. 219.

72. Shaw, *Terrestrial-Cosmological Correlations in Evolutionary Processes*, p. 2.

73. Ibid. p. 231.

74. G.W.F Hegel, *Phenomenology of Spirit*, Oxford 1977, pp. 7–9. Responding to a draft of this article, Shaw answers my criticism by pointing out that "unique frequencies [read: linear cause-and-effect] are illusions in natural phenomena—rather there are multifractal spectra of very wide range that have power peaks in certain frequency ranges in different places and/or mechanisms, and at different times."

75. "Gaia Versus Shiva: Cosmic Effects on the Long-Term Evolution of the Terrestrial Biosphere," in Stephen Schnieder and Penelope Boston, eds., *Scientists on Gaia*, Cambridge, Mass. 1993, p. 388.

76. Richard Kerr, "Galileo Hits a Strange Spot on Jupiter," *Science*, no. 271, 2 February 1996, pp. 593–94.

77. Bob Holmes, "Probe Finds Jupiter Short of Water," *New Scientist*, 27 January 1996, p. 7; and Bob Holmes and Govert Schilling, "Hidden Helium Heats Jupiter from Within," *New Scientist*, 3 February 1996, p. 16.

78. B. Smith et al., *Science*, no. 204 (1979), p. 951.

79. Taylor, *Solar System Evolution*, p. xi.

80. Ibid., pp. 12–13, 289.

81. Ibid., p. 181.

82. Ibid., p. xii. Taylor stresses that even the impacters seemed to be recruited from diverse populations specific to different regions of the solar system. He thus concludes that "the accretion of the planets was largely a local affair, that mixing between the inner and outer reaches of the solar system was minimal and was perhaps localized even within the inner solar system..." (ibid., p. 175).

83. Ibid., pp. xi, 287–89. See also Taylor, "The Origin of the Earth," in Geoff Brown et al., eds., *Understanding the Earth*, Cambridge 1992.

84. The modern debate on terrestrial biological exceptionalism begins with Lawrence Henderson's landmark book *The Fitness of the Environment: An Inquiry into the Biological Significance of Matter*, New York 1913.

85. Michael Rampino and Ken Caldeira, "The Goldilocks Problem: Climatic Evolution and Long-Term Habitability of Terrestrial Planets," *Annu. Rev. Astron. Astrophys.*, no. 34 (1994), p. 83.

86. Michael Hart, "Atmospheric Evolution, the Drake Equation and DNA: Sparse Life in

an Infinite Universe," in Ben Zuckerman and Michael Hart, eds., *Extraterrestrials: Where Are They?* Cambridge 1995 (2d edn.), pp. 216–17. For slightly more optimistic views of the width of the CHZ, see Rampino and Caldeira, "The Goldilocks Problem," pp. 105–6; and G. Horneck, "Exobiology, the Study of the Origin, Evolution and Distribution of Life within the Context of Cosmic Evolution: A Review," *Planet. Space Sci.* 43, nos. 1–2 (1995), p. 195.

87.   P. Boston, M. Ivanov, and C. McKay, "On the Possibility of Chemosynthetic Ecosystems in Subsurface Habitats on Mars," *Icarus*, no. 95 (1992), pp. 300–8.

88.   "Hydrothermal systems are about the only environments where primitive life would have been protected against postulated meteorite impacts and partial vaporization of the ocean." See Nils Holm and Eva Andersson, "Abiotic Synthesis of Organic Compounds under the Conditions of Submarine Hydrothermal Systems: A Perspective," *Planet. Space Sci.*, nos. 1–2 (1995), p. 153.

89.   Jonathan Lunine, "The Frequency of Planetary Systems in the Galaxy," *Planet. Space. Sci.* 43, nos. 1–2 (1995), pp. 202–3.

90.   C. Chyba, T. Owen, and W. Ip, "Impact Delivery of Volatiles and Organic Molecules to Earth," in Gehrels, ed., *Hazards Due to Comets*, pp. 13–14, 43–44.

91.   See James Pollack, "Atmospheres of the Terrestrial Planets," in Beatty and Chaikin, eds., *The New Solar System*, pp. 91–106; and D. Brownlee, "The Origin and Early Evolution of the Earth," in Samuel Butcher et al., eds., *Global Biogeochemical Cycles*, London 1992, p. 18.

92.   Jacques Monod, *Chance and Necessity*, New York 1971, pp. 145–46.

93.   "Preface to the Second Edition," Zuckerman and Hart, *Extraterrestrials: Where Are They?* p. xi.

94.   Popular accounts, for example, of the discovery of the planet 70 Vir B have emphasized its congenial temperature—possibly as high as 80°C—and the possible presence of water. But, as *New Scientist* points out, "claims that the planet may actually support life, however, are almost certainly wishful thinking because 70 Vir B is probably a gas giant, with no rocky surface on which life could evolve." See Gabrielle Walker, "Alien Worlds Boost Search for Life," *New Scientist*, 27 January 1996, p. 57.

95.   Christian de Duve, *Vital Dust: The Origin and Evolution of Life on Earth*, New York 1995, p. 292.

96.   Alexej Ghilarov, "Vernadsky's Biosphere Concept: An Historical Perspective," *The Quarterly Review of Biology* 70, no. 2 (1995), pp. 193–203. Together with the Norwegian V. M. Goldschmidt (1888–1947), Vernadsky founded the modern science of biogeochemistry.

97.   Rampino and Caldiera, "The Goldilocks Problem," p. 103.

98.   Ghilarov, "Vernadsky's Biosphere Concept."

99.   See Frank Golley, *A History of the Ecosystem Concept in Ecology*, New Haven, Conn. 1993, pp. 58–59.

100.   James Lovelock, *The Ages of Gaia*, New York 1995 (rev. edn.), pp. 7, 11.

101.   See the feast of ideas in Boston, *Scientists on Gaia*.

102. Again, early-twentieth-century Russian science—as in the cases of planetary theory and nonlinear dynamics—was in the forefront of symbiotic biology with such seminal, but forgotten figures as Famintsin, Mereschkovskii, and Kozo-Polianski. See Lynn Margulis, "Symbiogenesis and Symbionticism," in Lynn Margulis and Rene Fester, eds, *Symbiosis as a Source of Evolutionary Innovation: Speciation and Morphogenesis*, Cambridge, Mass. 1991, pp. 2–7.

103. Mark and Diana McMenamin, for example, have recently argued that mycorrhiza and other symbiotic fungi integrate all terrestrial flora into a single system—"Hypersea"—of nutrient circulation. See their *Hypersea: Life on Land*, New York 1994. For a brilliant defense, however, of the (competitive) Darwinian pathways to mutualism and the orchestration of biomes into a global biosphere, see E. G. Nisbet, "Archaean Ecology," in M. Coward and A. Ries, eds., *Early Precambrian Processes, Geological Society Special Publications*, no. 95, London 1995, pp. 46–47.

104. Ibid., p. 10.

105. Simon Conway Morris, "Ecology in Deep Time," *TREE* 10, no. 7 (1995), p. 292.

106. Niles Eldredge, *Reinventing Darwin: The Great Debate at the High Table of Evolutionary Theory*, New York 1995, p. 156.

107. "Comets are of special interest to exobiology because—among all celestial bodies—they contain the largest amount of organic molecules.... It has been argued that a soft landing of a cometary nucleus on the surface of a 'suitable' planet may provide all prerequisites for life to originate" (Horneck, *Exobiology*, p. 192).

108. Steel, *Rogue Asteroids*, p. 91; but Horneck estimates only 10,000 tons (*Exobiology*, p. 194).

109. Rampino, "Impact Cratering and Flood Basalt Volcanism," pp. 387–88. See also M. Rampino and B. Haggerty, "The 'Shiva Hypothesis': Impacts, Mass Extinctions and the Galaxy," *Earth, Moon, and Planets*, no. 72 (1996), pp. 441–60.

110. Sara Genuth, "Newton and the Ongoing Teleological Role of Comets," in Thrower, ed., *Standing on the Shoulders of Giants*, pp. 302–3. Halley's assertion—with which Newton apparently concurred—that the earth was "the wreck of a former world" caused consternation in Church of England circles and led to his loss of the Savilian chair in astronomy at Oxford. See Kubrin, in ibid., pp. 64–66.

111. Lewis, *Rain of Iron and Ice*, p. 557.

112. Edith Kristan-Tollmann and Alexander Tollmann, "The Youngest Big Impact on Earth Deduced from Geological and Historical Evidence," *Terra Nova*, no. 6 (1994), pp. 209–17.

113. Alexander Deutsch et al., "The Impact-Flood Connection: Does It Exist?" *Terra Nova*, no. 6, pp. 644–50. Ignatius Donnelly was the apocalyptic American populist whose *Ragnarok: The Age of Fire and Gravel* was a sensation of the 1880s; while Immanuel Velikovsky is, of course, the notorious author of *Worlds in Collision*, 1950.

114. For a recent example, see D. Allan and J. Delair, *When the Earth Nearly Died: Compelling Evidence of a Catastrophic World Change—9500 BC*, Bath 1995.

115. For a discussion of the dependence of the "decay constant" on crater size, see S. Yabushira, "Are Periodicities in Crater Formations and Maas Extinctions Related?" *Earth, Moon, and Planets,* no. 64 (1994), pp. 209–10.

116. D. Asher, S. Clube, W. Napier, and D. Steel, "Coherent Catastrophism," *Vistas in Astronomy,* no. 38 (1994), pp. 5, 20–21.

117. Ibid.; and S. Clube, "Evolution, Punctuational Crises and the Threat to Civilization," *Earth, Moon, and Planets,* no. 72, 1996, p. 437.

118. Ibid., pp. 81, 94, 104.

119. The small Arjunas—100 times more numerous than predicted by the power-law spectrum of true asteroids—may be "mostly fragments of decayed comets." See D. Rabinowitz, "The Flux of Small Asteroids Near the Earth," *Asteroids, Comets, Meteors 1991,* Lunar and Planetary Institute, Houston, Tex. 1992, p. 484.

120. V. Clube "The Catastrophic Role of Giant Comets," p. 101; and Asher et al., "Coherent Catastrophism," p. 5.

121. Clube and Napier, *The Cosmic Winter,* pp. 152, 157; Asher et al., "Coherent Catastrophism," pp. 7, 15; Steel, *Rogue Asteroids,* p. 203; and Tom Gehrels, "Collisions with Comets and Asteroids," *Scientific American,* March 1996, p. 59.

122. V. Clube, "The Catastrophic Role of Giant Comets," pp. 101–4; Steel, *Rogue Asteroids,* pp. 151–53. In addition, Chinese and Japanese meteor records, which Clube and Napier have mined, show flux maxima around 1000 AD and 1900 AD, which have been interpreted as Taurid in origin. See Ichiro Hasegawa, "Historical Variation in the Meteor Flux as Found in Chinese and Japanese Chronicles," *Celestial Mechanics and Dynamical Astronomy,* no. 54 (1992), pp. 129–42.

123. V. Clube, "The Catastrophic Role of Giant Comets," pp. 39, 103–4.

124. Clube and Napier, *Cosmic Winter,* p. 172.

125. *Cosmic Serpent,* pp. 157 passim and 254; Clube and Napier, *Cosmic Winter,* p. 172.

126. They frequently rely on the adventurous interpretation of cuneiform omens by Judith Bjorkman, "Meteors and Meteorites in the Ancient Near East," *Meteoritics* 8 (1973), pp. 91–132.

127. Clube and Napier, *Cosmic Serpent,* p. 196; Govert Schilling, "Stars Fell on Muggenburg," *New Scientist,* 16 December 1995, pp. 33–34. On the existence of similar "Taurid pits" at Stonehenge I, see Steel, *Rogue Asteroids,* pp. 148–49.

128. For the Clube-Napier view, see ch. 10, "1369 BC," in *The Cosmic Serpent,* pp. 224–72. On Thera, see P. LeMoreaux, "Worldwide Environmental Impacts from the Eruption of Thera," *Environmental Geology,* no. 26 (1995), pp. 172–75; and D. Hardy et al., eds., *Thera and the Aegean World III, Proceedings of the Third Congress* (Santorini), The Thera Foundation, London 1990.

129. See G. Valsecchi et al., "The Dynamics of Objects in Orbits Resembling That of P/Encke," *Icarus* 181, no. 1 (1995), pp. 177–79; and J. Klacka, "The Taurid Complex of Asteroids," *Astronomy and Astrophysics,* March 1995.

130. Other impact experts dispute the probability of an impact catastrophe in ancient

history. Morrison et al., for example, argue that the estimated dozen or more multiple-megaton events within human history probably involved sparsely inhabited areas like the Amazon or Siberia: "It is unlikely that Tunguska-like impact would destroy even one city in the entire 10-millennium span of human history" ("The Impact Hazard," p. 67).

131. K. Campbell, R. Grieve, J. Pacheco, and B. Garvin, "A Newly Discovered Probable Impact Structure in Amazonian Bolivia," *National Geographic Research* 5 (1989), pp. 495–99.

132. Patrick Huyghe, "Incident at Curuca," *The Sciences* (March/April 1995), p 16. The Curuca impact, if 1 megaton or larger, would suggest a higher frequency of Tunguska-type events, perhaps two or more per century, than previously believed (pp. 14–15).

133. P. Schultz and R. Lianza, "Recent Grazing Impacts on the Earth Recorded in the Rio Cuarto Crater Field," *Nature*, no. 355 (1992), pp. 234–37 (they estimate that the largest crater was the result of a 350-megaton impact); and Lewis, *Rain of Iron and Ice*, pp. 88, 99 (estimate of total megatonnage).

134. Lewis, *Rain of Iron and Ice*, p. 151.

135. The source of the Meteor Crater was an iron asteroid fragment thirty to fifty meters in diameter. "The earth collides with an object of this size or larger once in a century," Tom Gehrels, "Collisions with Comets and Asteroids," *Scientific American*, March 1995, p. 55. However, Neukum and Ivanov indicate a much lower cratering rate. See "Crater Size Distributions and Impact Probabilities," p. 411.

136. H. Melosh, "The Mechanics of Large Meteoroid Impacts in the Earth's Oceans," in Leon Silver and Peter Schultz, eds., *Geological Implications of Impacts of Large Asteroids and Comets on the Earth* (1985 Snowbird conference), GSA special paper 590, 1982, pp. 121–26.

137. Kerry Emanuel et al., "Hypercanes: A Possible Link in Global Extinction Scenarios," *Journal of Geophysical Research* 100, no. D7 (1995), pp. 13, 755–65.

138. Steel, *Rogue Asteroids*, p. 40.

139. Lewis, *Rain of Iron and Ice*, p. 157; S. Yabushita and N. Harra, "On the Possible Hazard to the Major Cities Caused by an Asteroid Impact in the Pacific Ocean," *Earth, Moon, and Planets*, no. 65 (1994), p. 7.

140. Lewis, *Rain of Iron and Ice*, p. 150.

141. J. Hills et al., "Tsunami Generated by Small Asteroid Impacts," in Gehrels, ed., *Hazards Due to Comets*, p. 788. Devonian megabreccias in the Las Vegas area recently have been identified as the signature of an asteroid impact in the palaeo-Pacific Ocean. See J. Warme and C. Sandberg, "Alamo Megabreccia: Record of a Late Devonian Impact in Southern Nevada," *GSA Today* 6, no. 1 (1994).

142. Stephen Jay Gould, *Time's Arrow, Time's Cycle: Myth and Metaphor in the Discovery of Geological Time*, Cambridge, Mass. 1987, p. 97.

143. Russian researchers also claim an end Eocene date for the 500-kilometer Popigai crater in Siberia. See Grieve and Pesonen, "Terrestrial Impact Craters," p. 367.

144. For the disintegration of the case for protracted extinction, see Peter Ward's account of the 1994 "Snowbird Three" conference: "The K/T Trial," *Paleobiology* 21, no. 3 (1995),

pp. 245–57.

145. M. Rampino and B. Haggerty, "Extraterrestrial Impacts and Mass Extinctions of Life," in Gehrels, ed., *Hazards Due to Comets*, p. 846. The Eocene palaeontologist Donald Prothero concedes the impacts but denies they played any significant role in the chaotic turnover of terrestrial and marine fauna. See *The Eocene-Oligocene Transition: Paradise Lost*, New York 1994, pp. 529–50.

146. See NASA Press Release 96: 55, "Chain of Impact Craters Suggested by Spaceborne Radar Images," 20 March 1996.

147. S. Clube and W. Napier, "Giant Comets and the Galaxy: Implications of the Terrestrial Record," in R. Smoluchowski, J. Bahcall, and M. Matthews, eds., *The Galaxy and the Solar System*, Tucson, Ariz. 1986, pp. 260–85. This is the most elegant of the diverse presentations of the "Clube-Napier Hypothesis," and it contains important clarifications of the difference between single impacts and protracted bombardment episodes (pp. 277–78), and the potential climate-shifting role of zodiacal "dustings" of the stratosphere (pp. 271–77).

148. D. Jablonski, "Mass Extinctions: Persistent Problems and New Directions," in *New Developments Regarding the KT Event*, p. 56.

149. Clube and Napier, "Giant Comets," pp. 277–78. In the six years since this article was published in *New Left Review* (May/June 1996), little evidence has emerged to support the case for impacts within a Holocene timeframe. The Taurid demons continue to be an unproven if hypnotic hypothesis. On the other hand, research has built an impressive dossier for extraterrestrial roles in major extinctions besides the K/T and the terminal Eocene. Most excitement has been generated by the recent publication of evidence for a series of impact-related isotopic anomalies in end-Permian sediments in southern China.

The end-Permian (some 151 million years ago) is the mother of extinctions: responsible for the disappearance of 90 percent of all marine species and 70 percent of all terrestrial vertebrates. Previously, the "Great Dying" had often been correlated to the Siberian Traps flood-basalt episode—the largest in the last 500 million years. But the Japanese team that studied the Chinese limestone found the chemical footprint of a "gigantic release of sulfur from the mantle" (leading to acid rain and a "Strangelove ocean") as well as impact-metamorphosed grains of iron-silicon-nickel (Kunio Kaiho et al., "End-Permian Catastrophe by Bolide Impact," *Geology* 29, no. 9 [September 2001], p. 815–18).

This provided corroboration for earlier claims by University of Hawai'i researchers that they had discovered an extraterrestrial signature in the isotopic composition of gasses trapped in "buckyballs" (fullerenes) in end-Permian sediments. The Siberian flood basalts, accordingly, may be themselves a consequence of a giant impact (as previously theorized for the coincidence of Chixculub and the Deccan Traps). Most of the leading scientists involved in this increasingly highstakes hunt for the Permian killer believe that the impact most likely took place in the oceans—thus the absence so far of shocked quartz or iridium spikes. An Australian geologist has recently claimed to have identified a huge impact structure—200 miles across and likely late or terminal Permian—offshore of northwestern

Australia. Is this the "smoking gun of the end-Permian? (Luann Becker et al., "Impact Event at the Permian-Triassic Border," *Science* 291 [23 February 2001], pp. 1530–33; and Luann Becker, "Repeated Blows," *Scientific American,* March 2002, pp. 78–83).

There is even more recent excitement. In spring 2002, Lamont-Doherty geologists identified iridium anomalies at the Triassic-Jurassic (T/J) boundary where other researchers had earlier found impact-shocked quartz grains and a fern spike: three key indicators of a global bolide catastrophe. The T/J transition corresponds to a major extinction of nondinosaur reptiles followed by a dramatic increase in the fossil record of meat-eating dinosaurs, suggesting that the domination of the dinosaurs, like their demise, was triggered by impact events (Richard Kerr, "Did an Impact Trigger the Dinosaurs' Rise?" *Science* 296 [17 May 2002], pp. 1215–16).

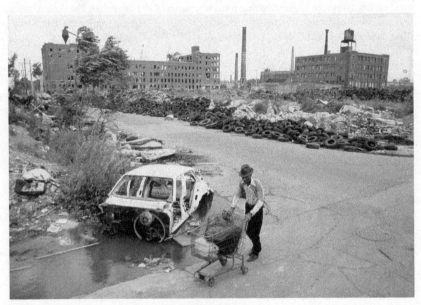

The Dead Zone

17

Dead Cities: A Natural History

Science in the Ruins

> For this marvellous city, of which such legends are related, was
> after all only of brick, and when the ivy grew over and trees
> and shrubs sprang up, and, last, the waters underneath burst
> in, this huge metropolis was soon overthrown.
>
> *Richard Jefferies*, After London: or, Wild England (1886)

From sometime in the late nineteenth century, the larger share of energy under the control of the human race has been devoted to the construction and maintenance of its urban habitats. Agriculture, for eight thousand years the primary locus of human and animal labor, is now secondary to the immense, literally "geological" drama of urbanization. Geologists calculate that the fossil energy currently expended in shaping the earth's surface to the needs of an exploding human population of city-dwellers is geomorphically equivalent—at least in the short run—to the work of the planet's primary tectonic engines: sea-floor spreading and mountain erosion. ("We have now become arguably the premier geomorphic agent sculpting the landscape," writes one expert on the history of human earth-moving activity.)[1] Even more alarmingly, the carbon metabolism of urban areas is transforming world climate, perhaps destroying in process the recent niche of moderate weather that has made super-urbanization possible.

Global change aside, there have always been compelling reasons to worry about the "sustainability" of big cities. The ability of a city's physical structure to organize and encode a stable social order depends on its capacity to master and manipulate nature. But cities are radically contingent artifacts whose "control of nature," as John McPhee famously pointed out, is ultimately illusory.[2] Nature is constantly straining against its chains: probing for weak points, cracks, faults, even a speck of rust. The forces at its command are of course as colossal as a hurricane and as invisible as bacilli. At either end of the scale, natural energies are capable of opening breaches that can quickly unravel the cultural order. Cities, accordingly, cannot afford to let flora or fauna, wind or water, run wild. Environmental control demands continuous investment and systematic maintenance: whether building a multi-billion-dollar flood control system or simply weeding the garden.

It is an inevitably Sisyphean labor. Even if we envision cities as "smart mountains," equipped with myriad human sensors to detect and counter erosion, fundamental interfaces with nature—the condition of the housing stock, the status of lifeline conduits of water supply and waste removal, the control of disease-carrying commensal species like rats and flies, and so on—are usually in disequilibrium. Environmental crisis is synonymous with expanding metropolitan scale. Rich cities, moreover, are not necessarily more stable than poor cities. "This grandly suspended, inorganic metropolis," Ernst Bloch wrote of New York or Berlin (see the preface), "must defend itself daily, hourly, against the elements as though against an enemy invasion."[3] Increasing infrastructural complexity, as Americans have become excruciatingly aware in the wake of 9/11, simply multiplies the critical nodes where catastrophic systems failure is possible.

Rich cities, however, do have greater ability to export their natural contradictions farther downstream. Los Angeles, for example, captures runoff and energy, and exports pollution, solid waste, and weekend recreation within a vast ambit of a dozen Western states and Baja California. The traditional natural condition of urbanization—the domination of a single large riverine watershed—has been transformed by the megalopolis into environmental imperialism of subcontinental scope. To cite other examples: New York City, which once simply stood astride the Hudson River, now extends its reach (via Québec hydropower) to Hudson's Bay. Tokyo, according to a 1998 Earth Council report, requires for its sustenance a biologically productive land area more than three times the size of Japan.[4]

Very large cities—those with a global not just regional environmental footprint—are thus the most dramatic end-product, in more than one sense, of human cultural evolution in the Holocene. Presumably they should be the subject of the most urgent and encompassing scientific inquiry. They are not. We know more about rainforest ecology than urban ecology. Moreover, the study of cities is one of the last bastions of linear problem analysis by mechanical decomposition. Engineering, for example, has always dealt with urban nature one problem at a time, as in the design of single-purpose infrastructures. In the same stubbornly monolithic spirit, the administrators of these lifeline technologies have little tradition of talking to one another, even where, as in the case of flood runoff, sewage, and aquifer water supply, they are dealing with aspects of the same integral natural system.

If in some areas—atmospheric chemistry or thermal meteorology—more holistic approaches dominate, we can still barely glimpse the distant form of a truly unified urban science. The most urgent need, perhaps, is for large-scale conceptual templates for understanding the city–nature dialectic. Here boldness may be a virtue. What would happen, for instance, if we simply erased from the blackboard all the differential equations (representing the "work" performed by humans on the environment) on the city side of the interaction. What would remain on the nature side? Indeed, what is "underlying" urban nature without human control? Would the city be gradually (or catastrophically) reclaimed by its "original" ecology, or by something else, possibly more like a chimera? "Dead cities," in other words, might tell us much about the dynamics of urban nature. But what forensic expert has ever examined as such the corpse of a great city? Who has ever put a microscope to the ruins of Metropolis?

In fact, two superb, self-trained naturalists—Richard Jefferies and George R. Stewart—conducted powerful thought-experiments in the guise of novels on the postmortem natural histories of the London and San Francisco Bay regions. They were, so to speak, Darwin and Wallace in the archipelago of the Apocalypse. Jefferies, the preeminent natural historian of late Victorian London and the Home Counties, wrote *After London* in 1886; Stewart, author as well of the Western environmental classics *Storm* and *Fire*, published *Earth Abides* in 1949. In both cases the fictional Last Man plots (ultimately derived from Cousin de Grainville's *Le Dernier Homme* of 1806) were overshadowed by brilliant depictions of natural reclamation and ecological succession. Jefferies's conjectures, moreover, were reexamined and updated in some detail by *The New Scientist* in 1996.

But dead cities are not fictional constructs alone. As the bumper sticker says: Apocalypse Happens. The strategic bombing of Europe and Japan during the Second World War, for example, inadvertently created numerous experimental stations for observing urban nature set free. In the bomb debris of Whitechapel, Altona, and Neukolln, botanists were able to empirically record the pioneer sequences of what Jefferies and Stewart could only surmise. War, they discovered, was the catalyst for the rapid expansion of previously rare alien species, resulting in the creation of a new urban flora sometimes referred to as "Nature II." Their work became the foundation for "ruderal ecology": the scientific study of urban margins and abandoned land.

Likewise, the destruction of US central city neighborhoods in the 1970s, in the wake of the "Second Civil War" (urban riots) of the late 1960s, produced eerily similar ruins to the aftermaths of thousand-bomber raids. If birdlovers and geologists have generally eschewed the desertified cores of the Bronx, Newark, and Detroit, photographer Camilo Vergara doggedly returned to document the same sites month after month, year after year. His time-lapse studies, published in part as *The New American Ghetto* (1995), are a unique archive for understanding dereliction as landscape process. Deborah and Rodrick Wallace, in turn, have used the formal tools of population ecology and mathematical epidemiology to link housing abandonment and neighborhood disinvestment with the "new plagues" of tuberculosis, HIV, rising infant mortality, and street violence. In particular, they have focused on the Frankenstein-like meddling of the Rand Institute with New York City's fire services during the 1970s. Their work is both a rich model of interdisciplinary urban science, and a warning about how little we still understand about the nonlinear dimensions of urban ecology.

The Toxic Metropolis

> The earth on which he walked, the black earth, leaving phos-
> phoric footmarks behind him, was composed of the mould-
> ered bodies of millions of men who had passed away in the
> centuries during which the city existed.
>
> *Jefferies*, After London

In February 1884 John Ruskin warned London audiences in a pair of notorious lectures (published as *The Storm-Cloud of the Nineteenth Century*) that their bourgeois world was on the brink of a supernatural catastrophe. A "poison cloud"—indeed, a "plague cloud"—now shrouded England, a symptom of a "miasmatic, progressive and apparent fatal infection of the sky." Skeptics, Ruskin admonished, need only to look out their windows at the darkening sky. His own blighted garden—"one miserable of weeds gone to seed, the roses in the higher garden putrefied into brown sponges"—provided additional evidence of irreversible degeneration:

> I will tell you this much: that had the weather when I was young been such as it
> is now, no book such as *Modern Painters* ever would or *could* have been written; for
> every argument, and every sentiment in that book, was rounded on the personal
> experience of the beauty and blessing of nature, all spring and summer long...,
> that harmony is now broken, and broken the world round ... month by month the
> darkness gains upon the day.[5]

Ruskin's state of mind, of course, was clearly delusional, yet, as Raymond Fitch has emphasized in a massive study, he also spoke with unique authority as the Victorian age's greatest student of the sky and connoisseur of clouds.[6] Moreover, his apprehensions were echoed later the same year by another expert observer: Richard Jefferies, renowned for his unique essays on the natural phenomena of London and its environs (for example, "The Pigeons at the British Museum,"

"Herbs at Kew Gardens," "A London Trout," etc.).[7] In a diary entry dated 21 July 1884, Jefferies wrote:

> Hyde Pk. Demonstration.
>
> Little Village. Mediaeval London. The Thames. Putrid black water, decomposed body under the paddle wheel. Deeds of darkness. The Body. Nine elms, sewn up in a sack. Children miserable, tortured—just the same. The tyranny of nobles now paralleled by the County Court. Machinery for extortion. The sewers system and the WC water. The ground prepared for the Cholera plague and fever, zymotic, killing as many as the plague. The 21 parishes of the Lower Thames Sewage Scheme without any drainage at all. The whole place prepared for disease and pestilence.
>
> This WC century.

A few weeks later, he added:

> The great hope of the future. The Revolutionist.[8]

Jefferies's extraordinary conjugation of unemployment, decomposed bodies in the Thames, venal officialdom, sewage, and revolution shared with Ruskin's poisoned sky the same relentless image of an all-engulfing miasma. Despite the new work of Koch and Pasteur, most educated people still believed, in the fashion of Sir Edwin Chadwick's famous 1842 *Report on the Sanitary Conditions of the Labouring Population*, that "sticky, miasmal atoms" spread disease as a literal corruption of the air. Indeed, "Chadwick and his collaborators," Carlo Cipolla observes, "acted and behaved not only as if all 'smell' was disease but also if all disease was 'smell.'"[9] If Ruskin now claimed that London's reek had fatally infected the sky, Jefferies saw evidence of a terminal urban crisis whose symbol was the new-fangled water closet flushing excrement into his beloved Thames.[10]

There is reason to believe that 1884 was indeed a year of many glooms and stenches. Nietzsche (or, rather, "Zarathustra's ape") also in that year was "nauseated by [the] great city ... where everything infirm, infamous, lustful, dusky, overmusty, cuntlike, and plotting putrefies together."[11] Air pollution, as Brimblecombe has shown, had reached its nineteenth-century apogee,[12] and stratospheric dust and sulphates from the eruption of Krakatau the previous year were producing anomalous weather across the globe. Most likely, Ruskin's weird "plague cloud" actually existed. Meanwhile, the continuing scandal of London's untreated sewage and contaminated water supply, four decades after Chadwick's identification of their lethal linkage, made the arrival of new pestilences almost inevitable. Equally, the growing desperation and anger of the East End, where tens of thousands had been thrown out of work by the world trade depression, raised spectres of Anarchy or even a London Commune.

If Ruskin turned from the demon skies in despair, Jefferies, whose own social stratum, the yeomanry of Wiltshire, had been decimated by the agricultural crisis of the previous decade, was more inclined to see glimpses of Arcadia beyond the Apocalypse.[13] Earlier in "Snowed Up," a 1876 story based on the famous blizzard of 1874, he depicted London's rapid collapse into barbarism as the metropolis was cut off from its vital imports of grain and coal and overwhelmed by rats and looters. Jefferies took undisguised pleasure in reminding readers that "but a thin, transparent sheet of brittle glass" stood between civilization and wilderness: a point underscored with a haunting image of an iceberg in the Thames.

After "Snowed Up" it was perhaps inevitable that Jefferies would return to the theme of Nature defeating and devouring the pestilential city. *After London; Or, Wild England* was clearly incubating in Jefferies's mind at the time of his 1884 diary entries; it was published in 1886 and has remained in print ever since. It is less a nightmare than a deep ecologist's dreamwish of wild powers re-enthroned. (William Morris reported that "absurd hopes curled around my heart as I read it.") The novel consists of two free-standing parts: "The Relapse into Barbarism" and "Wild England." The latter recounts the adventures of the archer-scholar Felix Aquilas, who survives various Darwinian perils (feral animals and devolved humans) as he crosses the medievalized landscape of postapocalyptic England in search of the "utterly extinct city of London." Jefferies cleanses the cultural as well as the natural landscape: every trace of the nineteenth century has vanished, although Roman and Greek classics somehow survive. It is a savage attack on Victorian civilization, although its style is anachronistic and "Wild England" now primarily interests critics as a literary forerunner to the kind of allegorical, political science fiction that H. G. Wells made famous.

"The Relapse into Barbarism," on the other hand, puts Jefferies's famous powers as a naturalist to the imaginative test of a mysterious catastrophe. His anonymous narrator, writing generations after the conflagrations that consumed the historical record, is as uncertain of the exact character of the Last Days as we are of the identity of the enigmatic plague and attendant disasters that depopulated Byzantium in the time of Polybius. Perhaps London and other cities were assassinated by a close encounter with an "enormous [cosmic] dark body which created gravitational chaos on earth"; on the other hand, the metropolis may simply have poisoned itself with its own pollution. "All that seems certain is, that when the event took place, the immense crowds collected in cities were most affected, and the richer and upper classes made use of their money to escape" (pp. 28–29). In contrast to the enigma of the Catastrophe, the narrator is able to describe the decomposition of the dead metropolis with forensic precision. "The Relapse into Barbarism" is worth summarizing at some length, not only for its speculations about the accelerated evolution of feral species, but above all for Jefferies's brilliant description of the natural forces reworking the urban landscape: first a weed explosion, then initial reforestation, followed by the vengeance of the Thames. It was one of the earliest, and certainly the most dramatic, descriptions of what was later termed "ecological succession."

In the first spring "after London" it "became green everywhere": wheat no longer sown grew wild, intermingling with couch grass and other weeds which also quickly covered footpaths. Mice in the millions—joined by sparrows, rooks, and pigeons—feasted on the fallen and over-ripe wheat in the fields, while vast armies of rats pillaged granaries and the cupboards of abandoned homes. Initially, predators made little headway against the rodents, but by winter their population explosion had reached its Malthusian limits. The mice and rats then cannibalized each other in desperate orgies. Winter storms beat down the last stalks of wheat and barley in the fields around London.

By summer of the second year, the once civilized grains had become almost indistinguishable in the wild jumble of docks, nestle, sorrel, wild carrots, thistles, oxeye daisies, and yellow charlock flowers. Later nettles and wild parsnips supplanted many of the pioneers, while briar and hawthorne followed bramble. (Jefferies knew that in Edward the Confessor's time, Westminster Abbey had been a mass of bramble: "Thorney Island.") Similarly, the hedges widened and began to narrow fields and lots until, after twenty years or so, they choked them completely. As fields, house sites, and roads were overrun, the saplings of new forests appeared. Elms, ashes, oaks, sycamores, and horse chestnuts thrived chaotically in the ruins while more disciplined copses of fir, beech, and nut trees relentlessly expanded their circumferences.

Meanwhile, kestrel hawks, owls, and especially weasels had brought the rodent irruption under control. As cats, now mostly grayish and longer in body than their domestic ancestors, recovered their ancestral competence in hunting, they chose fowl and poultry over mice. (Indeed, the detested "forest cat" sometimes even attacks travelers.) Forced to fend for themselves, the small lapdogs of the former middle classes (poodles, Maltese terriers, and Pomeranians) quickly became food items and perished. Bigger dogs—mastiffs, terriers, spaniels, and greyhounds—remained "faithful to man as ever" and followed their masters in the flight from the cities. A third class of canines, however, chose the freedom of the wild. After generations of natural selection they evolved into three new species which ceased to interbreed: black wood dogs (descendants of ancient sheep dogs, which hunt sheep and cattle in packs, but do not attack man); yellow wood dogs (smaller and devoted to the chase, pursuing hares and stags); and, pitifully, the white wood dog (a degenerate scavenger afraid to face even a tame cat).

Evolution also rapidly manufactured new species or subspecies out of other former domesticates. Jefferies delights in describing how after millennia of demeaning slavery to man, cattle have metamorphosed back into the fearsome gods of Minoan friezes. Indeed, the "very dangerous" white or dun bull is monarch of the new forest, although humans are also wary of black wild cattle as well. Formidable palisades defend farmsteads against the cattle and the four kinds of feral pigs as well as the large bush-horses which live in thickets near water. Thicker-set little hill-ponies share the Chalk Hills with two varieties of shaggy sheep. A third species of sheep has chosen island life as protection from wood dogs. Occasionally in calm weather, however, the dogs swim out to the islands and devour them. Initially survivors feared that the wild beasts from the London Zoo and various circuses would multiply in the new forests. Lions and bears had, in fact, roamed the fields for some years. But their progeny, together with those of escaped serpents, were gradually killed off by winter frosts. "In the castle yard at Longtover may still be seen the bones of an elephant which was found dying in the woods near that spot" (pp. 24–25).

Having thus decomposed and consumed the soft tissues of the city, the monstrous vegetative powers of feral nature begin a full-scale assault on London's brick, stone, and iron skeleton:

> By the thirtieth year there was not one single open place, the hills only excepted, where a man could walk, unless he followed the tracks of wild creatures or cut himself a path. The ditches, of course, had long since become full of leaves and dead branches, so that the water which should have run off down them stagnated, and presently spread out into the hollow places and by the corner of what had once been fields, forming marshes where the horsetails, flags and sedges hid the water. (p. 5)

As marsh recovered the floodplain, heavy rains flushed "vast quantities of timber, the wreckage of towns and bridges" downstream against the piers of the old, broken Thames bridges. Waterloo, London, and Tower bridges thus became dams, backing up waters and overspilling the remaining embankments. The hydraulic pressure of the flooded substratum of the city—underground passages, sewers, cellars, and drains—soon burst the foundations of homes and buildings, which in turn crumbled into rubble heaps, further impeding drainage.

Eventually, new vegetation and old debris completely blocked the Thames. Upstream, a 200-mile-long inland sea—The Lake—quickly formed. Its crystal waters—"exquisite to drink, abounding with fishes of every kind, and adorned with green islands"—symbolized the resurrection of a Green and Pleasant Land. (Hitler, in a curious coincidence, proposed to drown Moscow in a huge lake after conquest.) But downstream, where London was now most literally a Great Wen, all the toxicity of the Victorian age remained concentrated in the "green and rank" center of the swamp, exuding stinking vapors that mask the sun. "For all the rottenness of a thousand years and of many hundred millions of human beings is there, festering under the stagnant water, which has sunk down into and penetrated the earth, and floated up to the surface the contents of the buried cloacae" (p. 69). Jefferies's extinct London, in short, is a giant stopped-up toilet, threatening death as an "inevitable fate" to anyone foolish enough to expose themselves to its poisonous miasmas.

Jefferies Updated

> Some time, perhaps five hundred years after abandonment, the Great Leaning Tower of Canary Wharf finally crashes down.
> *New Scientist*

After London encouraged many sequels and imitations. I. F. Clarke has emphasized, for instance, the exceptional affinities between Jefferies and W. H. Hudson, the South American–born naturalist and, later, London bird expert, whose apocalyptic arcadia, *A Crystal Age*, appeared in 1887. "In violent and agreeably anticipative acts of destruction the two men called on nature to wipe out the infamy of urban civilization."[15] In his more celebrated variation on Jefferies, *News from Nowhere* (1890), William Morris does not completely destroy the "hideous town," but instead shrinks it down to the humane dimensions of the socialist garden city that he had first envisioned in *The Earthly Paradise* (1868): "London, small and white and clean/The clear Thames bordered by its gardens green."[16]

It took more than a century, however, for the natural history undergirding *After London* to be carefully scrutinized. In 1996, the *New Scientist* asked leading botanists, animal ethologists, material scientists, and engineers to reconsider Jefferies's "experiment."[17] The modern knowledges brought to bear on the problem of abandonment included a better understanding of the biogeochemistry of urban decay and the dynamics of reforestation as well as the fruit of a hundred years of research on the life-cycles of steel and ferro-concrete structures (the major innovations in urban materials, together with plastics, since the 1880s).

Within five years, the *New Scientist* found, weeds would indeed conquer the open spaces, pathways, and cracks of the city. Since Jefferies's time, however, herculean alien weeds—especially the formidable shrub *Buddleia davidii*—have established strongholds in London. If dandelions and other native weeds "only exploit existing weaknesses," buddleia has roots "powerful enough to penetrate bricks and motor to find moisture." A fast-growing, wind-disseminated native of the Himalayas—possibly imported as an ornamental as early as the 1880s but especially commonplace after the Blitz—buddleia is adapted to the monumental work of mountain erosion. Trafalgar Square is thus no problem. As one worried botantist told the *New Scientist*, "buddleia is already everywhere in London, poised to rid the city of its concrete and brick."

The city's disintegrating hardscape, however, is a poor, nitrogen-deficient diet for plants. Over the long run, nitrogen-fixers like clover and alder can fertilize the sandy detritus and create soils suitable for forest trees. Fire—which Jefferies oddly neglects—enormously accelerates this transition. Like almost all landscapes, London has a climatically regulated natural fire cycle and (in the view of one of the experts) "early autumn around five years after abandonment is a likely time for fire."

> The streets have built up a litter of grasses and fallen leaves. A dry spell and a lighting strike sets the city ablaze. Fire guts the buildings that still dominate the London landscape. As the houses burn and roofs come crashing down, nutrients are released from their timbers and from leaf litter, providing the fertiliser to speed London's return to its past.

Shrubs would flourish in the ash and rapidly build up the soil layer. Ivy tendrils would climb six stories high over abandoned townhouses and stores. Then trees—elder and birch saplings, as well as full-grown buddleia—would take over. Their roots have the power of hydraulic jackhammers. One of the scientists interviewed by the *New Scientist* was flabbergasted by the tree damage he had recently observed in the ghost city of Pripyat, near Chernobyl: "[T]he concrete paving stones have in one of the city's squares have been smashed and, in places, pushed up almost a metre by tree roots, as if a giant earthquake had struck."

Spring tides and flood surges, just as Jefferies had predicted, would meanwhile recycle much of central London into fen, bog, and swamp. The wild red deer, kingfishers, moorhens, and swallows—whose passing he had mourned in his London articles—would quickly return to the resurrected marshes.[18] Freed of its artificial constraints, the Thames is probably too wily to allow itself to be permanently dammed by debris as Jefferies imagined; instead ruined bridge piers would become weirs, ideal for spawning salmon. As the great river resumed its natural wander across a vast floodplain, the Isle of Dogs would revert to reeds and much of Southwark would again be a mudflat and migratory bird sanctuary. Likewise, former tributaries—London's famous "lost rivers" like the Westbourne under Sloane Square or the Fleet under Farringdon Road—would break through the surface and claim back their ancestral wetlands.

Everywhere, except on hilly islands like Hampstead and Highgate, rising groundwater, as Jefferies described, would rapidly undermine structures. Wooden constructions, a Japanese engineer explained to the *New Scientist*, "would be the first to vanish completely ... followed by the materials that glue a building together—partitions, insulation—materials that insects destroy by nesting in them." Unsuspected by Jefferies, birds will play a vital role in structural decay by introducing destructive insects. Modern steel and ferro-concrete structures will resist decay for a century or two but their eventual corrosion would produce a catastrophic denouement:

> While concrete remained alkaline, the steel bars that reinforce it held fast against corrosion. But carbon dioxide dissolved in rain has gradually carbonated the surface of the concrete and edged its way in, while acid from decaying organic matter in the ground has infiltrated concrete foundations.
>
> Once the steel corrodes, the end is swift. The corrosion products take up about three times the volume of the steel itself ... so as the steel rusts, it expands until its concrete covering spalls off.

By the time the Lloyd's Building begins to shed its rusting I-beams, the great Middlesex Forest would be extensively reestablished. There would, of course, be significant modifications: traces of the DNA of the abandoned city. Oak, followed by larch and spruce, would again predominate, but joined now by former street trees like sweet chestnut and foreign species like sycamore, Norway maple, and some conifers. In the understory, marsh, and meadow, many of the thousands of introduced house and garden plants would die off; but some, including hybrid "super-grasses," would flourish, perhaps even locally dominate.

Fauna in year 2556 might be more exotic. Although evolution among the higher animals doesn't work at the breakneck speed envisioned by Jefferies, escaped pets and new migrants might create a spectacular community of adapted species. The persistence for so many centuries of ruined skyscrapers, for example, would likely attract rough-legged buzzards from Scandinavia. Ring-necked parakeets (originally from Asia) would thrive in vast numbers, despite the attention of feral cats. And, "wolves, or wolf-like German shepherd hybrids, [would] roam the forests preying on roe, muntjac and sika deer and feral pigs descended from the stocks at London Zoo and city farms."

After Berkeley

> During thousands of years man had impressed himself upon
> the world. Now man was gone, certainly for a while, perhaps
> forever. Even if some survivors were left, they would be a long
> time in again obtaining supremacy. What would happen to the
> world and its creatures without man? *That* he was left to see!
> *George R. Stewart,* Earth Abides (1949)

Would cities decay differently in the compost of the New World? Jack London, not surprisingly, was first off the mark with an "After San Francisco" novel: *The Scarlet Plague* (1912). A crude acolyte of Spencer and Galton, he used the dead city by the Bay to illustrate the law of the survival of the fittest as well as the dangers of wanton race mixing of men and animals. His Nature, of course, is just as red in tooth and claw as the robber baron capitalism that it has overthrown. Dogs, for example, feed first upon the corpses of their masters, then upon each other. All small and weak types are quickly eliminated until a single breed of medium-sized wolf remains. Horses, on the other hand, "degenerate" into small, miserable mustangs: their vast herds trampling the former vineyards and farms of the San Joaquin Valley. Meanwhile, amongst the human survivors, a brutish "chauffer" claims a beautiful society woman as his "squaw" and imposes his appetites and will upon a "primitive horde." London's hero, a manly professor of literature at U.C. Berkeley, bears the lonely burden of promoting the "slow Aryan drift" back to civilization.[19]

The Scarlet Plague is a hysterical polemic about eugenics, not imaginative natural history. Both Jefferies and London, however, were rewritten in 1949 by a gifted amateur naturalist and Western historian, George R. Stewart. If his *Earth Abides* remains enshrined as a "science fiction classic," it is seldom cited, as it richly deserves, as a unique excursion in regional natural history. Indeed, in the pantheon of modern American environmental writers—Leopold, Stegner, Worster, Abbey, McPhee, and so on—Stewart is perhaps the most unfairly neglected major figure. During his four decades (from 1923) in U.C. Berkeley's English Department, he was part of an extraordinary community of savants that included Herbert Bolton, the father of comparative frontier history; Carl Sauer, the founder of the "Berkeley School" of cultural geography; Alfred Kroeber, the dominating figure in California anthropology; and Julian Stewart, the pioneer of "cultural ecology." In their different terrains of research, they gave similar priority to ecological interactions between humans and their natural region. Sauer's dialectical concept of the "cultural landscape" as the co-product of human praxis and natural process provided an unifying motif. "We are interested primarily," he wrote, "in cultures which grow with original vigor out of the lap of a maternal natural landscape, to which each is bound in the whole course of its existence."[20]

Stewart faithfully followed Sauer's prescription. A canonical New Deal regionalist, he authored seven novels and twenty-one nonfiction books, most of them about California or the West, including an authoritative biography of Bret Harte and a bestselling account of the ill-fated Donner Party's cannibal winter in the Sierras. The deep imprint of Berkeley cultural ecology is most evident in his quartet of "environmental novels"—*Storm* (1941), *Fire* (1948), *Earth Abides* (1949), and *Sheep Rock* (1951)—whose "heroes" are, respectively, a winter storm-system named "Maria," a wildfire called "Spitfire," California nature writ large, and a place known as "Sheep Rock."

As Wallace Stegner appreciated in an early review of *Storm*, Stewart's strategy of making nature the protagonist is not a return to Wordsworthian anthropomorphism and nature-worship, but a canny device that lets us look "at the mortar that holds a civilization together." "The storm is not the heroine. Maria is nothing but the crisis." Indeed, each novel of the quartet explores humanity's contingent control of nature from the vantage-point of a different crisis. Stewart, whose family had moved from Pennsylvania to the orange groves of Southern California (Azusa, then Pasadena) when he was twelve, found sublime drama in the Sisyphean struggle of Westerners to control their environment. Temporary human victories always yield in the end to the stubborn power of place. As Geoffrey Archer discovers in his battle to tame Sheep Rock: "Essentially he had not changed the place at all. It had kept its integrity again. It had conquered."

Trained as a biographer and historian, Stewart was truly a "poet and precisionist" (Stegner) in his painstaking accounts of environmental forces. Shortfalls in his scientific background were compensated by heroic and original research. Thus in the course of working on *Storm*, "Stewart drove [through two winters] the roads to Donner Pass during storms, rode on the snow plows of the Southern Pacific, observed the highway superintendent and his crews along US 40, and watched telephone and electric company linemen in action. He secured introductions to the staff of the Weather Bureau in San Francisco, visited them during storms, and learned to draw his own weather maps." Writing *Fire*, he visited fire lines and spent a week in a fire lookout at Sierra Buttes, while his intricate description of Nevada's Black Rock desert (the locale of *Sheep Rock*) was based on exploration in a desolate outback today famed for its neo-pagan "Burning Man" festival.[22]

Earth Abides, one reasonably expects, is distilled from all resources at hand, including Stewart's incomparable network of naturalist friends and scientific colleagues. If *After London* and *The Scarlet Plague* are compromised by lurid versions of Darwinism and reverse evolution, *Earth Abides* stands apart as the first novel to incorporate a sophisticated understanding of the young and still relatively obscure (in 1949) science of ecology. For example, Stewart explains human near-extinction at the beginning of the novel straightforwardly in terms of population ecology and microbial predation, without resort to the usual H-bombs, extraterrestrial monsters, or cosmic "dark bodies." As human numbers have exorbitantly overgrown the carrying capacities of their environment, new plagues have arisen to adjust the equilibrium. It is a human analogue to Aldo Leopold's famous deer population boom-and-bust cycle on Arizona's Kaibab Plateau in the 1920s. As Stewart knows, all exponentially increasing "r-selected" populations must ride a Malthusian rollercoaster—the "Lokta-Volterra curve"—of demographic peaks and cataclysmic collapses.[23]

Stewart's Survivor, a misanthropic and self-absorbed graduate student working on a thesis about the "ecology of the Black Creek area," is named Isherwood Williams or "Ish" for short: an obvious allusion to a *real* Last Man, Ishi, the sole Yahi Indian, whom Alfred Kroeber brought to San Francisco in 1911 as a "living fossil." Just as the tragic Ishi (who caught TB from Kroeber's wife and died in 1916) was a witness to the triumph of white, urban civilization, so Ish is the lonely scientific chronicler of its disintegration. Fortuitously inoculated against the plague by a rattlesnake bite while doing fieldwork in the same Sierra foothills where Ishi had once found refuge, Ish makes his way back to his home on San Lupo Drive in the Berkeley Hills. It is a grandstand seat for witnessing "the greatest of all dramas": Nature's majestic reclamation work on the Bay metropolis.

The new plague has worked so swiftly that, apart from looted liquor stores and a few localized fires, there is little physical destruction. Nor does the metropolitan infrastructure break down immediately. The great turbines in the Sierras assure hydropower for a few years; water continues to flow through aqueducts and taps for even longer. The absence of watering, of course, quickly kills temperate garden plants and lawns, while "fierce weeds press in to destroy [other] pampered nurslings of man." The hundreds of thousands of human cadavers meanwhile are devoured by the bigger dogs, while a glut of rotting foot provides a banquet for rats, roaches, ants, and flies, whose numbers temporarily soar toward infinity. Of domesticated and commensal species, Stewart asserts that only the three varieties of the human louse are immediately doomed by humanity's overthrow.

> At the funeral of *Homo sapiens* there will be few mourners. *Canis familiaris* as an individual will perhaps send up a few howls, but as a species, remembering all the kicks and curses, he will soon be comforted and run off to join his wild fellows. *Homo sapiens,* however, may take comfort from the thought that at his funeral there will be three wholly sincere mourners. (p. 59)

The first few years of the post-human era are a chaotic period of wild population shifts and merciless inter-species competition. Where Jefferies imagined a linear process of extinction, selection, and speciation, Stewart knows that modern ecology predicts nonlinear fluctuation. Predator/prey balance, it seems, can only be established through a sequence of reciprocal catastrophes. Thus after a few months, as corpses and accessible food are used up, ant and roach colonies collapse, while cats eat the rodents and are eaten, in turn, by the smaller dogs. But the dogs kill too many cats, and rats—from redoubts in grocery warehouses and grain elevators—make a spectacular resurgence. This second explosion of the rodent population brings bubonic plague in its wake: a new threat to the human survivors. Eventually, having consumed the remaining grain, the rats starve. In their frenzy, they attack even dogs, then eat each other.[24]

After this "secondary kill," there is a brief hiatus. Dogs (minus the "stupid overbred" varieties, which perished immediately) begin to sort out new pedigrees and build hunting packs. Cats, better adapted than dogs to feral survival, learn to surmount the canine menace and their population expands until it fatally collides with the encroaching domain of foothill bobcats (who prefer their feline cousins to rabbits).[25] In *year 2*, however, the countryside, in the form of hungry deer, rabbits, and wild cattle, invades the dead suburbs. They ravage much of the remaining garden flora and ornamental plantings.

Meanwhile, as one would expect in a Mediterranean climate, the natural engines of the landscape are working at much higher speeds than in Jefferies's Thames Valley. October rains have already initiated the erosion of neighborhoods: storm drains are blocked with debris, water ponds in the streets, homes are inundated with mud, and yards begin to gully. The summer of *year 3*, moreover, is extremely dry and ends with a lightning-ignited firestorm that reduces much of the East Bay to ash before rains extinguish it. The shrubs and grasses which grow from the burnt slopes the following spring provide rich fodder for the wild cattle. The bovine population, like that of the rats before it, overshoots any sustainable limit and then dies off *en masse* during the great drought of *year 6*. The mountain lions, which have followed the cattle into the city, feast on their carcasses and then, half mad with hunger, stalk all possible prey, including the surviving humans (who, in turn, quickly learn to become skilled lionhunters). Locusts are even more destructive campfollowers of the drought and with the terrible efficiency of myriads of nano-lawnmowers, they devour the stubble left by the starving cattle.

Much of the San Francisco area, as it previously did briefly during the cataclysmic 1860–63 drought, now looks like desert. But new rains, which gouge out huge arroyos and unleash thousands of landslides, turn the Bay hills green again. Long forgotten local springs—no longer slaves to an artificial water supply—reappear and spur more erosion. Riparian flora and fauna make spectacular comebacks. Humans are amazed in *year 10* to see schools of striped bass in San Francisco Bay and plentiful trout in the foothill streams. If the apocalypse has added new perils to human lives, it has also returned people to the state of biological grace enjoyed by their paleolithic ancestors before the agricultural revolution. The epidemic chain, which depends upon high densities of humans and commensal species, has been broken and survivors are freed from most infectious diseases. The human tribes also discover a new source of protein in *year 19* when elk dramatically appear in the Oakland Hills.

Meanwhile, fire-spared wooden structures—shotgun shacks in the Flats as well as magnificent Craftsman mansions in the Hills—begin an accelerated decay thanks to the collaboration of termites, rain, and rising groundwater. A major earthquake in *year 20* destroys thousands of these weakened structures. It also fractures concrete and asphalt, reducing their resistance to weeds and erosion. Much of the architecture on the U.C. campus, as in downtown San Francisco, is in an advanced stage of decrepitude, but the magnificent Bay and Golden Gate bridges, although rusted, are still structurally intact.

Over the next generation, the landscape and its life-worlds follow the same cycle of punctuated erosion: slow decay suddenly speeded up by extreme weather events, wildfires, and earthquakes. In *year 44*, for example, most of San Francisco burns down. Other fires destroy the old U.C. campus shortly afterwards. Then, in the final weeks of Ish's long tenure as "The Last American," a span of the Bay Bridge drops into the water. Saltwater corrosion has begun to dissolve the last overweening symbols of the old civilization.

Bomber Ecology

> The sign of our times is the ruins. They surround our lives.
> They line the streets of our cities. They are our reality. In
> their burned-out façades there blooms not the blue flower of
> romanticism but the daemonic spirit of destruction, decay, and
> apocalypse.
>
> *Hans Werner Richter*

There was, in fact, intense scientific curiosity about what would actually bloom
in the ruined cities of Europe. Blue flowers, daemonic flowers, or simply dandelions: botanists were unsure whether "potential natural vegetation" would reclaim
the rubble deserts of London's East End or Berlin's Neukolln district, or whether
alien weeds and escaped cultivars would become the occupying powers. Careful
observation of successional dynamics in the urban "dead zones" (a term coined
by Allied analysts of strategic bombing) might provide empirical answers to two
questions that taxed students of urban nature. First, how radically had urbanization (followed by "deurbanization" from the sky) altered the landscape's biophys-
ical template: soil and air chemistry, nutrient flows, hydrology, microclimates,
and genetic (pollen and seed) reservoirs. Second, did Clementsian paradigms of
regional plant ecology—an orderly succession of species "climaxing" in a peak
community optimally adapted to its environment—accurately describe population dynamics, or, as some prewar critics had argued, was "equilibrium" only
an illusion, and flux the reality?[26] The first published observations came from
London's Botanical Exchange Club while Werner von Braun's V-2 missiles were
still spreading their random terror; later, Berlin, during that "Age of Rubble" that
lasted until 1954,[27] became the primary laboratory for research in dead zone science.

In the case of London, there was already something of a tradition of natural
history in the aftermath of disaster. In the spring following the Great Fire of
1666, for example, the naturalist John Ray and other survivors were flabbergasted
by a spectacular and unexpected bloom of "fire flowers," the famous London
Rocket (*Sisymbrium irio*—a mustard).[28] Likewise, after the first Zeppelin attack

on London in spring 1915, the city's concerned birdlovers, led by W. H. Hudson, author of the magisterial 1895 *The Birds in London* (and, as we saw earlier, *A Crystal Age*), deployed to see whether bombing was scaring birds away from the city.[29] The eventual result of these observations was Sir Hugh Gladstone's odd monograph, *Birds and the War*, which concluded that only pigeons seemed unduly upset by falling bombs and anti-aircraft fire. Indeed "nightingales are well known for their indifference to gunfire, which indeed they often seem to regard as a particularly vigorous rival nightingale trying to muscle in on their territory: during a raid in May, 1918, for instance, one was singing loudly in a London suburb during heavy gunfire and bomb explosions."[30]

Goering's sleek bombers were incomparably better at enraging male nightingales than Count Zeppelin's lumbering airships. Although birdlovers continued their vigil through the Blitz (which, on balance, greatly favored rare species like the black redstart), the new wastelands were above all a scientific opportunity for London botanists, led by J. Lousley, R. Fitter, and E. Salisbury, the director of Kew Gardens. "Some of the sites now exposed to the foundations," Lousely enthused, "may have been continuously roofed over since Roman times, and one must go back to the years immediately following the Great Fire for the last opportunity of doing any extensive botanizing within the confines of the City."[31]

The botanical census of bomb sites in the City and the East End revealed a new pattern of urban vegetation adapted to fire, rubble, and open space. Uncommon natives and robust aliens dominated this unexpected "bomber ecology." The most successful colonist of blitzed sites, for example, was the formerly rare rosebay willowherb (*Epilobium angustifolium*), which in Jefferies's lifetime could be found only in Paddington Cemetery and on a few gravelly banks. Its tolerance for fire-baked soils, as well as its liking for light and its prodigious seed production, transformed its previous shyness into bullying aggressiveness, making it "probably the commonest plant in Central London" in 1943. Its chief allies were members of the groundsel family, especially the Oxford ragwort, which, despite its name, was actually a recent immigrant from Sicily, "where it frequents volcanic ash, so that, as Dr. Salisbury remarks, it may well find the site of a burnt-out building a congenial habitat."[32] Among the other aliens that flourished during the

Blitz were the Canadian fleabane, already a familiar plant on railway embankments, the redoubtable buddleia, previously described, and the Peruvian *Galin-
soga parviflora*, an escapee from Kew Gardens.[33]

These new "fire flowers," as O. Gilbert has more recently explained, were the
heralds of an irreversible revolution in the urban ecology of London and other
bombed cities:

> All these plants were at that time undergoing a period of steady expansion which
> was given renewed impetus by the sudden availability of the bombed site habitat.
> This acted like a catalyst. As populations built up they exerted a tremendous inoc-
> ulation pressure on the urban area and consequently were able to spread into new
> habitats. After the war they were found to have become permanent members of the
> urban flora in which most of our heavily bombed cities where previously they had
> often been rather rare.[34]

Gilbert emphasizes the fascinating generic overlap between the pioneer plant
species of London bombsites and the hardy weeds and shrubs that colonized
the terminal moraines of the last ice age. The Blitz, in several senses, turned the
ecological clock back 10,000 years. "The frequency of these genera [*Artemisia*,
Epilobium, etc.] suggests that conditions on the wasteground which include inter-
mittent disturbance, low grazing pressure, low competition and the presence of
unleached base-rich soils must have similarities to those existing just after the end
of the ice age: a number of the species may never had had it so good since."[35]

In central Europe, of course, even more of the late Holocene environment
was ruthlessly stripped away. The destruction inflicted upon urban Germany
by the Allies' "area bombing" campaign was a full order of magnitude greater
than the Luftwaffe's pounding of London and the Midlands. The Hamburg
Katastrophe of July 1943, when Bomber Command's "Operation Gomorrah"
opened the gates of hell with the first urban firestorm, spurred Churchill and
his advisors (with reluctant American participation) to launch an all-out air war
against German civilians. During the sustained bomber offensive from 1 August
1944 until Unconditional Surrender on 26 April 1945, the Allies conducted 205
great bomber raids, nearly half of which targeted Berlin. Two million tons of

high explosives and incendiaries killed an estimated 600,000 to 800,000 civilians (one-quarter of whom were slave-laborers or prisoners of war) and injured nearly a million more. (The Allies, in turn, lost 75 percent of their bomber crews: more than 100,000 men.) The most common street sign in many German cities became "Gruesome": warning of rotting cadavers in the rubble. "In more than 40 German cities the proportion of built-up area razed exceeded 50 percent" and 333 square kilometers was transformed from dense urban housing into debris.[36] Although no German city suffered the Carthaginian fate of Warsaw (700,000 dead, its ruins transformed by the SS into a vast minefield), millions of urban dwellers had been reduced to troglodytes: "cellar tribes ... crowded together in squalid basements, air-raid bunkers and subway tunnels eerily lit by flickering candles."[37]

As Niels Gutshow has shown, some hardcore Nazi ideologues actually welcomed the thousand-bomber raids and firestorms as a ritual cleansing of the "Jewish influence" of big city life and the beginning of a mystical regeneration of Aryan unity with nature. Thus, in the aftermath of the Hamburg, Cologne, and Kassel holocausts in 1943, the "eco-fascist" Max Karl Schwarz, who shared Ruskin's and Jefferies's aversion to the "toxicity" of big cities, proposed to "revitalize the landscape" by leveling the debris and planting trees. The old, dense cities would not be rebuilt; indeed, "only after the destroyed areas are animated through forest will they become a true urban landscape, that is, with houses and gardens." Authentically German garden cities would replace the decadent "Jewish" metropolis. After all, had not Zarathustra commanded his followers to "spit on this city of shopkeepers"?[38]

> It is the German fashion to derive spiritual abilities and physical strength from the connections with nature. However this source of vitality for the Germans has more and more been cut off by a pervasive alienation from Nature.... It is clear to me that multistorey buildings are an expression of the Jewish spirit, and with such buildings we have recently been spreading the paralysing idea that everything can be built only on the basis of mass, number, and weight.... The planned reconstruction is a genuine resettlement, a new rootedness, rooted in the land. That is no romantic idea, but rather it is truly the only basis for the future life of struggles that the now aged Europe has to carry out against the peoples of the East.[39]

In the event, weeds not linden trees grew over the corpses of slum and suburb alike. The poet Gottfried Benn wrote about the nettles "tall as men" that flourished everywhere in the "Mongolian border town provisionally still called Berlin."[40] In the hungry winters after the war (1945–49), Berliners became experts on the edible properties of bomb flora like dandelions and chickweed. In her study of the photodocumentation of Year Zero, Dagmar Barnow discusses a photograph taken by a city employee in the winter of 1945–46.

> At about the same time, Durniok photographed a group of haggard-looking Berliners participating in "Aktion Wildgemuse" ("wild vegetables," a euphemism for weeds) and for learning to distinguish edible from inedible plants. For the most part older men and women dressed respectably in coats and hats, they might have gathered in the overgrown neighborhood park or garden for an adult education course in botany, if it were not for their emaciated, anxious faces. Some of them are plucking weeds, others are looking on. They will take the weeds home and chop them up to make soup—if they have water and fuel.[41]

The winter of 1946–47 was the most terrible. Berliners felt themselves doomed members of a huge forgotten Donner party, and even in Charlottenburg, once the capital of the Weimar *haute monde,* there were verified reports of cannibalism. If wild rumors—of deliberate Allied starvation plots or of Hitler's escape by submarine to Antarctica—proliferated, so did wildlife. Occupied Berlin was suddenly as feral as Jefferies's dead London. Wild boar "ravaged" the city's outskirts in herds fifty-strong, and were avidly hunted by hungry civilians with bows and arrows or by bored GIs with burpguns. Likewise, British Tommies volunteered to help Berliners (now subsisting on less than 1000 calories per day) track the three species of starving deer that had taken refuge in the Spandau and Kopenick forests. Hard on the heels of the deer were their ancient predators. Signs were hastily posted on the autobahn, "BEWARE OF WOLVES!"[42]

While Berlin's "rubble women" toiled by hand to clear 100 million tons of bomb debris (eventually gathered into the three artificial mountains that allow contemporary Berliners to practice skiing), urban flora was undergoing a remarkable transfiguration. As in London but on a greater scale, the fired and alkaline substrate of the deadzone favored the propagation of previously exotic species like robinia, tree of heaven, traveller's joy, and butterfly bush. Botanists were particularly surprised by the rapid spread of *Ailanthus altissima* (tree of heaven), Berlin's equivalent to the London rocket of 1666. Imported from China in the era of Frederick the Great, it had never shown any capacity for spontaneous growth in some 200 years of cultivation in Berlin's gardens and parks. Then suddenly, thanks to Bomber Command and the Eighth Air Force, it became the avid colonist (with that other robust alien, *Robinia pseudacacia*) of calcaric bomb sites: "[T]he new habitat type enabled *Ailanthus* to establish new populations. This process was favored by its high reproductive potential including early and prolific flowering, which is usually converted into large quantities of seed, as well as quick growth." A thermophile, tree of heaven was subsequently nurtured by the postwar urban "heat island" (central Berlin is 3.2 degrees C warmer than its suburbs).[43]

The persistence into the 1980s of several uncleared bombsites in West Berlin (notably the Lutzowplatz in Tiergarten and the former Schoneberger Hafen in Kreuzberg), allowed ecologists led by Hans Sukopp and his colleagues to observe more than forty years of the successional dynamic. Indeed ruderal vegetation, and urban ecology as a whole, has probably been better studied in Berlin than anywhere else.[44] Research has confirmed the threshold role of the Second World War in naturalizing alien species and establishing unique urban biotopes, whose hypothetical stage of final succession and self-regulation is often referred to as "Nature II."[45] The floristic component of this war-mothered Second Nature is strikingly similar in most cities of Central and Western Europe, despite significant differences in climate. As in Britain, where Gilbert has emphasized the "unlikely mixtures" of tree species in ruderal copses, the mature flora is an unusual community of vascular plants of a southern European or American origin.[46] Sukopp has speculated that these alien-dominated ecologies, far from marginal, "may be the prevailing ecosystems of the future."[47]

Deadzone communities, moreover, are surprisingly species-rich. The old bomb site at Lutzowplatz in Berlin, for example, hosted over 100 different plant and more than 200 insect species in the early 1980s. The carefully tended parklands of the adjacent Tiergarten, by contrast, supported a mere quarter of this diversity. Such ruderal complexity has been discovered in the ruins of "Fordism" as well as Hitlerism. In recent years, environmentalists have belatedly begun to appreciate that the "brownfield" sites of post-industrial Europe are actually biological oases—"green islands"—whose species diversity typically exceeds not only the rest of the city but the surrounding factory-farmed, genetically modified countryside as well.[48] In the derelict colliery landscape of West Yorkshire, for instance, the *Guardian*'s environmental correspondent Pete Bowler marveled at the unexpected biodiversity:

> All of them had either common spotted orchids or bee orchids growing in large numbers. Some had both. Walkers on post-industrial sites frequently see brown hare, a species in sharp decline and which is included on the UK bio-diversity action plan list. I have found badger setts in disused airshafts, great crested newts in most old colliery and brickworks ponds. One pond has all three of our native newt species and both common toad and frog recorded.[49]

Ghetto Geomorphology

> Woe unto this great city! And I wish I already saw the pillar
> of fire in which it will be burned. For such pillars of fire must
> precede the great noon.
> *Nietzsche,* Thus Spake Zarathustra

In most of Western Europe (to the dismay of the friends of newts and endangered birds) urban-industrial brownfields, like the bombsites that preceded them, are eventually recycled into productive use, often after systematic surveys and planning. This is not the case in the United States, where many inner-city wastelands endure as seemingly permanent landscapes to the corresponding benefit of ruderal nature. Here urban dereliction has become the moral and natural-historical equivalent of war. In 1940–41, the Heinkel and Junkers bombers of the Luftwaffe destroyed 350,000 dwellings units and unhoused a million Londoners. In the 1970s, an equally savage "blitz" of landlord disinvestment, bank redlining, and federal "benign neglect" led to the destruction of 294,000 housing units in New York City alone.[50] In the course of the Nixon and Ford presidencies, much of the old urban core in the Midwest and Northeast began to look like the Ruhr in 1945. Cities like Detroit, St. Louis, and Paterson which formerly had the lowest percentage of vacant land per capita now had among the highest. St. Louis's housing stock diminished by almost a fifth in the 1960s and early 1970s, while some Chicago districts like North Lawndale lost more than half their homes to abandonment and arson during the 1970s. Nationally in 1980, according to Kevin Lynch, one of every twenty housing units in urban centers was boarded up. Dereliction in some cities exceeded park acreage.[51] No civilization—especially not one so rich and powerful—had ever tolerated such extensive physical destruction of its urban fabric in peacetime. And, even at the threshold of a new millennium, America's "age of rubble" was far from over. More than 20,000 housing units per year were still being abandoned in New York City as late as 1996, while pockmarked Philadelphia continued to grapple with the burden of 55,000 derelict buildings and vacant lots.[52]

Can environmental science add anything to our understanding of this catastrophe? Geomorphology, in fact, offers apposite epistemologies for sorting out the tangled causes of ghetto landscapes. Urban dead zones, first of all, illustrate a basic postulate that geomorphologists call *equifinality*: different processes producing essentially similar landscapes.[53] Indeed, process can never be simply read off from form. It would thus be a fallacy to assume from appearance alone that Newark and Detroit were actually carpet-bombed from 20,000 feet, or, for that matter, devastated by mega-earthquakes. In the same vein, geomorphology asserts that real landscapes are always the complex products of several or more processes (tectonic and erosional) operating at different tempos and scales. Presumably this should be as true of "ground zero" landscapes of the circa 1970s ghetto as of mountain ranges or sea terraces.

Here classical geomorphology provides essential distinctions between levels of analysis as well as an invaluable warning against collapsing them into one another. As Alistair Pitty explains:

No phrase bears more repetition than W. M. Davis's statement that landforms are a function of *structure, process* and *stage*. However, its actual implementation into specific geomorphological studies requires careful scrutiny. In particular, any emphasis on just one element of this trilogy encourages self-supporting interpretations, as enquiries then converge on that element. Contrasts in the element singled out may account for many differences of opinion in geomorphology. Dichotomies are manufactured since exclusive attention to one element allows quite separate conceptions of the element to run in parallel but without overlap.[54]

Davis's triple determinants of natural landscape can be analogized in the case of the economic and political erosion of inner-city neighborhoods. Thus *structure* in "ghetto geomorphology" is tantamount to the macro-economic determinants of inner-city decline operating on decadal frequencies: deindustrialization, white flight, housing and job discrimination, anti-urban (but pro-suburban) federal policies, capture of city revenues for corporate rather than neighborhood priorities, and so on.[55] Even at this scale, however, it would be erroneous to assume that the same story necessarily explains the background of housing destruction in Newark as well as Detroit, in Southside Chicago as well as Bedford-Stuyvesant. For example, in *The Assassination of New York* (1995), Robert Fitch has shown in compelling detail how private accumulation strategies (above all, the vast real estate calculations of the Rockefellers) manipulated public policy to help drive small manufacturing out of Manhattan.[56] This kind of purposeful deindustrialization, which greatly accelerated the dereliction of the city's blue-collar neighborhoods, was probably not the case in Chicago or Detroit.

Process, in the urban context, corresponds to the conjunctural forces that have translated the structural weakening of inner-city economies and public services into the actual abandonment of stores, factories, and housing. In the 1950s and 1960s the major engines of neighborhood destruction were urban renewal and, especially, freeway construction. Then, in the late 1960s, ghetto insurrections brought urban America to the brink of a "Second Civil War" and destroyed part of the commercial landscape of the inner city. But the rioters' bricks and molotov cocktails inflicted minor damage compared to the red pencils of mortgage lenders and insurance companies. The third and most cataclysmic phase of inner-city decay occurred during the 1970s and early 1980s. Although researchers still complain that there is "very little empirical analysis of housing abandonment" (especially specific city histories), the broad outlines of the process are clear enough.

Disinvestment in the older central cities was led by the banks, endorsed by federal policies, and reinforced by ensuing local fiscal crises and contraction of lifeline municipal services. Banks and S&Ls, first of all, pumped capital out of the inner city but refused to loan it back, especially to Black-majority neighborhoods. Instead they drained Northeastern savings to the Sunbelt, where they stoked a masssive speculative building boom. Local banks in Brooklyn in the 1970s, for example, committed less than 6 percent of mortgages to their home borough: fully 65 percent of local savings was exported to Florida and elsewhere. According to Richard Morris's influential 1978 study, this was nothing less than the "Mitchel-Nixon-Ford 'Southern Strategy'" in action.[57] As he shows in detail, the Nixon-era FHA was the mastermind of a policy of draining savings from regions with housing deficits toward regions with housing gluts. Instead of a firebreak against urban disinvestment, the FHA policies were gasoline on the flames.

> No longer was it an agency designed to shore up bank confidence in areas suffering from a shortage of mortgage capital. FHA became, instead, a mechanism to direct bank investment *away* from northeastern cities and toward the booming Sun Belt. No longer were mortgages insured in the inner city or in the Northeast; instead the bulk of FHA insurance commitments flowed south where it acted to attract mortgage money from all over the nation with the lure of a building boom supplemented by the attractions of federal insurance.[58]

Unable to sell buildings because banks refused to supply affordable mortgages, absentee owners began to walk away from the tax bills on their former rent plantations. "Redlining," write Jackie and Wilson, "usually set in motion a self-fulfilling prophecy of inevitable decline. In most cities, banks assessed neighborhoods negatively prior to the actual evidence of blight without reference to the specifics of resident credit ratings, housing conditions, community viability, or business solvency."[59] As property values collapsed, so did city revenues. In 1976—on the eve of New York's fiscal meltdown—fully half of the city's deficit consisted of uncollected property taxes from redlined neighborhoods. With their tax bases undermined, New York and other older cities were frozen out of the municipal bond market. The subsequent savage cuts to vital municipal services, including fire protection and building inspection, coincident with landlords' abdication of building maintenance, completed the vicious circle.

What happened at this point—when maintenance and vital services were withdrawn from inner-city residential neighborhoods—corresponds to the analysis of what Davis called *stage*. It is also where environmental forces come back into the picture. Curiously, in this period none of the Washington think-tanks, academic institutes, or government agencies traditionally committed to urban research seem to have paid any attention to how orphaned buildings and neighborhoods actually evolved into their terminal state of "bombed out" rubble. The ghetto of the 1970s and 1980s was intellectually as well as financially and fiscally abandoned. A major exception was the unsubsidized photo-observation and documentary journalism of Camilo Vergara. The Chilean-born writer/photographer has been derelict America's "Last Man"; like Ish in *Earth Abides*, its sole dedicated observer. Strange to say, without Vergara's thousands of time-consecutive photographs of individual buildings and neighborhoods, we would today possess almost no scientific or historical record of ghetto landscape processes.

Vergara's "New American Ghetto Archive"—the original of which is now owned by the Getty Research Institute in Los Angeles—consists of detailed time-studies of what might be called "canonical" urban decay. The dozen or so intensively documented neighborhoods include the South Bronx, Harlem, and most of north central Brooklyn, as well as comparable dead zones in Camden, Newark, and Detroit (especially the northwest side of downtown), as well as the major public-housing projects of Chicago. Vergara, like a good ecologist, deliberately focused on "disturbance" sites, including drug houses, homeless encampments, and areas of noxious landuse dumping as well as assorted margins and interstices "often lacking political representation or even a name." As the crack cocaine epidemic made street-level photography more risky, he began to use rooftops to generate panoramic bird's-eye views, which, in turn, revealed unsuspected landscape facets.

A good example of Vergara's tenacious methology is his case study of a once magnificent apartment complex at the corner of 178th Street and Vise Avenue, near the Bronx Zoo. When he first started visiting the "Castle" in the winter of

1980, the heating had failed and tenants were beginning to leave. Seemingly no resources were available to rehabilitate the building's faded glory. The next fall was fire season. Although one might assume that buildings burn and become derelict from the ground up, Vergara discovered that the opposite was true. The first of twelve apartment fires began in occupied units on the top floor. Subsequently scavengers looted pipes and radiators from the fire-damaged apartments, leading to flooding and water damage on floors below. Tenant flight accelerated. By January 1983 the complex was completely abandoned and efforts had been made to seal all the windows and entrances with cinder blocks. Scavengers, nonetheless, continued to find their way inside to "mine" the building of saleable materials. As the building continued to deteriorate over the next two years, it blighted the rest of the neighborhood: attracting crime, depressing property values, and encouraging more abandonment. Finally in 1985 it was bulldozed into oblivion.[60]

The Urban Pandemic

> With the prenatal clinic and AIDS-deaths data as indices, we can conclude that the destruction of low-income housing, through a variety of direct and indirect mechanisms, fueled the AIDS epidemic in New York City, concentrated it heavily in the poor minority areas, led to a much larger epidemic than would have otherwise occurred, and marked a generation as surely as did Pharoah's and Herod's murders of babies.
>
> *Deborah and Rodrick Wallace*, A Plague on Your Houses

Based on his unique familiarity with the history of specific sites and their residual populations, Vergara has published incisive analyses of the social processes of dereliction. In particular, he has hammered at misguided or hypocritical public policies, usually under the banner of "urban reconstruction" or "community reinvestment," that have targeted formerly devastated neighborhoods as receptacles for concentrating homeless shelters and related services. In New York, for example, Mayor Koch, followed by Giuliani, cleared the way for the corporate Times Square revival by shipping most of the homeless to the South Bronx. The result, Vergara fears, is hypersegregation and the emergence of the "New American

Ghetto" as a virtual prison for the city's redundant working classes. (Recently he has revised this thesis in light of the dynamic impact of Latino immigration.) It has, however, been left to others—particularly Rodrick and Deborah Wallace—to build a theoretical model of the processes captured in Vergara's archive.[61]

The couple—he, a mathematical epidemiologist, and she, a population ecologist—were members of Scientists and Engineers for Social and Political Action when in 1973 they were approached by New York City firefighters. Advised by the Santa Monica–based Rand Institute (whose normal business was nuclear war and Vietnam body counts), City Hall was proposing a downsizing of the Fire Department, closing fire stations in ghetto neighborhoods in favor of "automated" voice fire boxes. Although the Wallaces would later discover that the Rand Fire Project was only one element in a sweeping strategy of "planned shrinkage" of municipal services advocated by Housing Commissioner Roger Starr and other Downtown *intimati* with ties to the real estate industry,[62] Deborah Wallace's immediate attention was focused on the shoddy "science" of the Rand proposals.

> I am an ecologist and, at that time, was evaluating fish-population models in the Hudson river. … As I began to read through [the Rand reports], it slowly became clear that the data acquisition, analysis and interpretation, and the modeling methodology were much more primitive than those of the Hudson river fish population models. NYC Rand's level of science was inadequate for natural ecology and grossly inadequate for experiments on human populations.[63]

In 1973, of course, mysterious firestorms were sweeping across the ghetto neighborhoods of older cities. Although the South Bronx and inner Detroit (with its annual "Devil's Night") were the most notorious infernos, the fire epidemic spared few industrial cities of the Midwest and Northeast. Between 1970 to 1977, for example, more than one-third of the housing in Chicago's principal Puerto Rican *barrio*, the Division Street area, was destroyed, mostly by fire.[64] The view of law enforcement agencies, corroborated by a 1977 *Newsweek* investigation, was that the fires were a criminal form of urban redevelopment: a strategy for shifting the cost of urban decay from landlords to insurers. As *Newsweek* explained,

[B]uildings are sometimes torched for revenge, sometimes to cover other crimes such as murder, and sometimes just for kicks. But a great many deliberate fires, perhaps the majority, are the work of the arson industry—a shadow world of property owners, mortgage men, corrupt fire officials, insurance adjusters and mobsters.[65]

In their early articles, the Wallaces, without diminishing the importance of landlord arson, shifted the focus of the debate from fire events *per se* to the conditions that made them possible on such a rampant, epidemic scale. In their opinion, the ultimate arsonists were the powerful elite advocates (long before any fiscal emergency) of reduced and geographically redistributed municipal services. They pointed out that while structure fires had dramatically increased in the late 1960s, due to both housing abandonment and arson, the situation had been stabilized by the addition of new fire companies (partially funded by HUD) to the affected neighborhoods. Beginning in 1972, however, City Hall, advised by Rand operations management experts, began its restructuring of fire services: what the Wallaces denounce as "fire redlining." Thirty-five fire companies were ultimately withdrawn from poor, high-fire-incidence neighborhoods and overall NYFD strength was cut down from almost 15,000 personnel in 1970 to little more than 10,000 in 1976.[66]

The predictable result was the "South Bronx Firestorm" of 1974–77. The Wallaces employed sophisticated disease models to explore how the fire contagion, unleashed by "planned shrinkage" acting upon a preexisting housing crisis, set off a chain reaction ("highly unstable random walk") of pathologies:

Use of a mathematical model from epidemiology to explain the behavior of fires in New York City should distress the reader: such a model describes an ecological disaster in which stabilizing mechanisms needed to maintain a population have failed. An epidemic model of fires in New York City is, in spirit, like a model of the collapse of a fishery, the catastrophic destruction of habitat, or other ecological calamity.[67]

Indeed, what happened in the South Bronx (which, as they point out, "wasn't even listed as a poverty area in 1967") eerily resembled the primary and secondary kill dynamics described by Stewart in *Earth Abides*. First, as fire protection eroded, landlords *en masse* abdicated building maintenance, which, in turn, increased fire incidence. The rapid incineration of the core of the South Bronx produced a mass exodus into the West Bronx where overcrowding amid the continued reduction of fire and housing-related services punctually led to a second wave of fires, which also spread to Harlem and parts of Brooklyn. After peaking at 153,263 emergences in 1976 (when serious fires were triple their 1964 rate), the firestorm began to burn itself out. "The decline in structural fires after 1976," the Wallaces grimly point out, "represents not the ending of the crisis or the improvement of fire service, but rather the simple exhaustion of fuel in the primary areas of fire 'infection' such as the South Bronx, Bushwick, and so on."[68]

The Wallaces could easily have included housing inspection and rodent extermination, among other examples, in their discussion of vicious circles created by "planned shrinkage." At the beginning of the 1970s, New York was able to send 800 housing inspectors into the field at one time. From 1975, however, the numbers were progressively cut back until less than 200 inspectors were on the job—far too few to monitor landlord abuses and the deterioration of the housing stock. The Health Department likewise cut its rodent extermination force by two-thirds at the same time that it was banning old-fashioned, polluting incinerators. Their replacements, the huge steel trash-compactors that collect refuse from garbage shutes, quickly proved to be paradisical environments for rat reproduction. The burgeoning rodent population almost immediately overwhelmed the handful of remaining exterminators.[69]

The fires destroyed vital neighborhood-anchored social networks as well as homes. In their aftermath, homelessness and street violence soared to levels not seen since the Depression. The geography of poverty in New York City in the wake of the "desertification" of so much of the Bronx and parts of Brooklyn became "highly, almost explosively unstable" and white ethnics accelerated their flight to the 'burbs. As the Wallaces emphasize, all of these factors had profound implications for disease ecology and public health. Before 1970, for instance, heroin addiction in the Bronx was concentrated in a few stable nodes well known to public-health workers. By 1976, however, needle-users were dispersed throughout the borough, and thus more difficult to identify and treat. As a result, the epidemic spread of HIV among IV-drug users in the 1980s (as initially with TB in the 1990s) was impossible to contain. One-quarter of the emergency room admissions in some Bronx hospitals were tested as HIV-positive, a rate comparable to the AIDS death zones of Africa.[70]

In recent research, the Wallaces have continued to focus on the long-term downstream effects (infant mortality and low-birth-weight babies, homicide, cirrhosis, TB, and so on) of government retrenchment and the housing catastrophe of the 1970s. With other researchers, they have become worried about the role of collapsing and derelict neighborhoods as "virulence incubators." They argue that "cultural vectors," like Vergara's "new ghetto" with its hypersegregated poverty or, conversely, a nomadic and immune-suppressed homeless population, not only spread disease (like insect vectors) but increase its virulence. The emergence of multiple-drug-resistant TB among homeless needle-users is a frightening example of virulence evolution in conditions of high transmission rates. Only a massive, emergency effort by the US Department of Public Health in the 1990s brought TB in urban centers back under some semblance of public-health control.

Despite this success, the Wallaces are pessimistic that "the US urban-suburban system will long be able to restrict rising levels of ... increasingly virulent contagious disease within point-source minority epicenters." Indeed, their worst-case scenario—"the 'cascading supernova' model of [urban] collapse"—resembles nothing so much as Ruskin's and Jefferies's nightmare of the Metropolis killing itself with its own toxins. "The rapid, politically-driven physical and social implosion of minority communities produces 'delta functions' of increasingly high rates of increasingly virulent contagious disease which then explode into susceptible enclaves within surrounding affluent suburbs."[71] Needless to say, these emergent diseases and new plagues will pay little heed to the "Not In My Backyard" signs on suburban borders.

2001

Notes

1. Roger Hooke, "On the History of Humans as Geomorphic Agents," *Geology* 28, no. 9 (September 2000), p. 843.

2. John McPhee, *The Control of Nature*, New York 1990.

3. Ernst Bloch, "The Anxiety of the Engineer," in *Literary Essays*, Stanford, Calif. 1998, pp. 307.

4. Mathis Wackernagel et al., *Ecological Footprints of Nations*, Toronto 1998; and "The Ecological Footprints of Tokyo," www.soc.titech.ac.jp.

5. John Ruskin, *The Storm-Cloud of the Nineteenth Century*, Orpington 1884, pp. 137–38.

6. Raymond Fitch, *The Poison Sky: Myth and Apocalypse in Ruskin*, Athens, Ohio 1982. The magnificently Satanic London skies in the otherwise camp Hughes Brothers' film *From Hell* (2001) suggest that someone in the Art Department had read Ruskin.

7. His pieces on London natural history were anthologized by Samuel Looker in 1944 as *Richard Jefferies' London*.

8. Samuel Looker, ed., *The Nature Diaries and Note-Books of Richard Jefferies*, London 1948, pp. 180–81 and 188. There has been considerable debate about Jefferies's politics, but later entries in the *Note-Books* certainly evince an identification with Nihilism, if not with mystical Communism as in H. S. Salt's interpretation (see his *Richard Jefferies: A Study*, London 1894, pp. 72 and 86–87).

9. Carlo Cipolla, *Miasmas and Disease: Public Health and the Environment in the Pre-industrial Age*, New Haven 1992, pp. 4 and 7.

10. "… the steady pollution of the Thames by sewage converted a fine salmon river into a waterway that stank so much as to make it almost impossible to take tea on the terrace of the House of Commons on a hot summer's day" (in R. Fitter, *London's Natural History*, London 1945, p. 171).

11. Friedrich Nietzsche, *Thus Spake Zarathustra: A Book for All and None*, London 1978, pp. 176–78.

12. P. Brimblecombe, "London Air Pollution, 1500–1900," *Atmospheric Environment* 11 (1977), pp. 57–60. The Royal Society's study of the global effects of Krakatau, published in 1888, was a masterpiece of late-Victorian science: see G. Symons, Report of the Krakatau Committee, Royal Society, *The Eruption of Krakatau and Subsequent Phenomena*, London 1888.

13. "This is, in fact, a romance of the agricultural depression; country society has been deserted by the townsmen, and left to its fate": W. J. Keith, *Richard Jefferies*, London 1965, p. 117.

14. Jessica Maynard emphasizes the kinship between avenging nature in "Snowed Up" and Jefferies's essay, "Weeds and Waste." (See "A Marxist Reading of 'Snowed Up'," in Julian Wolfreys and William Baker, eds., *Literary Theories: A Case-Study in Critical Performance*, New York 1996, p. 154.)

15. Laura Spinney, "Return to Paradise," *New Scientist*, 20 July 1996, p 30.

16. William Morris, *Selected Writings and Designs*, London 1962, p. 68.

17. What follows is paraphrased or quoted from Spinney's exceptionally informative article.

18. See *Richard Jefferies' London*, pp. 124–25.

19. Jack London, *The Scarlet Plague*, New York 1912.

20. Sauer quoted in Garrett Eckbo, *Landscape for Living*, New York 1950, p. 31. Sauer's classic statement is *The Morphology of Landscape*, University of California Publications in Geography, Berkeley, Calif. 1929.

21. John Caldwell, *George R. Stewart*, Boise, Idaho 1981, pp. 30 (Stegner) and 41 (quote from *Sheep Rock*).

22. Ibid., p. 30.

23. Demographers and population ecologists contrast opportunistic R-selected or ruderal species, whose populations "explode" under favorable circumstances, and equilibrial or K-selected organisms. Rats are spectacular, even monstrous examples of "R-selection." Litters comprise up to a dozen pups and females are ready to go into heat within 48 hours of giving birth. Young rats can mate at two months.

24. (Missing?)

25. (Missing?)

26. Frederic E. Clements's *Research Methods in Ecology* (1905) and *Plant Succession* (1916) provided a ruling paradigm for plant ecology until the Second World War. Clements believed that a single plant community was best adapted to long-term equilibrium with each regional environment or biotope. He also maintained that communities were real co-evolved "superorganisms" and not just statistical artefacts.

27. Jeffry Diefendorf, *In the Wake of War: The Reconstruction of German Cities After World War II*, Oxford 1993, p. 30.

28. Fitter, p. 231.

29. Hudson's enduring popularity, however, is based on the posthumous success of his rainforest novel, *Green Mansions*, made into a famous film.

30. Fitter's summary, pp. 229–30, from Hugh Gladstone, *Birds and the War*, London 1919.

31. Edward Lousley, *The Flora of Bombed Sites in the City of London in 1944* (reprinted from *Report of the Botanical Exchange Club, 1943–44*), Arbroath 1946, p. 875.

32. Fitter, p. 232.

33. Ibid., pp. 232–33; and Gilbert, pp. 180–81.

34. O. L. Gilbert, *The Ecology of Urban Habitats*, London 1989, pp. 180–81.

35. Ibid., p. 72.

36. Kenneth Hewitt, "Place Annihilation: Area Bombing and the Fate of Urban Places," *Annals of the Association of American Geographers* 73, no. 2 (1983), p. 258.

37. Diefendorf, *In the Wake of War*, pp. 8–11 and 94; and Douglas Botting, *From the Ruins of the Reich: Germany, 1945–1949*, New York 1985, pp. 64 and 124.

38. Nietszche, *Thus Spake Zarathustra*, pp. 176–78.

39. Niels Gutshow, "Hamburg: The 'Catastrophe' of July 1943," in Jeffry Diefendorf, ed., *Rebuilding Europe's Bombed Cities*, pp. 115–19 and fn 14, p. 128 (quote).

40. Quoted in Wolfgang Schivelbusch, *In a Cold Crater*, Berkeley, Calif. 1998, p. 14.

41. Dagmar Barnow, *Germany 1945: Views of War and Violence*, Bloomington, Ind. 1995, p. 168.

42. Botting, *From the Ruins*, pp. 109, 136 and 147–50.

43. Ulrike Sachse, et al., "Synanthropic Woody Species in the Urban Area of Berlin (West)," in H. Sukopp and S. Hejay, eds., *Urban Ecology: Plants and Plant Communities in Urban Environments*, The Hague 1990, pp 239–42. Tree of heaven is also dominant in the derelict urban landscapes of New York and Philadelphia. Eric Darton in his "biography" of the World Trade Center marveled at plants "so hardy they will even grow in the soot and refuse of subway ventilation coffers" (*Divided We Stand: A Biography of New York's World Trade Center*, New York 1999, p. 221).

44. See, for example, H.-P. Blume et al. [with Sukopp], "Zur Okologie der Grosstadt unter besonderer Berucksichtigung von Berlin (West)," *Schriftenreihe des deutschen Rates fur Landespflege* 30 (1978). From the early postwar period, there were also important studies of the ecology of bombsites by W. Kreh in Stuttgart and R. Gutte in Leipzig (GDR).

45. Ingo Kowarik, "Some Responses of Flora and Vegetation to Urbanization in Central Europe," in Sukopp and Hejny, eds., *Urban Ecology*, pp. 57–58 (for the concepts of "Nature II" and the "hemeroby system" for scaling human unpact).

46. "These copses comprise unlikely mixtures of native ash, hawthorn, willows, broom and guelder rose growing alongside laburnum, domestic apple, Swedish whitebeam, cotoneasters, and garden privet" (Gilbert, p. 81). This chimerical deciduous woodland "climax" is typically achieved, according to Gilbert, after forty years or so, following successive Oxford ragwort, tall-herb, grassland, and scrub woodland stages—all of them dominated by hardy aliens (pp. 72–81).

47. Quoted in Kevin Anderson, "Marginal Nature: An Inquiry into the Meaning of Nature in the Margins of the Urban Landscape," Dept. of Geography, University of Texas, nd.

48. H. Sukopp, "Urban Ecology and Its Application in Europe," in Sukopp and Hejny, eds., *Urban Ecology*, pp. 3 and 10. A fundamental synthesis is H.-P. Blume et al. (with H. Sukopp), "Zur Okologie der Grosstadt unter besonderer Berucksichtigung von Berlin (West)," *Schriftenreihe des deutschen Rates fur Landespflege* 30 (1978).

49. A pioneering account is R. Gemmell, "The Origin and Botanical Importance of Industrial Habitats," in R. Bornkamm et al., eds., *Urban Ecology*, Oxford 1982.

50. *Guardian*, 15 August 2001.

51. John Jackie and David Wilson, *Derelict Landscapes: The Wasting of America's Built Environment*, Savage, Md. 1992, p. 176.

52. Kevin Lynch, *Wasting Away*, San Francisco 1981, p. 91; and Ray Northam, "Vacant Urban Land in the American City," *Land Economics*, pp. 352–53.

53. *USA Today*, 20 March 2000 (national survey of abandoned buildings; New York City refused to cooperate).

54. Alistair Pitty, *The Nature of Geomorphology*, London 1982, p. 90.

55. Ibid., p. 100. William Morris Davis was the counterpart of Frederic Clements in early-20th-century physical geography. In his view, landscapes evolved in orderly stages toward an end form, the peneplain, analogous to Clements's climax community. Although his vision

of landform development as a simple cycle of erosion and denudation is no longer accepted
by geomorphologists, his epistemological injunctions still retain much of their force.

56. Robert Fitch, *The Assassination of New York*, New York 1993. I disregard the rightwing
belief that rent control "killed" affordable housing in New York City and drove landlords
to abandon or torch their properties.

57. Richard Morris, *Bum Rap on America's Cities: The Real Causes of Urban Decay*, Englewood
Cliffs, N.J. 1978, pp. 73 and 80.

58. Ibid., p. 76. Jackie and Wilson also point to the disastrous impact in the Nixon era
of the FHA's Section 235 program, which theoretically supported housing rehabilitation
but in fact guaranteed speculators huge profits for picking the pockets of inner-city
homebuyers, who subsequently defaulted on their mortgages. "Urban ruin for profit," as
they call the program, was directly responsible for leaving 13 percent of Detroit's housing
stock vacant and boarded up; 40 percent of these units were quickly vandalized or torched
(*Derelict Landscapes*, pp. 170–71).

59. Ibid., p. 159.

60. Vergara's account in *The Liveable City* 15, no. 1 (March 1991), pp. 2–4.

61. Deborah and Rodrick Wallace, *A Plague on Your Houses: How New York Was Burned
Down and National Public Health Crumbled*, London 1998, p. 130.

62. The politico-intellectual justifications for Starr's attack on lifeline municipal services,
as the Wallaces point out, had already been created by Daniel Patrick Moynihan's advocacy
of "benign neglect" during his tenure in the Nixon Administration. Moynihan viewed
inner cities as "pathological communities."

63. Wallace and Wallace, *A Plague*, p. xii.

64. Felix Padilla, *Puerto Rican Chicago*, Notre Dame, Ind. 1987, p. 215.

65. *Newsweek*, 12 September 1977, p. 89.

66. Rodrick Wallace, "Fire Service Productivity and the New York City Fire Crisis: 1968–
1979," *Human Ecology* 9, no. 4 (1981), pp. 435–36. See also Rodrick and Deborah Wallace,
*Studies on the Collapse of Fire Service in New York City, 1972–76: The Impact of Pseudoscience on
Public Policy*, Washington, D.C. 1979.

67. Wallace, "Fire Service Productivity," p. 439.

68. Ibid.; 1976 figure from *New York Times*, 9 November 2001.

69. On rats, see *Los Angeles Times*, 14 July 1994.

70. Rodrick Wallace, "A Synergism of Plagues: 'Planned Shrinkage,' Contagious Housing
Destruction, and AIDS in the Bronx," *Environmental Research* 47 (1988), pp. 15 and 25.

71. Rodrick Wallace and Deborah Wallace, "Inner-City Disease and the Public Health of
the Suburbs: The Sociogeographic Dispersion of Point-Source Infection," *Environment and
Planning: A* 25 (1993), pp. 1709 and 1718.

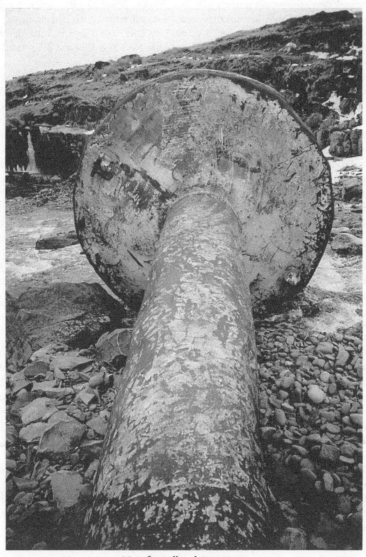

Newfoundland (2001)

18

Strange Times Begin

Year One

Last June (1998) *Science* published a short article about polar bears on the Arctic island of Spitzbergen in the Barents Sea. A team of Norwegian scientists has been systematically tranquilizing and tagging the animals in a long-term study of their population dynamics. They were shocked to discover that an anomalous number of the bears are hermaphrodites. They suspect that polychlorinated biphenyls (PCBs), which condense from the cold Arctic atmosphere and concentrate in the marine food chain, are the endocrine disrupters responsible for the bizarre sex-change.[1]

Amid all the environmental tremors of the recent past, including hemisphere-scale droughts and continent-size wildfires, it is the hermaphroditic polar bears that most haunt me. Like Central America's suddenly disappearing frog population a few years ago, the gender-bent bruins are a less immediately dramatic but ultimately more terrifying symptom of unspeakable tampering with the biosphere. They are also a small item in an extraordinary calendar of socio-natural catastrophe that may someday define the twelve months from June 1997 to July 1998 (the "weather year" as employed by meteorologists) as much or more a turning-point in world history as 1989.

More precisely, I think we may have just experienced the first year of life as it will be like in the early third millennium. As in a bad science-fiction novel overloaded with too many far-fetched scenes, the last two semesters have conjugated economic crisis, nuclear proliferation, ethnic cleansing, global warming, plague, drought, famine, and fire. If, for the moment, the OECD countries (Japan excepted) have been spared major shocks, and, indeed, still hallucinate themselves living luxuriously in the "end of history," the second and third tiers of the world market—the tigers and paupers—are reeling. The twentieth century is ending with a bang, not a whimper.

What the World Bank officially terms an "ASEAN depression" stokes religious and ethnic violence in Indonesia, Malaysia, and the Philippines.[2] Southeast Asia's hopes of recapitulating the Taiwanese and Korean paths of export-led industrialization have been sabotaged by the entry of China's immense, highly disciplined, ultra-low-wage workforce into the world market. The current meltdown of local currencies is mere lightning on the horizon, a warning of great storms yet to come. As it becomes fully integrated into GATT structures, China's staggering industrial capacity will eliminate the light-manufacturing and export-assembly niches that poorer Asian and Latin American countries have officially embraced as their engines of modernization. Their only recourse will be the accelerated export of primary products leading to even more deforestation, soil erosion, and water shortages.

Meanwhile, the emergent nuclear arms race that pits India against Pakistan and China has made regional atomic war between the world's most populous countries a terrifyingly real possibility. Chronic civil and ethnic warfare continues to spread across the Balkans, the Gulf of Guinea, the Horn of Africa, and parts of Latin America (with US military intervention against Colombian guerrillas now an open secret). And the popular belief that the HIV pandemic is under control was recently shattered by a UN study which reported that the potential death toll in Africa, where a quarter of the population in some countries (like Zimbabwe) is infected, will equal or exceed the mortality of the fourteenth-century Black Death or the influenza holocaust of 1918–19.

While Asian stock exchanges were collapsing and nuclear shockwaves were rattling seismographs, the most powerful El Niño of the century—in the context of the hottest decade in a thousand years—was orchestrating extreme climate events from Québec to Antarctica. Night after night the television news showed us a world on fire as more than ten thousand individual blazes consumed vast tracts of forest in Indonesia, Australia, Amazonia, Central America, southern Mexico, and Florida. Unnervingly, the seasons seemed to reverse order and regions seem to switch climates, as Boston sweltered in 90 degree heat in April and Los Angeles suddenly acquired the rainfall of Seattle. Deluges in the hyper-arid coastal deserts of Peru were matched by pigs and cattle dying of thirst in the jungles of Papua.

Confronted with James O'Connor's "second contradiction of capitalism" on such an epic scale conventional punditry has taken flight into absurd platonic distinctions between society and nature.[3] Although the mass media may occasionally concede some significant linkage between social and natural disorder—as between crony capitalism and rampant arson in Indonesian forests—the dominant strategy has been to reify El Niño as a virtually supernatural force that is exterior to history and whose human consequences have no social imprint.

In fact, as many researchers speculate, the present, intensified El Niño/Southern Oscillation (ENSO) cycle may well be the direct result—perhaps even the chief expression—of the anthropogenic warming of the troposphere. But regardless of ultimate mechanism, the El Niño events of the last year typically represent spikes in already chronic patterns of regional environmental decline resulting from breakneck industrialization and debt-enforced resource exploitation. To illustrate how global capitalism and climate change increasingly constitute a vicious circle, I offer thumbnail case studies of the contemporary drought syndrome in East Asia; the unexpectedly disastrous impact of extreme climate events on two major Western cities; and the looming biological catastrophe in the world's oceans and estuaries.

Depleted Futures

Consider, first, the persistent summer monsoon failures and severe droughts that have so troubled the humid as well as semi-arid parts of East Asia for most of the

Here climate change is colliding head-on with economic models based on unsustainable booms in commercial agriculture and forestry. From the standpoint of sacrificing priceless national ecological assets for short-term growth, "market socialism" in Vietnam and China is not entirely different from the "slash and burn" corporate capitalism practiced in Indonesia and the Philippines.

Although Vietnam figures little in environmental reportage and scarcely evokes images of Sahelian aridity, the current El Niño drought, following on the heels of crippling dry spells in 1992 and 1994, has caused huge damage to the coffee and rice crops that constitute its major exports and sources of foreign exchange. A staggering 50 percent of its rural work force was as a result estimated by government officials to be "basically unemployed" by the beginning of June 1998. Throughout the spring, moreover, acute water shortages caused regular power blackouts in Ho Chi Minh City and Hanoi, with further disruptions of production. A hundred major forest fires ravaged the highlands, and in some areas of central Vietnam like Thuan Hai province watertables fell so low after six years of drought that agronomists warned of eventual desertification.[5]

At a recent climate studies conference in London, the distinguished Vietnamese scientist Le Huy Ba attributed much of his nation's recent environmental distress to the drastic loss of upland and coastal forest-cover to war and agriculture over the last half-century (from 45 percent of the country's land area in 1945 to 19 percent in 1996). The increasing integration of Vietnam's economy into the world market has only accelerated deforestation. Whereas hillslope clearance has produced rapid soil erosion and catastrophic flooding, the destruction of mangrove swamps for speculative shrimp farming and firewood has led to acidification and biome loss. Forest destruction, he argues, is measurably and irreversibly changing the climate. "In the last twenty years, the average temperature has increased by .5°C while the rainfall has decreased by 100–150 mm."[6]

In a similar vein, World Watch Magazine published a disturbing account—borrowing heavily from a major US intelligence study based on spy satellite data—of China's growing water shortage. With barely 8 percent of the earth's freshwater resources, China feeds nearly a quarter of the world's population. Moreover, most of China's water is in the south while the majority of its cultivated land is in the north. This past year's El Niño drought, which pushed North Korea deeper into helpless famine by destroying 70 percent of its vital maize crop, also caused local agricultural emergencies throughout northern China. But as Lester Brown and Brian Halwell emphasize, China is becoming more vulnerable to drought primarily because it is extracting water from two of its five major river basins at twice the rate of natural replenishment. With 70 percent of its agricultural land dependent on irrigation, current overdraft levels are a recipe for national disaster, with unthinkable repercussions for the world food safety net if China is forced to import vast quantities of grain that otherwise might feed drought victims in Africa or the Near East.[7] The epicenter of the water crisis is in the cradle of Chinese civilization, the great plain of the Yellow River:

> In 1972, the water level fell so low that for the first time in China's long history it dried up before reaching the sea.... Since 1985, it has run dry each year, with the dry period becoming progressively longer. In 1996, it was dry for 133 days. In 1997... it failed to reach the sea for 226 days. For long stretches, it did not even reach Shandong Province, the last province it flows through en route to the sea. Shangdong, the source of one-fifth of China's corn and one-seventh of its wheat, depends on the Yellow River for half of its irrigation water.

Chronic drought, paradoxically, is accompanied by increasing flood danger. Even as it dries up in its lower reaches, the Yellow River continues to transfer fantastic quantities of silt from the deforested and eroded loess plateau region to the wheat-growing plains below. Siltation, by ever threatening to dam and divert the course of the river, requires ceaseless labors to rebuild embankments and levees as the river bed rises threateningly above the adjoining plain. This is, of course, a classic problem in China's history, which has seen millions die in Yellow River floods, but the danger has reached a modern apogee in the last few years. As one Chinese official recently explained, "Simply stated, embankment strengthening alone is doomed to failure in the race against channel siltation. Should the river break at its most dangerous locality, the resulting flood would directly affect an area with more than 150 million inhabitants. The only dependable remedy for this problem is soil conservation on the loess highlands."[8]

this problem is soil conservation on the loess highlands."[8]

But soil erosion in the loess region and elsewhere has skyrocketed since the decline of the communes and the introduction of market agriculture. As a result, the "average area classified as 'easily flooded and drought damaged' in China can be seen to rise to nearly 40 million hectares during the period from 1985 to 1990 from less than 30 million hectares in the early 1980s." More than 200,000 square kilometers of former crop or grassland were degraded into desert in the same period.[9] Beijing and even Seoul are now regularly shrouded by loess and sand blown out of China's growing Dust Bowl. These towering dust plumes quickly cross the Pacific where they merge with local pollution to form a continuous front of ochre haze from Arizona to Alberta. It is the startling signature of China's new state capitalism.

For a deeper analysis of the relationship between environmental degradation and China's "reformed" agrarian social structure, it is necessary to go beyond the mere alarmism of the World Watch Institute's and the National Intelligence Council's reports. The radical geographer Joshua Muldavin, who has worked as an advisor for many years in rural China, recently published a penetrating account of the ecological consequences of decollectivization in Heilongjiang province (former Manchuria). Market socialism, he argues, has abandoned sustainable agronomic practices and is ruthlessly "mining" the communal assets, built up so heroically by mass labor in 1950s and 1960s, that are the principal shock-absorbers of environmental stress. The dismantling of the commune system after 1978 led to the destructive expropriation by the richest or most influential peasant households of formerly common forest and grasslands resources. Although none of the local ecosystems involved in this transformation have the celebrity of tropical rainforests, the inevitable devastation is similar in process and scale. Treated as short-sighted windfalls, forests have been rapidly cut down while grasslands have been converted into temporary croplands, their productivity artificially maintained for a few seasons by increasing quantities of chemical fertilizers. Faced with the enclosure of their pasture lands, herders in turn attempt to stave off pauperization by crowding as many animals as possible into the remaining open grasslands. Rapidly industrializing towns, for their part, divert water resources from agriculture, then dump their untreated sewage and toxic wastes into the nearest stream or river.[10]

Meanwhile, "throughout China, villagers have been unable [under the new regime] to organize the necessary labor and capital investment for agricultural infrastructure. Reservoirs, dikes, irrigation canals, tube wells, erosion control, tree planting—all critical to sustaining and increasing production—receive little investment for maintenance, let alone improvement or expansion, and are in a state of serious disrepair." The central government has refused to make up the shortfall in local public-works spending, and state investment in agriculture in general fell precipitously (from 13 percent of the total budget to under 5 percent) during the Deng years. In Heilongjiang, the permanent local fiscal crisis, together with the privatization of the commons and the reliance on nonrenewable inputs like chemical fertilizers, is producing an environmental nightmare of barren, eroding hillsides; desertified grasslands; salinized fields; silt-filled reservoirs; broken and unrepaired dikes; polluted and unpotable streams; and more frequent "natural" disasters. As Muldavin somberly observes, "A national policy that relies on a raid of natural systems structures economic growth on ecological decline."

Godzilla

Most of us in the First World take for granted that our centrally heated and air-conditioned cities will shelter us from the seasonal discomforts of the weather—even as extreme climate incidents become more common. It is the misfortune, rather, of Third World shanty-town dwellers to face the full brunt of scorching heat, torrential rains, and merciless dust storms. All the more shocking, then, that the 1997–98 El Niño chose to act like Godzilla and bring two rich metropolises—Montréal and Auckland—to their knees in January and February of 1998. In so doing, it exposed the extraordinary vulnerability of even the most advanced urban and regional infrastructures to primordial forces of ice and heat.

At 4 a.m. on the morning of 5 January, while Montreal was asleep, a radio announcer warned that a powerful low-pressure system bearing down on Québec from the Gulf of St. Lawrence might bring freezing rain and ice. Ice storms are a familiar hazard in eastern Canada, and the previous November ice had collapsed high-voltage lines at the great hydro-electric complex at Churchill Falls, Labrador, which supplies backup power to New England and New York. Sturdy Montréalers and their neighbors, therefore, were braced for "ordinary" winter disaster. What they woke up to, however, on the morning of 5 January, and endured miserably for the next month, was the most destructive climate event in Québec history. Thanks to an El Niño–diverted jet stream (according to Environment Canada meteorologists) a vast amount of warm moist air was hijacked from the southern United States and lofted over a cool bottom layer covering the St. Lawrence and Ottawa Valleys. As rain fell through the cold layer it was super-cooled and transformed into an icy glaze weighing millions of tons that completely covered Québec as well as parts of Ontario and New England. Slender structures like trees, telephone poles, and transmission towers toppled under the weight of ice layers three or four inches thick. (Hydro-Québec in an ill-fated cost-saving move in the early 1970s had reduced the ice-bearing structural standards for its pylons and towers.)[11]

No strategic bombing campaign could have inflicted such comprehensive damage on Hydro-Québec and other public utilities as the *Le Grand Verglas*, which brought down *72,000 miles* of power and telephone line, in addition to millions of trees (including a large fraction of Québec's famous maples). The nation whose future is firmly mortgaged to its role as eastern North America's electricity reservoir—with billions of dollars of recent investment in ecologically destructive dams and artificial lakes—suddenly found itself without power for nearly a month. As the first deaths from hypothermia were announced, more than 100,000 residents of Montréal and its southern suburbs were moved into emergency shelters. (Residents of the most hard-hit part of the metropolitan region—the 89 municipalities of the so-called "Triangle noir"—did not have power restored until the end of the first week of February.) Economic life ground to a halt, and, for the first time since the "October Crisis" of 1970, the army was called out to patrol the darkened streets.

As Louis Francoeur, one of the Montréal's best-known progressive journalists, later pointed out, the heroism of Hydro-Québec's linemen and blue-collar employees, who worked day after day without sleep to rebuild 30,000 toppled wracked by recent financial scandals, has long been the principal enemy of Québec environmentalists and native peoples. *Le Grand Verglas*, on the contrary, provided the emergency raison d'être for the Québec Parliament to set aside normal debate and ratify Hydro-Québec's new strategic plan to boost electricity and gas sales in deregulated US markets. Public controversy was instead siphoned into angry squabbling over how the municipalities and the province would divide the several-billion-dollar cost of the great ice storm.[12]

Winter in Québec, of course, is summer in the Antipodes. While Montréalers were still assessing ice damage in late February, the New Zealand government was weighing the unprecedented option of sending in the army to evacuate the population of Auckland's central business district. In the face of a prolonged El Niño heat wave, the city's recently privatized electric grid had died in a dramatic month-long series of rolling blackouts. Each power failure had been punctually followed by smiling reassurances from Mercury Energy Ltd. that "the crisis was over and power would be shortly restored." In fact, as a later ministerial inquiry established, the private utility had neglected routine maintenance on the four underground cables that feed power to 8500 businesses and 75,000 people in the center of New Zealand's largest city. In a drive to increase profits, it had also laid off half its workforce as "unnecessary."[13]

Despite an improvised hookup with a power ship anchored in the harbor, the blackouts continued through the end of May. Surgeons operated by flashlight and New Zealand's High Court was forced to move sessions first to a church crypt, then to a racetrack. One hundred and twenty blocks of downtown Auckland became a ghost town. As the city's mayor confessed to angry protester, "We have had people trapped in elevators, overcome by fumes, there have been fires … people have been hurt." Eventually, scores of businesses were driven into bankruptcy and thousands of residents to the suburbs.[14] (In the meantime, the Kansas City–based holding company UtiliCorp, which already had a one-third interest in Power New Zealand, exploited the Auckland crisis to launch a hostile takeover bid against Mercury Energy.)[15] In effect, this was inadvertent urban renewal, thanks to a strange alliance of ENSO and the politics of privatization.

Dead Seas

Last fall the bodies of thousands of dead sea lions and hundreds of thousands of dead sea birds began to wash up on the beaches of Peru and Ecuador. In November, one naturalist discovered 500 dead sea lion pups on a single stretch of beach near Punta San Juan de Marcona in Peru.[16] As in 1972–73 and 1982–83, El Niño was decimating marine species throughout the western Pacific. Its warm waters suppress the cold upwelling of nutrients that nourish the huge biomass of anchoveta in the Humboldt Current off Peru. As the anchoveta population crashes, it brings down with it an entire food chain of larger fish, seabirds, and marine mammals. In the most extreme ENSO events, 80 to 90 percent of the interdependent populations perish. And ecosystem turmoil off the Peruvian coast can ricochet thousands of miles up and down the western edge of the Pacific: with migrating jack mackerel, for example, killing millions of juvenile salmon off British Columbia.[17]

Since the 12 million metric ton anchoveta harvest is normally the earth's single largest source of fish protein, its disappearance also sends shock waves through the global food-security system. Soya and beef futures soar on the Chicago commodity exchange while belts are tightened throughout the protein-deficient Third World. Although the ENSO-anchoveta cycle has existed for at least five thousand years, the recovery of anchoveta stocks and the other species that feed on them has been compromised by overfishing and coastal pollution. Climate change, again, is only exacerbating an already existing manmade crisis—this time with global dimensions. The 1950s–60s dream of the oceans easily feeding a hungry world has proven to be a cruel disappointment.[18]

Indeed, the fastest growing biological deserts on earth are the seven seas. As the *Washington Post* recently warned, "The oceans, long thought to hold unlimited bounty, are emptying. From Iceland to India, from Namibia to Norway fish catches are decreasing every year.... About 60 percent of the fish types tracked by the UN Food and Agriculture Organization are categorized as fully exploited, overexploited or depleted." Official figures, moreover, probably understate the crisis since "five low-value species—only one of them eaten by humans, the rest used for animal feed—have accounted for the entire growth in the marine catch since 1983."[19]

The world fishing fleets—dominated by 300-foot-long, high-tech supertrawlers that literally plow the seas with kilometer-long nets that can pull in 60,000 pounds of fish and hapless dolphins in a single catch—are heavily overcapitalized and depend on an estimated $54 billion per year in national subsidies.[20] Moreover, as a BBC editorial recently pointed out, "Even as fisheries decline, governments are continuing to subsidise an increase in fishing capacity ... particularly giant factory trawlers and purse-seiners—which [can] venture further afield, principally into southern waters where stocks were not as heavily fished."[21] It is the sheer overcapacity of First World fishing fleets, in other words, rather than the population ecology of the fish, that drives the annual harvest, usually to the detriment of poor countries whose fish stocks are raided so that multinational corporations can sell the protein back to them in the form of fishmeal and other products.

These factory fleets are, of course, highly efficient in wiping out entire marine ecologies. "Industrial" fishing for fishmeal to feed poultry and pigs (now about one-third of the world catch) is particularly insidious in destroying small fish populations and thus starving the sportfish, birds, and mammals that feed on them. As surface fisheries for cod, salmon, red snapper, mackerel, herring, tuna, and so on have been depleted, the fleets have moved into deep fishing for species like orange roughy which live a mile or more below, on the edge of the continental shelf. As marine biologists have futilely pointed out, these are not renewable stocks. "Orange roughy takes up to 30 years to reproduce and some specimens on the fishmongers slab are thought to have been alive when Queen Victoria was on the throne."[22]

Meanwhile, highly touted schemes to introduce sustainable-yield management into the major fisheries have been costly debacles. Over the last twenty years, for example, the world's coastal fisheries have been divided up into nationally exclusive 200-mile economic zones. Justified as a solution to overfishing by foreign trawler fleets, the zones have simply encouraged further government-subsidized overcapitalization of local seafood industries (or, in the case of poor countries, a lucrative, often corrupt market in selling fishing rights to foreign corporations at the expense of local fishing communities). In the US case, eight regional fisheries councils were established to set catch limits in accord with the best scientific estimates. But in classic fashion, the political control of the councils was captured by the large commercial fleets that they were designed to regulate. As a result, the smaller, community-based fishing vessels have been driven to the wall while the big boys have depleted 82 percent of the commercial stocks. Attempts in the United States and elsewhere to rectify the situation with Harvard Business School gimmickry like "individual transferable fishing quotas" have led only to the quotas being gobbled up by such odd anglers as the Caterpillar Corp., the huge KMPG accountancy conglomerate, and the National Westminster Bank.[23]

As marine populations are threatened with extinction so are the traditional fishing communities that have depended on them for centuries.[24] The collapse of the Grand Banks cod fishery (once the most productive in the world) in 1992, for example, has been an economic death sentence on the tiny "outport" communities of Newfoundland, where local fishermen had long protested in vain against the Ministry of Fisheries' connivance with corporate interests to inflate stock assessments of the cod population and thus justify the subsidies that Ottawa was giving to big companies to enlarge their fleets and processing plants. The subsidies to a handful of vertically integrated seafood monopolies (Fisheries Products International, National Sea Products, and so on) have continued while 40,000 unemployed Newfoundlanders are regularly denounced in Canada's Parliament as "welfare spongers" and "parasites." Most pathetically, demagogic politicians, led by Fisheries Minister Brian Tobin, are now claiming that it was harp seals, not corporate greed, that destroyed the cod, and are demanding an expansion of the controversial annual seal pup slaughter. "Save Our Cod, Eat a Seal" buttons are being distributed by the government-funded Canadian Sealers Association.[25]

Several thousand miles down coast from the dying fishing villages of Newfoundland, a deadly biological crisis, which could be radically aggravated by a warming climate, is rapidly growing out of control in the polluted estuarine and coastal waters of the Mid-Atlantic and Southern states. Every summer, vast algal blooms, fed by riverine runoff of nitrogen-rich waste from Iowa corn farms, Arkansas poultry processing plants, and North Carolina hog confinement lots, eutrophy and kill shallow-water biota. One example is the notorious "Dead Zone"—currently 7000 square miles and stretching from the mouth of the Mississippi to the Texas border—that emerges annually off the coast of Louisiana. In the spring the Mississippi fertilizes the Gulf with the agricultural runoff and municipal sewage from thirty upstream states. As the summer sun warms this polluted surface layer, algae blooms explosively then rapidly dies, decays, and removes oxygen from the water column. The resulting hypoxic water—thanks to Archer Midlands Daniel and the rest of agribusiness—is incapable of sustaining marine life.[26]

To prevent increasing eutrophication of the Gulf, Midwestern and Southern commercial agriculture would have to be radically modified to reduce its dependence on chemical fertilizers and pesticides. Elaborate grassland and wetland borders would have to be created to capture and dilute farm runoff before it reached the Mississippi or its principal tributaries. Since such a radical modification of American agriculture is almost impossible to imagine, the Dead Zone will continue to grow at the expense of marine life and the 200,000 Gulf area fishermen and processing workers who depend on the health of coastal ecosystems. At the same time, the witches' brew of petrochemical waste, livestock and poultry manure, artificial fertilizers, heavy metals, and exotic chemicals in urban runoff is producing nightmarish downstream effects in scores of coastal counties.[27]

In Carolina estuaries and the Chesapeake Bay, "the protozoan from hell," the incredibly sinister dinoflagellate *Pfiesteria piscidia*—which flourishes in manure-contaminated waters—has been killing millions of fish and inflicting a strange illness on local fishermen whose symptoms resemble acute multiple sclerosis or Alzheimer's disease. The courageous crusade by a North Carolina biologist, Dr. JoAnn Burkholder, against the official coverup of the *Pfiesteria* epidemic forms the suspenseful plot of Rodney Barker's 1997 bestseller *And the Waters Turned to Blood*. As the research of Burkholder and her colleagues indicates, the "microbialization" of the Chesapeake and other coastal waters seems to be a collaboration between nutrient dumping and overfishing of oysters and other efficient plankton feeders. *Pfiesteria* is only the latest, if most frightening, stage in a process of eutrophication that can be traced back to the impact of mechanical harvesting (first introduced in the 1930s) on the Chesapeake's formerly vast oyster reefs.[28]

Further south, along the Gulf Coast, the immediate and present danger seems to arise primarily from the toxicants and endocrine disrupters that have entered the marine food chain. Hormonal damage to wildlife is already assuming science-fiction proportions as gulls are born with thyroid glands so big they eventually explode, and male alligators with stunted sexual organs that prevent reproduction. Such horror stories, of course, bring us full circle to the PCB-polluted Arctic Ocean and the astounding prevalence of hermaphroditic polar bears....

1998

Postscript: Four Years Later

El Niño and his delinquent sister, La Niña, have come and gone, but strange times persist. It was the second warmest year (after 1998) in a thousand years. In March 2002, while crews were still excavating the remains of dead firefighters and stockbrokers from the crater that was once the World Trade Center, the Larsen B ice shelf in Antarctic suddenly collapsed. Sixty stories thick and covering more than 1300 square miles, the shelf had survived every warm pulse since the end of the Pleistocene. Yet in a mere few weeks, it pulverized—like a window hit by a cannonball—into thousands of iceberg shards. Where white ice had been locked solid for perhaps 50,000 years, there was now the angry velvet swell of the Southern Ocean.

The deep blue sea was an equally disturbing sight 10,000 miles northward, where, for the first time in the communal memory of the Inuit of Russia's Chukota Peninsula, the Bering Strait failed to freeze. "Strange portents are everywhere," reported an American visitor. "Thunder and lightning, once rare, have become commonplace. An eerie warm wind now blows in from the south. Hunters who pride themselves on their ability to read the sky say they can no longer predict the sudden blizzards. 'The Earth,' one hunter concluded, 'is turning faster.'"[29]

Indeed. Parts of the Arctic have warmed ten degrees in the last generation and climatologists predict that the Arctic Ocean may be open to summer navigation by the middle of this century. Those magnificent predators of the sea ice, the polar bears, won't fret about their sexual identity since they will be extinct. So too will be the seal species that depend on pack ice for nativity. The notorious seal cull in the Gulf of St. Lawrence had to be called off this year because there was so little ice. Instead the beaches of Newfoundland were macabrely littered with tens of thousands of dead harp seal pups who had drowned when their mothers couldn't find sea ice to give birth on.

Meanwhile, winter went AWOL across much of the temperate latitudes. New Yorkers walked in shirtsleeves on Christmas Day and turned on their air conditioners in mid April. Geese stayed home in balmy Wisconsin instead of migrating to southern climes, while rats and ticks rioted in the suburbs of Boston. Beijing, normally racked by icy Siberian winds, basked in 84 degree heat in March, while Indian scientists pondered the probability of catastrophic flooding from melting mountain glaciers in the Himalayas. Auckland put out a welcome mat for pioneer contingents of refugees from Tuvalu: the first island nation to begin planned evacuation in the face of rising sea levels.

None of this, of course, has had the slightest impact on the (future tropical) city on the Potomac. Even if the West Lawn turned into a sand dune or monkeys jabbered in the galleries of Congress, energy industry lobbyists—echoed by Rush Limbaugh—would still decry global warming as science fiction. Although it may be theoretically possible to imagine a "Green" global capitalism without rampant fossil fuel dependency, the actual outcome is dirty environmental counterrevolution. Bush's controversial election in 2000 was, first and above all, a coup d'état on behalf of the chief carbon dioxide producers: the auto and energy industries. Subsequent US interventions in Afghanistan and Colombia, as well as the foiled Washington-sponsored coup in Venezuela, have brazenly followed the (existing or proposed) routes of oil pipelines. Although the academy may still favor the esoteric relativity of postmodern textualism, vulgar economic determinism—which begins and ends with the superprofits of the energy sector—currently holds the real seats of power. We don't need Derrida to know which the way the wind blows or why the pack ice is disappearing.

2002

Notes

1. "Polar Bears and PCBs," *Science* 280 (26 June 1998), p. 2053.

2. See the pessimistic interview ("this depression may be very long lasting") with the World Bank's vice-president for Asia, Jean-Michel Severino, in the *Sidney Morning Herald*, 17 June 1998.

3. The "second contradiction" arises from individual capitals' externalization of social and environmental costs which, like an army of ghosts, return to haunt Capital as a whole. Although local communities of labor are the first victims of environmental degradation, O'Connor argues that sprawl, pollution, deforestation, and global warming eventually become constraints on global profitability. See "The Second Contradiction of Capitalism: Causes and Consequences" in O'Connor, *Conference Papers* (CNS Pamphlet 1), Santa Cruz, Calif. 1991.

4. There is debate as to whether the 1994 heat wave and drought—Southeast Asia's worst in a half-century—were caused by the lingering 1990–92 El Niño or by unusual synoptic conditions over the Tibetan Plateau. See Chung-Kyu Park and Siegfried Schubert, "On the Nature of the 1994 East Asian Summer Drought," *Journal of Climate* 10 (May 1997), pp. 1056–57.

5. Cf. "Global Warming and Vietnam," *Tiempo* (online magazine of the University of East Anglia Climate Research Group); "Drought Damage to Vietnam Coffee Seen Increasing," *Financial Express* (Singapore), 17 April 1998; "Parched: El Niño Is Blamed," *Los Angeles Times*, 17 June 1998.

6. Le Huy Ba, "The change of regional climate and environment caused by deforestation, irrational landuse and urbanization, biotic responses," poster at PAGES Open Science Meeting, University of London, 20–23 1998.

7. Lester Brown and Brian Halwell, "China's Water Shortage Could Shake World Food Security," *World Watch Magazine* (July/August 1998).

8. Huang Bingwei, quoted in Ian Douglas et al., "Water Resources and Environmental Problems of China's Great Rivers," in Denis Dwyer, ed., *China: The Next Decades*, London 1994, p. 192.

9. Erosion and deforestation increase drought as well as flood loss by silting up irrigation systems and reducing ground absorption of rainfall. In Guangxi province, for example, erosion has destroyed one-eighth of the vital irrigation infrastructure, with corresponding declines in grain production. See Scott Rozelle et al., "The Impact of Environmental Degradation on Grain Production in China, 1975–1990," *Economic Geography* 73, no. 1 (January 1997), pp. 52–53.

10. Joshua Muldavin, "Environmental Degradation in Heilongjiang: Policy Reform and Agrarian Dynamics in China's New Hybrid Economy," *Annals of the AAG* 87, no. 1 (1997), pp. 579–613.

11. This account of Montréal and Québec's ordeal is based on Louis Francoeur, "Le Québec `Sonne' par le Grand Verglas," in *Le Monde Diplomatique*, May 1998, pp. 20–21; and

various press releases and background briefings (including "Is There a Link to El Niño?") from Environment Canada, Ottawa 1998.

12. Ibid.
13. BBC News, 20 March 1998; and *The Press* (Christchurch, New Zealand), 11 May 1998.
14. Associated Press wire, 26 February 1998.
15. "Blackouts in New Zealand a Real Turn-on for UtiliCorp," *Kansas City Business Journal*, 16 March 1998.
16. From the international bulletin board of ENSO events at *The 1997 El Niño/Southern Oscillation* (www.darwin.bio.uci.edu).
17. See Michael Glantz, *Currents of Change: El Niño's Impact on Climate and Society*, Cambridge 1996, esp. pp. 23–32.
18. Early estimates that humans exploited only 2 percent of the oceans' primary production have been supplanted by studies that indicate that the real catch in prime fishing grounds often approaches 40 percent of theoretical potential. (See the account of the work of fisheries biologist Daniel Pauly in *Science* 296 [19 April 2002].)
19. Anne Swardson, "Once Bountiful, the Seas Slowly Empty," *Washington Post*, 3 October 1994.
20. From the award-winning *New Orleans Times-Picayne* series "Oceans of Trouble," 24–30 March, 1995.
21. Kieran Mulvaney, "The Fish Is Off," *BBC Wildlife Magazine*, October 1994.
22. "The Fisheries Effect," briefing from the World Wide Fund for Nature, 1997.
23. Dick Russell, "Vacuming the Sea," *E Magazine* 7, no. 4 (July/August 1996).
24. As fish stocks and family incomes collapse in unison, local fishermen are compelled to seek out potential catches in any sea, any weather. Sebastian Junger's *The Perfect Storm* (New York 1997) is a brilliant account—indeed, class analysis—of the almost suicidal risks now routinely taken by New England swordfishing boats.
25. Cf. Alan Finlayson and Bonnie McCay, "The Political Ecology of Crisis and Institutional Change: The Case of Northern Cod," in Fiknet Berkes et al., eds., *Linking Social and Ecological Systems: Management Practices and Social Mechanisms for Building Resilience*, Cambridge 1998; Christopher Chipello, "Nothing But Net," *Wall Street Journal*, 19 May 1998; and David Lavigne, "Seals and Fisheries, Science and Politics" (paper read at the Eleventh Biennial Conference on the Biology of Marine Mammals, Orlando, Fl., December 1995).
26. "Oceans of Trouble," ibid.
27. Ibid.
28. C. B. Officer et al., *Science* 223, no. 22 (1995).
29. Usha MacFarling in the *Seattle Times*, 15 April 2002.

Acknowledgments

Original publication only: *New Left Review* (preface and chapters 2, 13, and 16), *Grand Street* (chapter 3), *The Nation* (chapters 9, 12, and 14), *Socialist Review* (chapter 10), *International Socialism* (chapter 11), and *Capital, Nature, Socialism* (chapter 18). Chapter 7 first appeared in Mark Shiel and Tony Fitzmaurice, eds., *Cinema and the City*, London 2001; chapter 8 in Diane Ghirardo, ed., *Out of Site*, Seattle 1991; and chapter 15 in Lars Nittve et al., *Sunshine and Noir*, New Orleans 1998.

Thanks to Colin Robinson, Steve Hiatt, Samita Sinha and the rest of the New Press gang.

Photo Credits

The photographs that introduce chapters 8, 10, 14, and 15 are from the extraordinary portfolio of Diego Cardoso. Born in Ecuador, Diego (a leading city planner and civic activist) has for years kept a magical-realist photo diary of the streets of Los Angeles. His uncanny images have profoundly influenced my own perception of the city.

Preface, *Los Muertos*, by José Clemente Orozco, courtesy of the Orozco family, Centro Nacional de Conservación, and Museo de Carrillo Gil, Mexico City; chapter 1, courtesy of the Southwest Museum, Los Angeles, photo #P.36657; chapter

ter 1, courtesy of the Southwest Museum, Los Angeles, photo #P.36657; chapter
2, *Dead Animals #327*, © 1987 by Richard Misrach, courtesy Robert Mann Gallery;
chapter 3, United States Army; chapter 6, © Premium Stock/CORBIS; chapter
7, US Works Progress Administration; chapter 9, © Robert Yager; chapter 11, *Los
Angeles Herald-Examiner;* chapter 12, © Ted Soqui; chapter 16, © Premium Stock/
CORBIS; chapter 17, ©1988 by Camilo Vergara. Other photos: MD.

Index

Adamic, Louis, 122, 131

Adams, Ansel, 39

Adams, Robert, 38

African Americans:
civil rights and, 199, 208, 212, 234, 265
middle class, 231–32, 280, 282, 288
police and, 199, 214, 231–32, 266, 277–78 rebellions among, 230–34, 388
relations with Latinos, 242, 276–77, 281–303
struggle against racism, 193, 199, 214–15, 221, 279–80
youth, 214–15, 221, 289, 293–96, 300

After London (Jeffries), 364, 365, 367–71

AIDS, 391, 394–95, 402

Alatorre, Richard, 169, 171

Aldrich, Robert
Kiss Me Deadly, 136–38

Alhambra, Calif., 218–19

Alvarez, Walter, 315, 336, 351 n 45

American Ground Zero (Gallagher), 42–46

American Notes (Dickens), 129–31

And the Rivers Turned to Blood (Barker), 413

Angel's Flight, 131, 133, 152

Angel's Flight (Sheets), 133

Angelus Temple, 121–23

Another World (Grandville), 9

Antarctica, 414

anti-Communism, 207–9, 212, 211

anti-Semitism, 68

"Anxiety of the Engineer, The" (Bloch), 7–8

Arapahoe people, 28, 29

Arctic, 401, 414–15

arson, 394–398

Asian Americans, 204, 233

Assassination of New York, The (Fitch), 388

asteroids, 307, 309, 317–19, 325–29, 337–44

astronomy, 349 n 28
  discoveries, 308, 347 n 4, 354 n 94

astrophysics, 323

Atomic Energy Commission, 43, 44, 45

Atomic Photographers Guild, 39, 40, 42

At Work in the Fields of the Bomb (Del Tredici) 42

Auckland, 409

Ballard, J.G., 38

Baltz, Lewis, 38

Banham, Reyner, 127

Barker, Rodney
  And the Rivers Turned to Blood, 413

Barrett, Wayne 10

Beggar's Opera, The (Gay), 128

Bell, Calif., 196, 203

Bell Gardens, Calif., 196–97, 198, 199

Benjamin, Walter, 3

Bernardi, Ernani, 153, 165

Berlin
  bombing of, 65–78, 380, 382–83
  ruderal ecology of, 380–86

biological warfare, 51–53, 76

biosphere concept, 334–35

Birds of London, The (Hudson), 381

Black Panther Party, 221, 223

Bloch, Ernst, 7–9, 362
  "The Anxiety of the Engineer," 7–8
  The Principle of Hope, 9

bombing
  deurbanization and, 380–81
  of Germany, 380, 382–83
  incendiary, 65–80
  London Blitz, 69, 381
  precision vs. area, 69–72

Booker, Claude, 201

Bradley, Omar, 282–83

Bradley, Tom, 153, 154, 159, 162, 171

Brecht, Bertolt, and Kurt Weill
  The Threepenny Opera, 127, 128

Bretz, Harlan, 41

Brittain, Vera, 70
  Massacre by Bombing, 70

Brown, Jerry, 238

Brown, Willie, 164

Boyle Heights (Los Angeles), 200

Bunker Hill (Los Angeles), 127–29, 131–32, 135–36, 138–40, 147, 151–53, 157, 160, 162

Burkholder, JoAnn, 413

Bush Administration, 234, 239–40, 243–45, 272 n 10

Bush, George, 243

California
  budget crisis in 265–69
  environment, 196, 374–80
  immigration, 191–93, 199, 227–28
  recession in, 235
  taxes in, 202, 262, 273 n 64
  youth culture, 207–23

Canaries on the Rim (Ward) 63 n 77

capitalism, 41, 402, 403, 404, 406, 409, 411–12

Carpenter, John
  They Live!, 138–40

Carter Administration, 246, 247
Catholic Church, and redevelopment,
    165, 175
causality, in science, 325–29, 346
Central Business District Association
    (CBDA), 146, 153, 184
Central City Committee, 149
Central City West Assoc. (CCWA),
    166–68
Century City (Los Angeles), 150
Chance and Necessity (Monod), 339
Chandler, Raymond, 131
chaos theory, 309, 317–18, 321, 347 n 8
Cherokee people, 24
Chesapeake Bay, 413–14
Cheyenne people, 28, 29
Chicxulub impact, 315, 320, 336
China
    astronomy in, 341
    economy of, 402
    foreign investment, 194, 196
    scientific tradition in, 322
    water shortages in, 404–7
Churchill, Winston, 69, 70, 73, 77
cities
    disinvestment in, 390–95
    downtowns, 11–12, 143–76
    ethnic population shifts, 191–93,
        199, 252–55
    federal policy toward, 240–52, 389
    gambling as means of finance,
        202–4, 281
    health problems in, 391–95
    nature and, 8, 367–74, 377–86
    religion and, 119–27, 128
    scientific study of, 363–64, 371–74
    sustainability of, 362–63

technology and, 8, 361–62
    urban renewal, 143–76, 202, 242,
        281, 388–95
City of Commerce, Calif., 153, 194,
    196, 203
civil rights movement, 208, 212, 223,
    234
Clements, Frederic E., 380
    Plant Succession, 397 n 25
    Research Methods in Ecology, 397 n 25
climate, impactors' affects on, 345–46
    changes in, 402–5, 407–8, 414–15
Clinton Administration, 49, 240, 257–
    59, 272 n 46, 307
Clube, Victor, 310, 313, 320, 321, 338–
    46, 358 n 149
    The Cosmic Serpent, 340
    The Cosmic Winter, 340
    The Origin of Comets, 340
coherent catastrophism, 339
Cold War, 34–35, 207, 211–20, 309, 311
comets, 340
    fear of, 308, 314, 341
    observation of, 308
    role of, in history, 337, 341,
        356 n 122,
        358 n 149
Committee for Central City Planning,
    Inc. (CCP), 150–54
Community Redevelopment Agency
    (CRA), 149. 151, 153–54, 158–60,
    162, 164, 168–71
Compton, Calif.
    Black politicians, 279, 281
    corruption in, 277, 281
    economic decline, 280
    gangs in, 275–76

Police Dept., 275–79
politics, 279–83
white flight, 280
Cooper, James Fenimore
The Last of the Mohicans, 128
Corwin, Bruce, 164
Cosmic Serpent, The (Clube and Napier), 340
Cosmic Winter, The (Clube and Napier), 340
Craters, Cosmos, and Chronicles (Shaw), 309
Cretaceous/Tertiary boundary, 315, 317, 319, 322, 345
Criss Cross (Siodmak), 133–36
Crystal Age, A (Hudson), 371
Crystal Cathedral, 123–24
Cudahy, Calif., 196, 197, 202, 203

Darman, Richard, 245
Darwin, Charles, 313, 335–36
Darwinism, 374–75
Dawkins, Richard, 335
Deal, Joel, 39
de Duve, Christian, 334
Vital Dust, 334
deficit, federal, politics of 258–62
deindustrialization, 158, 159, 193–94, 235
Del Tredici, Robert, 39
At Work in the Fields of the Bomb, 42
Democratic Party, 199–201, 240, 245, 246, 247, 253, 247, 256, 266–69
Desert Cantos (Misrach) 36, 38
Dickens, Charles, 129–31
American Notes, 129–31
Disney Corporation, 163

Dollarhide, Douglas, 279
Dos Passos, John
Manhattan Transfer, 9–10
dot-com boom/crash, 236
Downwinders, Inc., 52–53, 54
Dresden, bombing of (1945), 76–78
drought
China, 404–7
New Zealand, 409
Vietnam, 404
Dugway Proving Ground, 51–53, 65–69
Dutton, Clarence, 39

Earth Abides (Stewart), 364, 375–79
Earthly Paradise, The (Morris), 371
Ecocide in the USSR (Friendly), 34
ecology, 361–86, 397n25, 401–15
"edge cities," 252–55
education, 198–99, 228, 243, 290–91
El Cajon, Calif., 210–11
El Niño, 403–7, 408, 410, 414
environmental degradation, 33–38, 196, 365–66, 401
Erickson, Steve 52, 53
eutrophication, 412–14
evolution, theories of, 335–36, 346
extinctions, 319, 322, 325, 332, 335–36, 345–46, 358n149

Fanon, Franz, 14
FBI, 17, 244, 297
fear
economic exploitation of, 12–13
social study of, 4–7
fertilizers, overuse of 412–13
film industry, 124–25

film noir, 121–32
Fire (Stewart), 364, 375
fish, depletion of, 410–13
Fitch, Robert
   The Assassination of New York, 388
Five Points (New York), 128–29
floods, 406
Freud, Sigmund, 6
Friendly, Al
   Ecocide in the USSR, 34–35

Gaia controversy, 333–35
Gallagher, Carole, 39
   American Ground Zero, 42–46
Galmond, Mary, 120
Galveston, Tex., 213
gambling, 202–4, 281
gangs, 198–99
   and 1992 riots, 232–34
   in Compton, 275–76
   fear of, 236, 244
   L.A. gang truce, 230–31, 236,
      293, 296
   warfare, 285–87, 290–303
Gampel, Yolanda, 6
García Lorca, Federico, 3
   "Dance of Death," 3
Garden Grove, Calif., 123
geomorphology, 40–41
gentrification, 116, 163–64, 167, 174,
   300, 303
geology
   catastrophism, 308, 309, 319–331,
      338–45
   discoveries, 310, 340, 343, 345,
      348 n 12, 358 n 149
   uniformitarianism, 308, 310–13, 318

geophysics, 323, 328
Ghilarov, Alexej, 334
Ghost Dance religion, 23–31
Gilbert, Carl Grove, 40–41, 316
Gitlin, Todd, 222, 224 n 42
Giuliani, Rudolph, 10–11, 391
glaciers, melting of, 414
Glassner, Barry, 5–6
Glendale, Calif., 159
Goebbels, Joseph, 74–75
Goin, Peter, 39
   Nuclear Landscapes, 42
González, Eusebio Joaquín, 122–23
Gosiute people, 27, 50, 51, 58–59
Gould, Stephen Jay, 310, 319, 322,
   351 n 46
Grainville, Jean-Baptist Cousin de
   Le Dernier Homme, 364
Grandville, Jean-Ignace de
   Another World, 9
Graves, Michael, 163
Great Basin, 35–41
Greater Los Angeles Plans, Inc.,
   140, 152
Green, Mark, 14
Griffith Park (Los Angeles), 214–15,
   220
Grieve, R.A.F., 321, 351 n 45, 352 n 66

Haagen, Alexander, 235
Halley, Edmund, 337
Harlins, Latasha, 232
Harris, Sir Arthur, 73–75, 77
Hawaiian Gardens, Calif., 297
Hayden, Tom, 185, 189
Healing Global Wounds, 48, 49, 53
Hernandez, Rosa Maria, 201

Heym, George, 7
High School Sex Club, 212
Hilo, Hawai'i, 107–16
Hitler, Adolph, 78
Hollywood, Calif., 150, 172, 183–84, 188
Holmes, Arthur, 319
Holmes, Dewayne, 293–94
Holmes, Dick, 37
homeless population, 391, 395
housing, 67–69, 147–49, 161, 163, 197–98
   abandonment, 388–93
Hudson, W. H., 371, 381, 397 n 28
   The Birds of London, 381
   A Crystal Age, 371
Huggett, Richard, 326
Huntington Park, Calif., 196, 197, 202–3
Hutchinson, C. Evelyn, 334

ice storms, 407–09
immigration, 197–198, 199, 227–30, 250–51
Indians (Native Americans), 23–31, 33, 50, 51, 58–59
Institute for Creative Technology, 5
Ireland, 24, 25
Islam, 15

Japan
   bombing of, 79–80, 83 n 57
   investment in US, 155–56
Japanese Americans, 286
Jefferies, Richard, 367
   After London, 364, 365–71
   "Snowed Up," 367

Johnson Administration, 242
Johnson, Philip, 123
Judson, E.Z.C.
   Mysteries and Miseries of New York, 130

Kazakhstan, 33
Kemp, Jack, 244, 252
Kennedy, Edward, 270 n 15
Kiowa, 24
Kirchner, Ernst, 7
Kiss Me Deadly (Aldrich), 136–38
Klett, Mark, 39, 40
Korean Americans, 232, 235, 290
Krakatau, 343, 366, 396 n 12
Kristan-Tollmann, Edith, 337–39
K/T debate, 332, 334, 336, 345
Ku Klux Klan, 123, 214, 299

Lakota people, 24, 25, 29
Last of the Mohicans, The (Cooper), 128
Latinos
   immigrants, 191–93, 199–200, 227–30, 252–54
   in Los Angeles, 154, 169–70, 193, 199–201, 214, 220, 227–30
   politics and, 199–201, 279–83
   relations with African Americans, 242, 276–77, 279, 281–303
Le May, Curtis, 71, 79, 80
Lewis, John 337, 343
life, conditions for, 331–35
Little Tokyo, 120, 157, 162, 163, 164
London
   fictional environmental collapse, 367–71

pollution, 365–66
ruderal ecology, 371–74, 380–82
underworld of, 129
London, Jack
    The Scarlet Plague, 374–75, 376
Long Beach, Calif., 213–14
Los Angeles
    bohemia in, 131–32
    Centropolis plan, 149–51
    Downtown, 132, 143–76, 185
    environment and, 362
    as film set, 127–28, 130–31, 137–40
    as financial center, 143, 152–56
    foreign investment in, 155–56, 194
    freeways in, 146, 167
    politics in, 147, 153–56, 161, 166,
        169–70, 172, 175–76, 247–48
    racism in, 123, 146
    religion in, 119–25
    riots in, 150
    Silverbook plan, 151–58, 163
    Skid Row in 149, 150, 152, 154, 156,
        160, 164
    transportation, 146, 167, 177–83
    urban renewal in, 146–50
Los Angeles Bus Riders' Union, 184,
    186–89
Los Angeles Community
    Redevelopment Agency, 147
Los Angeles County Sheriff's Dept.,
    213, 216, 218, 231, 293–296
Los Angeles Examiner, 209, 220
Los Angeles Police Dept. (LAPD), 150,
    214–15, 217, 229, 231, 236, 243, 292,
    302
Los Angeles Times, 147, 216, 219, 234,
    239–40, 267

Lovelock, James, 335
Luz del Mundo, La, 124–25
Lyell, Charles, 310, 312, 313–14, 323

MacArthur Park (Los Angeles), 230–29
MacPherson, Aimee Semple, 121–23
    This Is That, 122
Manhattan Transfer (Dos Passos), 9–10
Mann, Eric, 181
Marc, Franz, 7
Marshall, George, 71, 76
Marvin, Ursula, 320, 321, 350 n 39,
    351 n 50
Massacre by Bombing (Brittain), 68
mass transit
    buses, 180–81, 183
    light rail, 161–62
    subways, 146, 152, 157–58, 183–89
    streetcars, 146
    terminals, 153, 169, 171
Maywood, Calif., 196
Mencken, H. L., 131, 132
Mendelsohn, Eric, 66, 67, 68
Metroplolitan Transportation
    Authority (MTA), 183–89
Mexican Americans, 252, 289
Mexico, 124–25, 196
Mighty Uncle (test), 46–49
Milestones (Qutb), 14
millennialism, 26–31, 119–20
Misrach, Richard, 35–38
    Desert Cantos, 36
    Violent Legacies, 37, 47
Molina, Gloria, 166, 168, 171
Monod, Jacques, 333
    Chance and Necessity, 333
Montréal, 407–9

Mooney, James, 24–26, 28–29
Mormons, 45, 46, 48, 50, 52, 53
Morris, William, 367
   The Earthly Paradise, 371
   News from Nowhere, 371
Muertos, Los (Orozco), 3
Mysteries and Miseries of New York, The
   (Judson), 130
Mysteries de Paris, Les (Sue), 128

Nagatani, Patrick, 39
Napier, William, 310, 313, 320, 321,
   338–46, 351 n 53, 358 n 149
   The Cosmic Serpent, 340
   The Cosmic Winter, 340
   The Origin of Comets, 340
Naples, 8
"national sacrifice zones"
   in US, 36–59
   in USSR, 33–35
Navajo people, 27
Nazi ideology and ecology, 383
Near-Earth Objects (NEOs), 308,
   317–19, 337–44
Newton, Sir Isaac, 313, 314, 320,
   348 n 17
New Zealand, 409
Nevada, 23–24, 26–27, 29, 35–38,
   48–49
Nevada Nuclear Test Site, 29, 43, 45,
   49
Nevada-Semipalatinsk Movement, 48
New American Ghetto, The (Vergara),
   364
New Scientist, 354 n 94, 371–74
News from Nowhere (Morris), 371
New Topographics, 38–39

New York
   9/11 attack on, 4–6, 9, 362
   environment and, 362, 390
   fictional attacks on, 1–6
   as financial center 11–12
   fiscal crisis of 13–14, 248, 389–90
   racism in 13–14, 388–95
   underworld, 128–30
   urban renewal in, 9–11, 388–95
Nietzsche, Friedrich, 366
Nisbet, E. G., 314
Nixon Administration, 270 n 10, 389,
   399 n 58
Nuclear Landscapes (Goin) 42
nuclear testing, protests / campaigns
   against, 46–50

oceanography, 310–12, 315, 316
O'Sullivan, Timothy, 39, 40
Olvera Street (Los Angeles), 149,
   169–70
Origin of Comets, The (Clube and
   Napier), 340
Orozco, José Clemente, 3
   Los Muertos, 3

Paiute people, 26–27, 33, 48
Pallan, Pedro, 282
Parker, William, 147, 214–15, 243
Pasadena, Calif., 218
Pasadena Star-News, 218
Pawnee people, 28
Pentecostals, 119–25
Perot movement, 259–62
Pershing Square (Los Angeles), 157
Peterson, D.J.
   Troubled Lands, 34–35

Pitchess, Peter, 216, 220, 293
Plague on Your Houses, A (Wallace and
Wallace), 391–95
planetology, 308, 309, 316, 318–19
Plant Succession (Clements) 397 n 25
plate tectonics, 308, 309, 312–14
Poincaré, Henri, 350 n 35
police brutality, 275–79
political realignment, 258, 260
pollution, 196, 198, 401, 412–14
Pool, Robert Jr., 245
postmodernism, 309
Powell, John Wesley, 24, 40, 41
*Report on the Lands of the Arid
Region*, 41
Principle of Hope (Bloch), 9
privatization, 243
Proposition 13, 262, 266
punctuated equilibrium theory, 337

Quayle, Dan, 240, 245, 270 n 15
Québec, 407–9
Qutb, Sayyid, 14
Milestones, 14

racism, 15–16, 58, 120, 122–23, 150,
159, 186–87, 199, 214–15, 221, 231,
237, 256, 374–75
political language and, 255–56
religion and, 120, 122–23
segregation, 122, 252–56, 288, 299
Rampino, Michael, 319–20, 321, 336–
37, 345, 352 n 67
Rand Institute, 392–93
Raup, David, 322, 323, 350 n 39
Reagan Administration, 243–46,
262 n 10, 264 n 46

Reason Institute, 245
Rebuild L.A., 235, 241
redlining, 389–95
"reindustrialization," 194–95
Rephotographic Survey Project, 39
Report on the Lands of the Arid Region
(Powell), 41
Republican Party, 201, 240, 243–48,
250–51, 266–69
Research Methods in Ecology (Clements)
397 n 25
Reviving the American Dream (Rivlin),
264 n 46
Riel, Louis, 25
Riordan, Richard, 159, 188, 236
Rivlin, Alice
Reviving the American Dream,
272 n 46
Rodney King riots (1992), 227–31, 276,
291–92
response to 239–43
Roosevelt, Franklin, 75, 76, 83 n 42
Royal Air Force, 72–78, 381–82
ruderal ecology, 364, 380–84, 398 n 45
Ruskin, John
*The Storm Cloud of the Nineteenth
Century*, 365

San Diego, Calif., 207–12
San Diego Union, 208, 209, 211, 212
San Fernando Valley (Los Angeles),
146, 198
secession movement in, 236–37
Salvadoran Americans, 227
San Francisco Bay Area, 255
fictional environmental collapse,
374–79

Sauer, Carl, 375
Savas, Emmanuel, 246
Scarlet Plague, The (London), 374–75, 376
Schabarum, Peter, 201
Schuller, Robert A., 123–24
Schwartz, Fred, 207–8, 212 223
security services/technology, 12–14, 17–19, 152
Seymour, William Joseph, 119–21, 125
Shaw, Herbert, 353 n 74
    Craters, Cosmos, and Chronicles, 309, 323–29
Sheep Rock (Steart) 375, 376
Sheets, Millard
    Angel's Flight, 133
Shiva impacts, 319, 321, 322, 337
Shoemaker, Carolyn, 337, 350 n 42
Shoemaker, Eugene, 317, 319, 350 n 42, 352 n 66
Shoshone people, 27, 29, 33, 46, 48, 54
Siodmak, Robert
    Criss Cross, 133–36
sixties, interpretations of, 220–23
Skull Valley, Utah, 50–59
skyscrapers, 12, 143, 152, 155, 158, 168, 173
Smithsonian Institution, 24
"Snowed Up" (Jefferies), 367
Solar System Evolution (Taylor), 309–10
South Central L.A., 156
South Gate, Calif., 196, 197
South Park (Los Angeles), 162–63
Spillane, Mickey, 136, 137
states (US), and budget cutbacks, 262–65
Steel, Duncan, 338, 340, 344

Stegner, Wallace, 375
Stewart, George R., 375–76
    Earth Abides, 364, 375–79
    Fire, 364
    Sheep Rock, 375, 376
    Storm, 366, 375, 376
Stimson, Henry, 71
stochastic catastrophism, 338–45
Stockman, David, 246
Storm (Stewart), 364, 375, 376
suburbs
    federal subsidies to, 251
    growth of, 24, 251–52
    political power of, 242, 247, 255–62
    segregation and, 252–55
Sue, Eugene
    Les Mysteries de Paris, 128

Taiwan, 194
Taylor, Stuart Ross, 330–31, 353 n 82
    Solar System Evolution, 309
They Live! (Carpenter), 138–40
This Is That (MacPherson), 122
Threepenny Opera, The (Brecht and Weill), 127, 128
Tokyo
    bombing of, 78–80
    environment and, 362
Tollmann, Alexander, 337–38
tourism, 11, 107, 116
trade unions, 193, 201
    in Hawai'i, 109
Trakl, Georg, 7
Troubled Lands (Peterson), 34–35
tsunamis, 108–9, 112–15, 344
Tunguska impact, 317

Ueberroth, Peter, 241, 268
Universal City, Calif., 185
University of California, Berkeley, 375
University of Southern California, 5, 151, 162, 170
urban ecology, 8, 362–64
urban renewal, 143–76, 195, 202, 281, 388–95
    opposition to, 165–68
US Air Force, 68–72
US Dept. of Defense, 38, 40, 49–55, 307
US Geological Survey, 24
USSR, 33–35, 48
Utah, 42–48, 50–59
Ute people, 27, 28
utopias
    "black," 8, 9
    bourgeois, 8

Van Hoddis, Jakob, 7
    "World's End," 7
Vergara, Camilo, 364
    The New American Ghetto, 364, 390–91
Vernadsky, Vladimir, 310, 321, 334–35
Vernon, Calif., 191, 194–96
Village Voice, 10
Violent Legacies (Misrach), 36, 47
Vital Dust (de Duve), 334
Volk, Tyler, 350 n 67

Wallace, Rodrick and Deborah, 364–65
    A Plague on Your Houses, 391–95
Ward, Chip, 57–58
    Canaries on the Rim, 63 n 77

War in the Air, The (Wells), 1–2
"War of the Worlds," (Welles) 5
"War on Drugs," 251, 263
Washington, D.C., 252
Washington Post, 17
Washoo people, 27
Watts rebellion (1965), 150, 221, 231, 279
"Weed and Seed" program, 240, 244
welfare programs, cuts in, 263–64, 267–69
Welles, Orson
    "War of the Worlds," 5
Wells, H. G., 1–2
    The War in the Air, 1–2
Westlake (Los Angeles), 200
Westside (Los Angeles), 150, 154, 288
whites
    flight from cities, 252, 279
    nativism among, 230, 236
    racism among, 193, 279
wildlife, 369, 374, 377–79, 381, 384, 401, 411–13
Wilson, Pete, 266–68
Wilson, J. Tuzo, 312
working class
    Berlin, 67, 68, 70, 73–75
    Hawai'i, 109, 111–12, 116
    housing of, 67, 68, 70, 147–49, 197–198, 388–93
    Los Angeles, 191–93
    retraining of, 249–51
    youth, 208, 210, 214–16, 221–23
World's Columbian Exposition, 25
"World's End," (Van Hoddis) 7
World Trade Center 3, 6, 9
Wounded Knee massacre, 25, 26

Wovoka, 26–31

Yaroslavsky, Zev, 165
Yorty, Sam, 154, 219
youth
   gangs and, 283–87, 290–92, 296–98,
      300–3

jails and, 293–96
police harassment of, 198–99,
   214–15, 232
rebelliousness, 207–27
Yowell, Raymond, 46–47

Zoroastrian Society, 69–72, 78

About the Authors

Mike Davis (1946-2022) was a writer, political activist, urban theorist, and historian. He is best known for his investigations of power and class in works such as *City of Quartz*, *Late Victorian Holocausts*, and *Planet of Slums*. His last two nonfiction books are *Set the Night on Fire: L.A. in the Sixties*, coauthored by Jon Wiener, and *The Monster Enters: COVID-19, Avian Flu, and the Plagues of Capitalism*. He was the recipient of the MacArthur Fellowship and the Lannan Literary Award.

Rebecca Solnit is the author of more than twenty books on feminism, western and urban history, popular power, social change and insurrection, wandering and walking, hope and catastrophe. Her books include *Orwell's Roses*; *Recollections of My Nonexistence*; *Hope in the Dark*; *Men Explain Things to Me*; *A Paradise Built in Hell: The Extraordinary Communities that Arise in Disaster*; and *A Field Guide to Getting Lost*. A product of the California public education system from kindergarten to graduate school, she writes regularly for the *Guardian*, serves on the board of the climate group Oil Change International, and recently launched the climate project Not Too Late (nottoolateclimate.com).

About Haymarket Books

Haymarket Books is a radical, independent, nonprofit book publisher based in Chicago. Our mission is to publish books that contribute to struggles for social and economic justice. We strive to make our books a vibrant and organic part of social movements and the education and development of a critical, engaged, and internationalist Left.

We take inspiration and courage from our namesakes, the Haymarket Martyrs, who gave their lives fighting for a better world. Their 1886 struggle for the eight-hour day—which gave us May Day, the international workers' holiday—reminds workers around the world that ordinary people can organize and struggle for their own liberation. These struggles—against oppression, exploitation, environmental devastation, and war—continue today across the globe.

Since our founding in 2001, Haymarket has published more than nine hundred titles. Radically independent, we seek to drive a wedge into the risk-averse world of corporate book publishing. Our authors include Angela Y. Davis, Arundhati Roy, Keeanga-Yamahtta Taylor, Eve Ewing, Aja Monet, Mariame Kaba, Naomi Klein, Rebecca Solnit, Olúfẹ́mi O. Táíwò, Mohammed El-Kurd, José Olivarez, Noam Chomsky, Winona LaDuke, Robyn Maynard, Leanne Betasamosake Simpson, Howard Zinn, Mike Davis, Marc Lamont Hill, Dave Zirin, Astra Taylor, and Amy Goodman, among many other leading writers of our time. We are also the trade publishers of the acclaimed Historical Materialism Book Series.

Haymarket also manages a vibrant community organizing and event space in Chicago, Haymarket House, the popular Haymarket Books Live event series and podcast, and the annual Socialism Conference.